PAPER PROFITS

POLLUTION IN THE PULP AND PAPER INDUSTRY

The MIT Press
Cambridge, Massachusetts, and London, England

PAPER PROFITS

POLLUTION IN THE PULP AND PAPER INDUSTRY

researched and written by
Leslie Allan
Eileen Kohl Kaufman
Joanna Underwood

with the assistance of
Howard Dykoff
Abby Friedman
Holly Miller
Constance Peters
Michael Taylor

The Council on Economic Priorities
New York, New York

Typed with an IBM Executive

by Mrs. Inge Calci

printed on Oxford Earthtone (recycled paper)

by Halliday Lithograph Corp.

and bound in Interlaken AL1-975 matte

by Halliday Lithograph Corp.

in the United States of America.

Library of Congress Cataloging in Publication Data

Council on Economic Priorities.
    Paper profits.

    Bibliography: p.
    1. Wood-pulp industry--United States--Waste disposal.
2. Pollution--United States. I. Allan, Leslie.
II. Kaufman, Eileen Kohl. III. Underwood, Joanna.
IV. Title.
TD899.W65C65 1972   338.4'7'676   72-2471
ISBN 0-262-03045-4

CONTENTS

Contents

Environmental problems are a perplexing mixture of broad, global generali-
zations and minute, local details. Anyone who has been assaulted by the smell
of a pulp mill is intensely aware of its impact on the local environment. What
is less obvious is that the pulp mill is associated with two much larger systems; indeed, it is the link between them. One of these is the ecosystem that
feeds logs into the pulp mill and receives waste in return. The other is the
economic system, which governs the mill's appetite for logs and its ability to
control its wastes. To appreciate fully the import of the details about pulp and
paper mills which this book offers the reader, it is useful to set them in their
ecological and economic contexts.

The basic unit in an ecosystem is a cycle—a circular system of inter-
linked events. Two such systems are particularly relevent to paper and pulp
mills. One of these operates in the soil, where plants (for example, trees)
produce, as they grow, organic materials (complex carbon-containing compounds); these eventually fall to the ground, where they are converted by microbial action into humus—the soil's store of organic matter; in turn humus
is degraded by further microbial action, yielding inorganic products (such as
carbon dioxide, nitrate and phosphate); the latter, taken up by the plant's
roots, support its growth and are converted into organic matter—thus completing the cycle. There is a similar cycle in surface waters. Fish excrete
organic wastes; these are converted by the bacteria of decay into inorganic
products, oxygen being consumed in the process; the inorganic materials
(again, carbon dioxide, nitrate and phosphate) nourish the microscopic green
plants (algae) which convert them into organic matter; the algae are eaten by
the fish, and the cycle is complete. There is an important difference between
these two systems: the soil is rich in organic matter, while the water contains rather little. Indeed, the aquatic system can be readily overburdened if
extra organic matter (such as sewage)—beyond that produced by the cycle itself—enters it, for so much oxygen may then be consumed as to asphyxiate
the bacteria of decay and bring the entire self-sustained cycle to a halt.

The paper and pulp industry exploits a particular organic product of plant
growth—cellulose. Cellulose is made up of very long fibrous molecules, each
composed of thousands of simple sugar (glucose) molecules attached end to
end. To organize the requisite carbon, oxygen, and hydrogen atoms into a
glucose molecule and to link the latter into a cellulose chain requires energy.
This is derived by the plant from sunlight, through the process of photosynthesis. This process is responsible for the synthesis, from inorganic raw
materials, of all the organic matter which supports life on the earth. Cellulose plays a vital role in the plant; its strong, fibrous character is a major
source of the structural strength that holds the plant aloft, where its leaves
can catch the sun.

In nature, cellulose, like the rest of the plant substance, is sooner or
later brought to the soil, as leaves fall, plants die, and trees topple. Here it
is attacked by molds and bacteria; the constituent glucose molecules are
freed, to become energy-yielding food for microorganisms that on their death
add to the deposit of humus. As the humus is gradually converted to inorganic

materials that are then taken up by plants, cellulose is formed again. Here nothing is lost, nothing is misplaced, and the cycle remains in balance.

Enter the pulp mill. Now the trees are removed from the land and taken into the mill, where, by a variety of processes, the cellulose fibers are separated and made into paper or one of the other numerous products of cellulose pulp.

How can we judge the ecological impact of a paper and pulp mill? First, since the mill removes from the soil a major constituent of the soil ecosystem—trees—this operation in itself may be a hazard. A harshly logged mountainside, lacking the soil-binding tree roots, may erode rapidly so that the soil system loses much of its biological capability. Even if logging is sound, the soil system may nevertheless be stressed, for the mere removal of the logs from the soil cycle disrupts the circular processes which in nature sustain the level of essential elements, particularly nitrogen and phosphorus, in the soil. If logs are taken from the soil, and nothing is returned to compensate for the nitrogen and phosphorus which they contain, the amounts of these essential elements in the soil will inevitably decline. (It should be noted that purified cellulose usually contains rather little nitrogen or phosphorus; these are largely in the wastes of the pulping operation.)

Once sold, the pulp and paper products—and the cellulose which they contain—are removed from the ecological cycle which produced them, and to which they must be returned if the cycle is to remain in balance. However, in principle nothing prevents the ultimate return of the cellulose to the soil cycle; the cellulose in most paper products remains accessible to the attack of soil microorganisms, and could readily be reintroduced into the soil system. (A simple experiment can establish this fact: insert a bit of paper in the soil of a potted plant and observe its condition after a few weeks.)

It follows that, in principle at least, a pulp mill which converts the trees' cellulose into paper products need not seriously disrupt the ecosystem, so long as the soil is properly sustained and unused organic matter in the logs, and eventually the cellulose itself, is returned to it. It is sometimes suggested that restoration of soil fertility can be accomplished by returning the missing chemical elements themselves to the soil, in the form of inorganic fertilizer. This is not necessarily the case, for unless the organic content of the soil is maintained, the efficiency of nutrient uptake by the plant roots (which depends, among other things, upon the aerating effect of humus in the soil) may decline, so that nutrients leach from the soil into surface waters—which thereby become polluted. In any case, as described thus far, a pulp mill is not a great ecological hazard. Indeed, as indicated in Chapter 2, the oldest pulping method—groundwood pulping—which simply breaks up the wood by mechanical means and converts nearly all of it into pulp without the intervention of chemicals, has a minimal impact on the environment. The only negative effect is due to the misplaced "disposal" of waste organic matter from the groundwood pulping process into surface waters. Here the added organic matter increases the rate of oxygen utilization by the bacteria of decay, which may become so rapid as to threaten the latter with asphyxiation and thus dis-

rupt the self-purifying ecological cycle. If the wastes of the groundwood pulping method, and ultimately the paper product itself, were returned to the soil,
the process would have almost no harmful effect on the environment.

However, for high-quality paper products, it is necessary (given present
technology) to remove certain noncellulose constituents of the wood pulp by
chemical means. This greatly intensifies the environmental impact, for the
reagents represent an <u>artificial</u> stress on the ecosystem into which they intrude when released into the air or water as wastes. In contrast, organic
waste from the groundwood process is natural and therefore can be readily
accommodated if returned to the <u>appropriate</u> ecosystem (i.e., the soil rather
than water). However <u>no</u> ecosystem can accommodate sulfite and other chemical pulping wastes; since these are not part of the natural ecological cycle,
they are not readily tolerated by the organisms which are, and the resultant
toxic effects become serious ecological threats.

Again, <u>in principle</u>, there is no reason why even the chemical processes
need to stress the environment—all that is required is to accomplish the <u>total</u>
recovery of the added chemicals, so that none are released into the environment. As clearly revealed by this study, in practice the environmental impact of a chemical pulping plant simply reflects the effort made to recover
the added chemicals (and the organic wastes from the wood itself) from the
mill wastes. And what determines this effort is the willingness of the opera-
tor to meet the costs of chemical recovery. There are, in effect, no natural
or technological carriers to essentially complete control of pollution from
such pulping plants.

This conclusion is, of course, basic to the evaluation offered in the CEP
study, for in each case it is assumed that the pollutants <u>can</u> be controlled by
a given level of effort, and that the difference between the actual effort and
the required one can therefore be computed. The validity of this assumption
and the accuracy of the computations of needed control effort proposed in this
volume has—quite fortuitously—been tested very recently. It will be noted
from Chapter 29 that, according to the CEP analyses, the Weyerhaeuser Corporation has been operating a number of plants which are effectively controlled
with respect to environmental pollution, and a few which are essentially uncontrolled. One of the latter is the Weyerhaeuser sulfite mill in Everett,
Washington. According to the CEP study (see Table 29.9) an expenditure of
about $11,500,000 would be needed to accomplish the necessary control of
pollutants from this plant (now largely discharged untreated, with only some
dilution, into the waters of Puget Sound).

The Weyerhaeuser Company has itself provided the test of the CEP estimate of what it would take to correct the plant's environmental transgressions.
The company has recently announced that, in view of its conclusion that about
$10,000,000 would be needed to install the control equipment required to meet
federal pollution standards, it has decided to close down the Everett plant.

Here, then, we find the pulp mill linked to the second complex system of
which it is a part—the economic system. Clearly the Weyerhaeuser decision
to close down the Everett plant, rather than to meet the cost of installing pol-

lution equipment (or for that matter the much larger cost of constructing a wholly new plant at Everett) must have been based on economic considerations. These must have included the plant's competitive position relative to other pulp operations in the Everett area, changing labor costs, prices and demand for the product, and so forth. As we trace the origins of these economic factors, external to the operation of the mill itself, that influence the capability of the company to meet the costs of needed pollution control, matters become increasingly complex. For example, in part, the company's economic position must depend on the economic effect of competition from paper substitutes, such as plastics. In turn this requires that we consider the trade-off between more or less comparable paper and plastic products with respect to their environmental impacts. Insofar as its production and use pollutes the environment, either type of product involves a cost which is met, not by the producer or the user, but by society generally—or, more particularly, by that segment most affected by the pollutant. To make a socially rational choice between the alternative products, each has to be given a kind of "pollution price tag." At this point the implications become very broad indeed. For there is a fundamental ecological difference between otherwise comparable paper and pulp products: cellulose, being a natural product, is in principle capable of being produced and used without ecological stress, so long as the technological process is a loop in the closed ecological cycle of which, in nature, cellulose is a part. In contrast, no synthetic plastic is part of a natural cycle. It is produced at the expense of a nonrenewable resource, such as petroleum. The necessary chemical reactions, unlike those which enable the tree to synthesize cellulose, occur at high temperatures, producing combustion products (such as nitrogen oxides) that are unnatural and therefore pollute. Finally, unlike a natural substance such as cellulose, the synthetic plastic is incapable of being degraded biologically and either accumulates in the environment or is burned—often leading to air pollution. On these grounds, any plastic substitute for paper is likely to involve a more serious environmental impact.

These are some of the specific economic interactions which may determine the degree to which a paper product or a synthetic substitute may stress the environment. Beyond these factors is the overall pattern of the economic system in which basic operational decisions—for example that made by the Weyerhaeuser Company with respect to the Everett plant—are governed by the rate of return or profit, a factor which is itself determined by fundamental economic parameters, such as interest rate, or the return expected from alternative investments. In this connection, a recent study based on Paper Profits points out that those companies which have made the stronger efforts to control pollution have yielded higher profits. (See Appendix 2.) It suggests that the more farsighted managements, in the habit of anticipating a wide range of financial and production problems, have designed plants over the years that were both efficient and clean and are in a better position today to meet mounting demands for rapid curtailment of pollution. However, even a company with a generally good record like Weyerhaeuser is not problem free, and the story at its Everett plant suggests the complexity of this matter. Con-

sider the decision to close down this plant on the company's record in pollution control and profit-making. With the heavily polluting Everett plant closed,
the company can clearly improve its ratio of product to pollutants; it can claim
to have enhanced its already "good" (according to CEP) pollution-control record. The same decision will presumably also enhance the company's economic
position if we credit it with operating on the basis of valid cost accounting.
Yet, while this action seems to have had a parallel, beneficial effect on both
the environment and the company's finances it has involved a cost that has
been borne elsewhere in society: The closing of the Everett plant wiped out
hundreds of jobs.

Suppose, then, that for the sake of maintaining employment, the company
were required to improve its environmental operation by investing the $10-11
million needed to bring the Everett plant up to standard. In this case, the social needs for a livable environment and employment would be served; but on
the grounds given above, the company would presumably suffer financially.
Thus, if—as is customary in the economics of the private enterprise system—
the company endeavors to keep its financial position intact and maintain its
rate of profit, then the social cost represented by environmental pollution is
simply exchanged for the social cost represented by unemployment.

It should be pointed out as well that the apparent parallelism between
good environmental practice and an industry's economic condition may collapse when the effect of synthetic substitutes is taken into account. For there
is considerable evidence that the new synthetic industries, which are more
polluting than the natural ones that they tend to displace, yield the higher
rates of profit. (See Barry Commoner, The Closing Circle, Knopf, New York,
1971, Chapter 12.)

Clearly we need to know a great deal more about all of these complex and
interconnected factors, if we hope to achieve a rational use of our resources—
and thereby preserve the environment from the catastrophic outcome of the
present assaults upon it. This book is a valuable contribution to this effort.
Apart from its explicit, substantative contribution, the book makes a particularly valuable, albeit implicit, point: that despite the vast resources available to United States officialdom, and the serious need for the type of analysis offered by this study, it was made not by a government agency, but by a
few determined individuals. There is a lesson here. If we would save the environment which is our habitat, we shall have to do it ourselves.

Barry Commoner
January 1972

In boardrooms, executive offices, sales departments, and at plant sites, thousands of decisions are made daily by U.S. corporations. Many of these decisions vitally affect not only corporate growth and business but the quality of society as a whole. Decisions concerning environmental policies and their implementation may determine how clean our air and water remain; those concerning employment may affect the degree to which opportunities for jobs and advancement are open to minority group members and women; decisions regarding the types of research, products and services to be supplied to the Department of Defense may influence the scope of our military commitments; those regarding overseas investments and how foreign resources and personnel are utilized may alter national as well as corporate images abroad.

The Council on Economic Priorities was founded two years ago to research the practices resulting from such decisions, to develop criteria for evaluating socially responsible performance, and to define in detail which corporations have the most responsible records. With such information available, concerned Americans could evaluate the impact of industries and individual companies on their own lives and provide pressure or encouragement — in their roles as consumers, stockholders, business men and women, or students — for a greater degree of social responsibility.

Council research is published primarily in bi-monthly Economic Priorities Reports: some have dealt with comparing corporate practices (GM, Ford, Chrysler in South Africa); some with a broader analysis of socially relevant areas (Corporate Advertising and the Environment); and some with a review of major campaigns for corporate responsibility (Corporate Challengers: 1971). Research is also available in longer periodic in-depth studies. Council publications are now read regularly by over 3,500 subscribers (see the last page of this book for subscription information).

Paper Profits is the Council's first in-depth study in the environmental area. It required eight months of research on the practices of the 24 largest pulp and paper makers, the pollution problems they face, and the pollution control steps they have taken. From the research, distinct and strikingly contrasting portraits of each company emerged. Pollution control performance ranged from that of Owens-Illinois with four mills carefully equipped to minimize discharge of soot and gas into the air and solids and chemicals into the water, to that of St. Regis, with nine large and often poorly controlled mills; corporate attitudes on disclosure varied from that of the giant International Paper Company, which refused to share even the most basic information on its operations and environmental efforts, to that of Georgia-Pacific, whose technical director took many hours to explain each control system and how it performed at each location.

The portraits reflect many years of cumulative effort — or lack of effort — to control pollution as well as the underlying corporate philosophies about the public's right to information. However, the portraits also reflect the paper industry at a given point in time. Since December, 1970, changes within each company have gradually occurred; new equipment has been installed, operating procedures have been revised, new mills have opened and small old (and

often more polluting) ones have shut down. To trace the developments that
have taken place, each company in the original study was re-interviewed in
December, 1971. The follow-up study will be found in Appendix 3.

Paper Profits was prepared when the Council was still quite a new organization; it has changed and grown considerably since then. A year ago Council
members numbered barely a dozen while now there is a full-time staff of 23,
and a Board chaired by Dr. Robert Heilbroner.

Though the staff has greatly expanded, no council study could be done by
staff alone. Our work relies heavily on the voluntary input and involvement of
many technical experts, public interest groups, and legal authorities. For
this book we owe special thanks to many: to the executives of the American
Paper Institute and of the 24 companies under study, to representatives of the
fifteen state pollution control agencies, and to numerous environmentalists
across the country, especially to Helen Bird of the Southeast Environmental
Council in Jacksonville, Florida. Without their help, the material in Paper
Profits could not have been compiled; without their explanations and sugges-
tions, it could not have been adequately evaluated. We are also deeply in-
debted to Timothy Collins of Collins Securities Corporation and to John
Westergaard of Equity Research Associates, Inc. Before we had an office of
our own in New York, they shared their offices and facilities with us. During
the many months of work on this project, their staffs showed an indescribable
amount of patience and the most encouraging interest in our research; they
offered their assistance whenever possible.

Alice Tepper Marlin and Joanna Underwood
co-directors

BOARD OF DIRECTORS
Robert Heilbroner, Chairman
Thomas R. Asher
Peter Bernstein
Timothy Collins
Hazel Henderson
Edgar Lockwood
Alice Tepper Marlin
Leon S. Reed
William Samuels, Jr.
Stanley K. Scheinbaum
Ed Swan, Jr.
Joanna Underwood

Part I

PAPERMAKING AND POLLUTION

Chapter 1

THE POLLUTANTS OF PULPING

"A skunk and her kittens were travelling west
When dawn drew nigh and she stopped to rest
She found a place close by the town
Where she fixed a bed and they all lay down.

As the paper mill odor filled the air
Like rotten fish and burning hair
She awoke and crawled right out of bed
(You can't appreciate this unless you've
smelled the damned thing).

She turned to the kittens all seated there
Saying 'Bow your heads, it's time for prayer
Lord, you gave us a weapon for self-defense
It was an awful odor, both potent and dense.

'I'm a sow pole-cat of ripe old age
A cat that nobody keeps in a cage
But I can always recognize defeat
Compared with it, my smell is sweet.

'The extent of the difference I realize
But I think it's partly a matter of size
So, Lord, I ask, if it be thy will
Make me as big as the kraft pulp mill.' "

Narcissus of Tuscaloosa

Many would argue that the sentiments expressed in this poem are simply an emotional response on the part of one mill town resident, and not of much significance in approaching so important a question as pollution in the pulp and paper industry. The odor produced by pulp mills is, in fact, only one aspect of the problem, and its importance in any terms but human comfort is debatable. Other sources of pollution stemming from the pulping process are no longer the subject of debate.

In May, 1968, one of three underground wells used by the Hammermill Paper Company to store "spent" pulping liquor burst from accumulated pressure and spewed a geyser of "brown, putrid broth" twenty feet in the air. During the three weeks it took the company to stem the flow, two million gallons of raw chemical and organic waste streamed into nearby Lake Erie. [1]

At the mouth of the Mobile River, all shellfish beds have been closed because the bacteria level is often as much as two thousand times higher than
the Public Health Department standard. Over forty-seven percent of the coliform pollution load is from one International Paper mill, while thirty-nine
percent comes from the untreated wastes of the residents of downtown Mobile.
[2]

A professional shrimp fisherman from Jacksonville, Florida, Linton
Cook, described his first-hand experience with pulping wastes (primarily
from the Rayonier and St. Regis mills): "That damned mill yonder is just
putting the fishermen out of business. Not only shrimp fishermen but fellows
that's crabbing, the oyster beds, and the whole damned business. I've caught
fish and you couldn't eat them for the turpentine in them." [3]

On November 28th and 30th, 1970, outside of Canton, North Carolina,
massive, multi-car collisions (involving at least thirty-four vehicles) caused
by the smog from the U.S. Plywood-Champion Paper mill, led to the injury
of at least twenty-five people. ... The smog is so continuously dense near
this mill that highway signs read, "Caution: Dense Fog Likely Next Four
Miles." [4]

These are not isolated examples, but rather part of an all too general
pattern of neglected care for the environment on the part of the pulp and paper
industry. Statistics from as recently as 1964 show that the industry was responsible for 23.5 percent of the industrial effluent discharge in the Northeast, 26.4 percent in the Southwest, and 16.7 percent in the Pacific Northwest. [5] The pollution streaming from the stacks and sewers is as much a
product of the industry as are newsprint, brown paper bags and the finest
writing papers.

An obvious reason for the environmental impact of the pulp and paper industry is sheer size. "Pulp and paper" ranks as the third largest nondurable
industry in the United States; in 1969 it had sales of $ 20.5 billion and produced
53.8 million tons of paper, or 576 pounds for every man, woman and child in
the country. [6] In compiling this impressive record, the industry "borrowed"
over two trillion gallons of water and consumed sixty million cords of pulp-
wood. [7]

Production and consumption are not the only important measures of size.
The industry's real estate holdings are at least as imposing. For example:

International Paper, the largest company in the industry, ranks second only
to the Federal Government in total acreage across the country.

Paper companies control about 52 percent of the land area of the State of
Maine.

Boise Cascade (in a joint corporate venture) is the largest land owner in the
State of Louisiana.

The estimated total land holdings of all pulp and paper companies amounts to fifty million acres, an area equal to all of New England and New Jersey.

With the economic power represented by the assets and sales of the pulp and paper companies must go responsibility, for it is estimated that in 1969 the industry discharged 15 percent of the total industrial effluent of the United States. [8] The pollution contained in that effluent affects everyone. It is discharged into the streams, rivers, lakes, and oceans which are our only source of water. It is carried in the air all over our atmosphere.

Air Pollution

Until recently, residents of many mill towns rationalized the nearly unbearable atmosphere around them as something that had to be tolerated in the name of economic prosperity. Reactions such as, "we need the jobs," or, "that smell is the smell of bacon and eggs" were typical. Lately, however, a countertheme of complaints has begun to be heard: "the town stinks constantly of sulfur from that damn mill"; "you can't get away from the smell, the haze, the feeling that your lungs are swallowing in some vaporous disaster." [9]

The amount of air pollutants emitted from a pulp mill depends upon what method of pulping is being used and what fuel is burned to produce steam for generating electrical power and heating various stages of the production process. Almost all mills burn some sort of fuel (coal, oil, natural gas) or waste (bark, unusable wood scraps) and thus emit some amount of carbon ash, soot, and dust. All mills using chemical pulping methods (kraft, sulfite, or NSSC, to be described in Chapter 2) cook wood chips in a solution containing sulfur. This process causes these mills to emit sulfurous gases, mists, and water vapor. Many mills burn the "spent" cooking solutions after the pulp has been extracted to recover chemicals for reuse and to generate further heat for the production process. These mills commonly emit still greater amounts of chemical mists and gases, as well as salts and dust particles.

The sheer quantity of air pollutants is staggering. The 120 United States mills that use the kraft process annually emit 630,000 tons of particulates, 2.6 million tons of carbon monoxide gas, and 84,000 tons of gaseous sulfur oxides. [10]

Particulate emissions, i.e., fine particles of dust, chemicals, and carbons, are a serious nuisance. In high concentrations they can aggravate respiratory and lung diseases such as emphysema and asthma. They can cause such a visibility problem that cars must drive with their headlights on in broad daylight because "a heavy blanket of foul smelling smoke from a kraft plant effectively blots out the sun." [11] This specific example relates to the beautiful Missoula Valley in Montana. The Hoerner-Waldorf mill there emits 11,000 pounds of black soot and ash each day from the boiler stacks alone. [12] Particulates and corrosive gases formed and emitted from the chemical pulping and recovery processes are so damaging to car paint that many kraft mills, including those of International Paper in Panama City, Florida, and

Weyerhaeuser in Everett, Washington, provide a car rinsing service for their
employees to drive through each day after work.

Particulate emissions have a curious side effect on nature. They provide
a nucleus around which water particles may condense, thus increasing rain-
fall. A study in the state of Washington has shown that "annual precipitation
has increased since 1946 more than 30 percent in some cases in regions adjacent to or downwind of some of the larger pulp and paper mills in the state."
[13]

The most annoying pollutants in terms of human comfort are <u>hydrogen
sulfide</u> and the <u>methyl mercaptans</u>, produced by chemical reactions in the
kraft process. These gases are characterized by an "ugly, pervading, penetrating, sickening stench," which can spread for thirty miles in a strong wind.
According to a study on noxious odors, done by the Copley Corp., this kraft
mill odor is the third most serious, affects the second largest area per
source, and is the longest lasting of all our modern odors. [14]

In addition to its noxious qualities, hydrogen sulfide is highly toxic in
large quantities. The safety coordinator of the International Paper Company
in Mobile, Alabama, described the effects on employees who came into close
contact with high concentrations of the gas:

"Unconsciousness comes immediately, breathing either stops or is so light
that artificial respiration has to be administered immediately... [In the case
of several workers during incidents in 1966 and 1968] the unconsciousness
lasted from one week to 10 days ... other effects such as hallucination, loss
of memory, etc. were present, and the patients were unable to concentrate
normally ... a lack of coordination was noticed ... The symptoms in both
brain hemorrhage and hydrogen sulfide contact are very much the same." [15]

Another sulfur gas formed in chemical pulping is <u>sulfur dioxide</u>, a highly
irritating toxic compound. Edmund Faltermayer, in an article in <u>Fortune</u>,
wrote:

"Sulfur dioxide is the most worrisome of the major pollutants ... It has been
implicated in most of the famous air pollution disasters, such as the 1948 one
at Donora, Pennsylvania (twenty dead) and the 400 "excess deaths" recorded
during a fifteen-day smog episode in New York City early in 1963. While not
toxic to man in the concentrations ordinarily found in the atmosphere, it can
cause acute crop damage in relatively small concentrations. In industrial regions it causes nickel to corrode twenty-five times as fast as in rural air,
and copper five times as fast. And under certain conditions it kills people.
One of its derivitives, sulfuric acid mist, can get past the body's natural filtration system and penetrate deep into the lungs, causing severe damage." [16]

Plant life can also be damaged by sulfur dioxide pulp mill emissions. In
a pretrial report on the Hoerner-Waldorf mill in Missoula, Montana, it was
shown that tissues of plants evidenced diseases caused by these gases as far
as 15 miles away from the mill site. [17] In a more extreme case, the moun-

tainside rising next to a no longer operative Westvaco mill in Luke, Maryland, were virtually stripped of all foliage decades ago; they have never fully recovered.

In Jacksonville, Florida, which has six paper and pulp mills within a fifty mile radius, the acid in the air is so damaging that a warranty certificate for a car paint job in the area reads:

"If your car is constantly exposed to extreme deterioration such as that caused by acid from a paper mill, or sitting directly on the beach, then our warranty must not be considered to be in effect since even a new car will rust under such conditions." [18]

The owner of this warranty amended it with the declaration: "Corrosion is four times higher in downtown Jax than on Neptune Beach!"

Obviously the effects of these pollutants are multiplied when they are prevented from rapidly dispersing through the atmosphere. A temperature inversion can often trap polluted air in a local area. In an inversion, warm air sits on the cooler surface air below, preventing it from rising. It acts as a lid, holding down the air and any pollutants it contains. Contrary to popular belief, inversions are quite frequent occurrences. In some valley areas, such as Lewiston, Idaho, the condition occurs as much as 80 percent of the time, [19] while even in nonvalley systems, the air may be inverted 40 to 50 percent of the time. In cities such as New York and Los Angeles, the serious consequences of long periods of inversion are becoming increasingly alarming. As Edmund Faltermayer so aptly put it: "The air above us is not a boundless ocean. Much of the time it is a shallow stagnant pond, and we are the fish at the bottom." [20]

Water Pollution

All types of pulp mills use vast amounts of water. "As a material for use in the manufacture of paper and pulp, water is second in importance only to the fiber itself." In 1967, the paper and allied products industry used nearly 2.1 trillion gallons and discharged 1.9 trillion gallons. Only 34 percent of this discharged water received any treatment to alleviate its chemical and physical adulteration. [21]

Depending on the process, pulp mills use a maximum of 34,650 gallons of water per ton of kraft pulp and 62,700 gallons per ton of sulfite pulp. [22] Logs may be debarked by high pressure hydraulic jets; the pumps and machines themselves are cooled by a constant flow of thousands of gallons of water. Water is also used in cooking, washing, and as a method of transporting fibers throughout the processing. Between each of the phases of the pulping process, the fibers are diluted with up to 99 percent water and then reconstituted.

The St. Regis mill in Jacksonville, Florida, pumps over 18 million gallons of fresh water each day from an underground well, and uses an additional 60 million gallons from the Broward River to cool the pumps and machines. This 80 million gallons is discharged, untreated, into the St. John's River.

In Everett, Washington, one of the largest pulp producing cities on the West Coast, four pulp and paper mills and 60,000 residents use more water than the entire city of Seattle with 600,000 residents.

Although the industry constantly reiterates that most of its water is not "consumed," but only "used" and then returned to the waterways, a great amount and variety of waste matter is discharged in these trillions of gallons. There are heavy solids like fibers, bark, uncooked wood chips, and dirt; there are dissolved solids like carbohydrates and soluble wood matter; and there are the cooking and bleaching chemicals. Although they are not pathogenic, these pulping pollutants have disastrous effects on our waters. The pulp and paper industry has long admitted that water pollution is a major problem, although, according to J. Alfred Hall, "the tradition of using the river both as water supply and sewer has persisted." [23]

Solid wastes sink to the bottom of slowly moving streams and lakes, where they form beds of sludge that are not only unpleasant but also ruin the quality of the water and aquatic life. They destroy the natural bottom habitat of fish, they blanket existing clam and oyster beds, and as they decay, they use up the oxygen in the water and release toxic gases. A 1963 report on the Kennebec River in Maine indicated that one calcium sulfite and tissue mill complex discharging 400 tons of these solid wastes each day created a potential deposit of 35,000 cubic feet per day on the bottom of the river. [24] For the last 100 years, paper mills on Lake Champlain (now primarily International Paper at Ticonderoga) have dumped so much of this waste into the lake that "300 acres of lake bottom are coated with guck in depths of up to 12 feet." [25]

Many of these wastes are in mill effluents because of spills, dumps, and improper maintenance of screening equipment. Eighty percent of them could be filtered out by primary treatment facilities such as sedimentation tanks, screens, or basins, and could be disposed of by burning or in land fill areas.

Carbohydrates dissolved from wood in the cooking process and drained from the pulp are even more polluting than solid wastes. These materials (sugars and sugar derivatives) are decomposed by the microorganisms in the water as long as there is oxygen present. However, if the concentrations of waste are too high, bacteria overfeed, overpopulate, and consume the available oxygen in the stream, often diminishing it to a level insufficient to support aquatic life. This biological decomposition process is further hampered by the chemicals dumped, leaked, or spilled from mills. Often waste chemicals react directly with the oxygen in the water to form new chemical compounds. This reduces the amount of oxygen available for bacterial breakdown of organic mill wastes. The quantity of oxygen consumed in reactions not mediated by bacteria or other organisms is called the chemical oxygen demand (COD).

The amount of waste material dumped from a pulp mill is measured in terms of biochemical oxygen demand (BOD). The BOD is the weight of oxygen needed by bacteria to degrade and stabilize a discharge of waste materials. The average waste load of a mill varies according to the process. From groundwood pulp, it is 30-60 pounds per ton of pulp; from unbleached kraft,

25-50 lb/ton; from unbleached sulfite, 550-750 lb/ton. Bleaching raises the
BOD poundage by 25-250 lb/ton. [26]

The BOD is the primary measurement of water pollution. If it is too high,
it adversely affects the complicated life chain in natural waters from bacteria
to plankton to plants to fish life. The importance of this measurement can be
seen by comparing human and industrial discharges. While an estimated 0.17
pounds of oxygen is required by bacteria to decompose a daily human discharge of waste, nearly 1,000 pounds of oxygen are required for the decomposition of untreated wastes from one ton of bleached sulfite pulp. Thus an
average sized sulfite mill that produces 200 tons of pulp each day, can create
a BOD load which is equivalent to that of raw wastes from one million people.
Describing the BOD of the waste output of specific mills in terms of population equivalents makes the measure more graphic. This equivalency however
refers only to oxygen demand since industrial wastes do not contain the pathogenic bacteria, nutrients etc. present in domestic wastes. For example, the
combined output of seven mills on the Columbia River impose as much BOD
as would 4.5 million people.[27] In 1966, the mills on the Fox River in Wisconsin dumped wastes measuring 438,000 pounds of BOD daily, the equivalent
of the wastes of 2.6 million people, or 8 1/2 times as much as the sewage of
the actual population discharge.[28]

Often pulp and paper mills are the main contributor of BOD to a waterway.
According to the Wisconsin Resource Conservation Council, pulp and paper
manufacture accounts for 82 percent of the total BOD in the state's waters; [29]
in the Willamette Basin in Oregon, 80 percent of the total population equivalent
in terms of BOD is contributed by mill wastes. [30] In 1964, the industry discharged 5.9 billion pounds of BOD into the nation's waters. [31]

Two other water pollutants, foam and discoloring matter, are more annoying than damaging. The foam results from small amounts of resin, fatty
acids, and chemicals in the effluent. It bubbles up from mill sewer outfalls,
spreading brownish suds over the water. In the South, where highly resinous
woods are common, foam in rivers can build up to a thickness of several feet.
The dark color in mill effluent can be both distressing and harmful, for as it
spreads over the surface of rivers and lakes, it can block out sunlight, thereby reducing the rate of photosynthesis and upsetting the ecological balance.

Until recently, mercury was another pollutant from pulp and paper manufacture. It was used as a fungicide to preserve pulp or chips in warm weather,
as a slimicide, and as a catalyst in the manufacture of bleaching chemicals.
According to a Bureau of Mines estimate, the paper industry used 46,000
pounds of mercury in 1968. [32] However, since the "mercury scare" in mid-
1970, the industry has almost totally discontinued using and discharging mercury.

Industry Efforts

The president of the Technical Association of the Pulp and Paper Industry
(TAPPI), has called upon the industry to try to stop "hiding behind the door

as polluters" and, instead, "proudly open the door as depolluters." [33] This is a bit premature, inasmuch as there are vast quantities of untreated wastes discharged by pulp and paper manufacturers. Still, an effort is being made. In the past 25 years, according to the American Paper Institute, over $800 million has been spent to reduce air and water pollution. [34] Of this, over $180 million was spent in 1969 alone on pollution control capital expenditures, operating costs, and research and development. The National Industrial Conference Board has determined that, of the 17 major industries studied, the paper and pulp and allied products industry spent the largest percentage of its capital expenditures, 6.72 percent, for water pollution abatement facilities in 1968. [35]

A recently presented position paper, "The Paper Industry's Part in Protecting the Environment," claims that these expenditures have gone far toward reducing the industry's pollution. "Today, 80 percent of the mills, which provide 90 percent of total national paper and paperboard production, have waste treatment facilities in operation . . . " [36] If all of these facilities provided adequate primary treatment, removing, say, 90 percent of the suspended organic matter, this achievement would indeed be laudable. However, included in the 80 percent is such equipment as screens, savealls and diffusers on outfall pipes, i.e., the barest minimal treatment still authorized by some of the less enlightened state authorities.

The industry's efforts to reduce the BOD have been more productive. In the past twenty years, the industry-wide average BOD discharge has been reduced from 140 pounds to 68 pounds per ton of paper produced. [37] Thus, in a period during which total paper production has risen 120 percent, the total BOD discharged into the water has risen only 7 percent.

Of course, there is no way to determine how much of this effort was a direct result of public pressure and governmental pressure (as at Covington, Virginia, and at Missoula, Montana), or how much was simply the expression of the industry's realization of its responsibility toward the nation's resources and their preservation.

One industry achievement that gives grounds for hope, at least, is in the management and multiple use of forest lands for growing and harvesting trees, protecting watersheds, providing recreational areas, and preserving the habitats of animals and birds. Until 50 or 75 years ago, the life span of a forest depended solely on the speed of the sawmill. Around 1800, "a frantic wood rush began that would strip most of our forests." [38] The lumbermen leveled them in a hundred years, moving from New England to New York to Pennsylvania, then to the lake States, the South, and the West. By the 1930s, there was little left in the entire eastern half of the United States but hardwoods, which were not then suitable for pulping.

Only then, when the raw material was exhausted by carelessness and misuse, did the lumber and paper industries resort to conservation practices, making an all-out effort to recreate ecologically sound, well-managed forests. In the past 20 years, over one billion dollars has been spent to repair the damage, reforest the stripped lands, and reestablish the natural balance. [39] Since 1960, over 3 billion new trees have been planted. [40] Individual com-

panies have established over 50 million acres of forest preserves, 90 percent
of them open to the public, where they plant, nurture, and harvest selectively
bred trees. Constant research is done on tree genetics and chemistry, ecology,
fertilizers, planting and harvesting techniques, and other forestry practices.

This achievement illustrates a sad truth: conservation and ecology become prime considerations only when an industry is faced with collapse because of its past practices. Perhaps there will be a similar all-out effort to
reclaim and preserve our waters when they become so clogged with sewage
and industrial wastes that this too threatens industry's ability to produce.

A second positive achievement by the paper industry is the increased use
of lumber wastes in pulp mills. Sawdust, chips, and shavings which were once
burned by lumber mills are now used in many pulp mills. In fact, a significant
portion of the increased paper production since World War II has been made
by using these residues, which now account for nearly 25 percent of all the
fibers used in paper. [41] The Hoerner-Waldorf Corporation, which operates
3 pulp mills in the Midwest, uses only small reject logs, lumber wastes, and
recycled paper as raw materials. In Oregon and Washington, as of 1968,
around 8 million tons of this lumber waste were used for pulp, a 200 percent
increase from the 4 million tons used in 1962. [42]

Unfortunately, in another tree-saving area, the paper industry is lagging
far behind its own potential and the actual achievements made by other nations.
The American Paper Institute calculates that each ton of recycled paper saves
17 trees, yet in the United States only 20 percent of all paper is recycled. [43]
In 1969 this amounted to about 11.4 million tons, representing a saving of
nearly 200 million trees. [44] One-tenth of this was collected by one company,
Container Corporation of America.)

A 20 percent recycling is not very high. The National Academy of Sciences
has recommended that the industry attain a 35 percent level by 1985. Japan
now recycles 38 percent of its paper, Germany recycles 31 percent [45], and
the United States recycled 40 percent of its total paper production during
World War II. But, it has been explained, in these instances, "recycling has
an economic necessity." [46]

While the blame for the enormous waste paper problem in the United
States must be shared by the entire population,* the paper industry could do
more to help solve it by making a greater effort to recycle paper. Since 1930,
annual paper consumption has risen from 75 pounds per person to today's

----

*Consumers, for example, continue to pay for several layers of wrapping
around each packaged commodity and stimulate the increased use of "disposable" linens, plates, cups, diapers, and even clothing. They do not react,
for example, to the fact that every 1/3 ounce of "Pillsbury Space Food Stick"
is wrapped in another 1/3 ounce of cardboard and tinfoil. In a forthcoming
study, Ralph Nader's Task Force reveals that the United States is the only
nation in the world which annually consumes over 500 pounds of paper per
person, and that the U.S. population, less than 6 percent of the world's total,
consumes almost 45 percent of the world's paper. [48]

level of 576 pounds per person [47]; there has certainly not been a comparable effort to increase the rate of reuse.

References for Chapter 1

1. "Waste Disposal Wells, Once Considered Safe, Now Seen As Polluters," by Richard D. James, Wall Street Journal, May 21, 1970.

2. Pollution Affecting Shellfish Harvesting in Mobile Bay, Alabama, Washington, D.C.: U.S. Government Printing Office, 1970, p. 3.

3. WJXT News Special Report, by Charles Thompson, aired on April 13, 1970. Reprint sent to CEP by the Southeastern Environmental Council, Inc., P.O. Box 31278, Yukon Branch, Jacksonville, Florida, 32230.

4. The Waynesville Mountaineer, Waynesville, North Carolina, November 30, 1970, p. 1.

5. Cost of Clean Water and Its Economic Impact, Vol. II, U.S. Government Printing Office, January 10, 1969, pp. 63-65.

6. How the American Paper Industry Performed in 1969, prepared by the American Paper Institute, p. 2.

7. The Paper Industry's Part in Protecting the Environment, American Paper Institute, November, 1970, p. 29.

8. How the American Paper Industry Performed in 1969, p. 11.

9. "Pollution and Power in a Small Mill Town," by Sheldon Coffey III, Potomac, January 26, 1969.

10. National Emissions Standards Study, Report of the Secretary of Health, Education, and Welfare to the United States Congress, March, 1970, p. 82.

11. "The Effluent of the Affluent," by Creighton Peet, American Forests Magazine, May, 1969, p. 19.

12. "Hoerner Waldorf's Clean Air Proposals," The Sunday Missoulian, November 9, 1969, p. 4.

13. Pulp and Paper, August, 1970, p. 63.

14. Study prepared by the Copley Corporation, La Jolla, California, for the National Air Pollution Control Administration, February, 1970, p. 14.

15. "H$_2$S: not enough is known about it," by Harris K. Williams, Pulp and Paper, May, 1969, p. 140.

16. "A Fortune Proposition: We Can Afford Clean Air," by Edmund K. Faltermeyer, Fortune, November, 1965, p. 5.

17. "Hoerner Waldorf Pre-Trial Report to Garlington," September, 1969, by C. C. Gordon, p. 9.

18. Letter to the CEP.

19. Lewiston, Idaho - Clarkston, Washington Air Pollution Abatement Activity, U.S. Department of Health, Education, and Welfare, Public Health Service, February, 1967, p. 34.

20. A Fortune Proposition: We Can Afford Clean Air, p. 4.

21. The Pulp and Paper Industry and the Northwest, by J. Alfred Hall, Pacific Northwest Forest and Range Experimentation Station, Forest Service, U.S. Department of Agriculture, 1969, p. 34.

22. Ibid. p. 34.

23. Ibid. p. 33.

24. "The 70s Beckon: Will We Enter Them with a Half-Hearted Anti-Pollution Plan?" by Sheldon Brooks, Pulp and Paper, January 1970, p. 97.

25. "Pollution by Pulp Mills Stirs More Localities to Press for Curbs," by Stanford Sesser, Wall Street Journal, June 25, 1970, p. 1.

26. The Cost of Clean Water and Its Economic Impact, Federal Water Pollution Control Administration, 1968, Vol. III.

27. The Pulp and Paper Industry and the Northwest, p. 33.

28. Letter to the CEP from the Wisconsin Ecological Society, June 4, 1970, quoting the 1966 report from the Federal Water Pollution Control Administration to the Lake Michigan Pollution Conference.

29. Letter to the CEP from the Wisconsin Resource Conservation Council, May 3, 1970.

30. The Pulp and Paper Industry and the Northwest, p. 36.

31. Cost of Clean Water and Its Economic Impact, p. 63.

32. "A New Pollution Problem," Sheldon Novick, Environment Magazine, May 4, 1969, p. 5.

33. Paper Age, May, 1970, p. 1.

34. "How the American Paper Industry Performed in 1969," p. 9.

35. "Clean Water Report," March, 1970, p. 25.

36. The Paper Industry's Part in Protecting the Environment, p. 13.

37. Ibid. p. 14.

38. The Quiet Crisis, by Stuart Udall, Holt, Kinehart & Winston, 1963.

39. The Paper Industry's Part in Protecting the Environment, p. 32.

40. Ibid. p. 30.

41. Ibid. p. 33.

42. The Pulp and Paper Industry and the Northwest.

43. Testimony of Edwin A. Locke, Jr., President, API, on the Resource Recovery Act of 1969 before the Senate Subcommittee on Air and Water Pollution, February 26, 1970.

44. Ibid.

45. "Recycling Waste Paper into Paper and Paperboard Packaging," remarks by Henry G. Van der Erb, President, Container Corporation of America, at the Waste Paper Recycling Seminar of the API, Washington, D.C., October 16, 1970, p. 4.

46. Ibid.

47. "The Water Lords," draft of a forthcoming study by Ralph Nader's Study Group, to be published by Grossman Publishers, March, 1971.

48. Ibid.

"So _that's_ where it goes! Well, I'd like to thank you fellows for bringing this to my attention." Drawing by Stevenson; © The New Yorker Magazine, Inc.

Chapter 2

THE PULPING PROCESSES

In order to understand how potential pollutants may be identified and removed rather than allowed to enter the environment, it is necessary to know where and how they are formed. Paper, in all forms from newsprint to the finest writing paper, is basically wood which has been reduced to fibers by chemical or mechanical means and rearranged into a sheet. There are four basic pulping methods which achieve this reduction: groundwood, semichemical, sulfite, and kraft.

Groundwood Pulping

The oldest and simplest pulping method, called groundwood, was developed around 1840 in Germany. In this strictly mechanical process, bolts of barked wood or steam-softened wood chips are pressed against large rotating corrugated stones or discs under a flow of water. The abrasion pulverizes and tears the wood apart by physical force until it is reduced to fiber bunches. These are flushed away, pressed, matted, and dried into a sheet of pulp.

Groundwood is the cleanest, cheapest, and least harmful pulping process. The mills are generally small, there is no odor problem, and there are no chemicals to discharge. (Some mills use a mild sulfur liquor to presoak chips, but this liquor is reused and not discharged.) The only materials rinsed into the river are unscreened suspended woodwastes and water soluble wood compounds. The BOD load is normally 30 to 60 pounds per ton of pulp, depending on the effectiveness of the pulping process and the quantity of water soluble material in the wood used. Finally, because none of the wood is dissolved by chemical action, the pulp yield is from 93 percent to 98 percent of the weight of the original materials.

From the standpoint of pollution control, groundwood is the optimal process. However, because mechanical grinding tends to bruise and rupture the fibers, and does not remove noncellulosic materials, the paper produced by this method is brittle, nondurable, and easily discolored by contact with sunlight and air. Despite the fact that it has excellent printing qualities, it is used mainly in combination with other pulps or for papers where strength, quality, and brightness are not required, such as newsprint, corrugated paper, molded packaging materials, and hardboards.

Chemical Pulping

Wood is basically made of cellulose fibers, lignin (a gluelike substance that binds the cellulosic fibers together) and tannins, resins, and oils. While the groundwood process simply pulverizes the wood into fiber-sized pieces, in the chemical pulping methods, acid or alkaline sulfur-containing solutions fill the void spaces in and around the fibers and dissolve the noncellulosic materi-

als. As much as half the tree is dissolved and, literally, washes down the
drain or goes up in smoke.

Chemical pulping is much more complex than mechanical pulping. It produces a wider range of paper types, qualities, and strengths, from tissue
paper to shipping cartons, and it does so from a wider variety of trees. However, it also has a much greater potential for creating water and air pollution.
Before being pulped, the logs must be debarked, chipped to expose the fiber
ends to the chemicals, and often steam softened. This preparation creates
wastes of wood fibers and bark, sand, and grit. Drastic cooking in the acid or
alkaline solutions results in mill effluents that affect the color, temperature,
acidity or alkalinity, chemical content, and foaming tendencies of the rivers
or waterways into which they are discharged. In addition, a large percentage
of the pulp produced by chemical methods is bleached to achieve whiteness
and prevent yellowing. Chlorine, caustic, and other compounds remove residual coloring materials and lignin left in the pulp, greatly increasing the amount
of organic and chemical wastes and creating a serious color problem in the
waters that receive the mill effluent.

Neutral Sulfite Semichemical Pulping

Neutral sulfite semichemical (NSSC) pulping is a combination of mechanical
and chemical methods. Pulp produced by this process is used mainly to make
the wavy corrugated paper that is glued between the two outer layers of cardboard in boxes and cartons. Wood chips are cooked at a high temperature in
a mild solution of sodium sulfite buffered with sodium carbonate to soften and
partially dissolve the lignin bonds. Mechanical grinding then completes the
breakdown to fibers. The pulp yield from this process ranges from 65 percent
to 85 percent of the weight of the wood, depending upon how much of the lignin
is dissolved out. Neutral sulfite semichemical mills face the difficult problem
of finding a way to dispose of their spent cooking liquor without dumping it into the mill effluent. Only a very few companies, e.g., Owens-Illinois, have
successfully evaporated the liquor to recover the chemicals in a recycleable
sodium sulfite form. It is also very difficult to dispose of the spent liquor by
incineration. Twelve years ago, Mead Corporation invested $3 million in a
system to burn the spent NSSC chemicals at its Lynchburg, Virginia, mill. A
new washer to thoroughly rinse the cooking liquor from the pulp, an evaporator to concentrate it, and a furnace to burn it were installed. However, in the
furnace, the combination of tremendous heat, high oxygen concentration, and
moisture caused the strong sulfur compounds to convert into the highly corrosive chemical, sulfuric acid. In a very few months, the Mead furnace was
completely eaten away by the acid and actually fell apart.

Because of the difficulties of chemical recovery or liquor burning, most
NSSC mills today have continued to empty all of their liquor into the waste
water. NSSC mills consequently are one of the greatest sources of water pollution, with a potential BOD discharge ranging from 175 to 400 pounds per ton
of pulp, depending upon the strength of the chemicals and the amount of dissolved wood matter.

There are two alternatives to dumping the effluent. The liquor may be concentrated and burned to recover the chemical sodium sulfate, which, although not usable in an NSSC mill, is the chemical used by kraft mills. Some companies have two mills (one NSSC, one kraft) on the same site, and the NSSC concentrated spent liquor is disposed of at the kraft mill in a process called "cross recovery." Other companies sell the concentrated liquor to nearby kraft mills.

The other option is to switch from sodium sulfite to another pulping liquor such as ammonium sulfite. This conversion causes a very slight downgrading in the quality and strength of the corrugated paper produced. Traces of ammonium compounds remaining in the pulp affect its strength, make it slightly more brittle, and keep it from holding its wavy shape as well. But the ammonia-based liquor can be burned easily, virtually eliminating the main source of water pollution. Mead Corporation, after the corrosion of its first furnace, did convert to ammonia. It invested another $1 million for a new furnace and now burns all its spent liquor. The company has since similarly converted a second NSSC mill to this base.

Acid Sulfite Pulping

The acid sulfite method, which is presently used to produce about 10 percent of the pulp in the United States, was first used commercially in 1882. Production is limited to woods with low resin contents (white firs, spruces, hemlocks) because the acid cooking liquor reacts with natural resins to form insoluble compounds. These compounds prevent the liquor from penetrating and softening the wood. Despite this limitation, sulfite pulping is widely used, particularly in the Pacific Northwest and Wisconsin. It produces soft, flexible, and easily bleached pulp used for fine writing papers, tissue, and other high quality papers where strength is not required. In 1970, sulfite pulp production from 41 mills was about 2.3 million tons.

In acid sulfite pulping, a giant pressure cooker, called a digester, may be filled with up to 40 tons of 3/4-inch wood chips and 50,000 gallons of strong acid cooking liquor. [1] This liquor is a solution of sulfurous acid and one of four sulfur-containing compounds: calcium bisulfite, sodium bisulfite, ammonium bisulfite, or magnesium bisulfite.

The mixture of chips and liquor is cooked at about 125 to 160 degrees centigrade for 6 to 12 hours, during which the liquor penetrates every hollow in the wood chips, dissolving the lignin and other noncellulosic materials. The digester is then partially depressurized and the soggy chips are "blown" into a huge pit twice the size of the digester. The sudden pressure drop and the impact against stainless steel "targets" disintegrates the chips into loose fibers, or pulp.

Most of the air pollution from sulfite pulping occurs in the blow pit. Although much of the sulfurous gas that forms during the cook is released into accumulators, stored, and reused to fortify the cooking liquor, a substantial amount is "blown" with the chips. These waste gases, containing large quantities of toxic, highly irritating sulfur dioxide, may go directly out into the atmosphere through stacks. The sulfur dioxide can combine with water vapor in the air to form sulfurous acid, a chemical that erodes limestone, corrodes metal, and aggravates respiratory ailments.

Once the chips are reduced to pulp, the mixture is washed and drained several times to completely remove the strong "spent" liquor containing the dissolved lignin, wood sugars, and noncellulosic matter. The pulp is then screened to remove knots and uncooked chips, dirt, and bark. The "spent" cooking liquor and screenings are often discharged. If the mill provides no treatment for this effluent, "a greater percentage of the tree leaves the mill by the sewer than is sold." [2] Over 55 percent of the weight of the wood and thousands of gallons of chemical solution can end up in the river, stream, or lake adjoining the mill. Untreated sulfite effluent has a BOD level of from 550 to 750 pounds per ton of pulp, the highest pollution load of any pulping process effluent. It causes discoloration of the water, buildup of sludge deposits, and serious depletion of the oxygen in the water.

This waste sulfite liquor has also been shown to be very harmful to shellfish in the early stages of their life cycle, before they have developed protective mechanisms against environmental conditions. The injuries from pollution can be so serious that the result is "the decline or disappearance of a species." [3] In Puget Sound, "pulping wastes, particularly from the Georgia-Pacific Corporation mill, have caused water quality throughout the largest part of the Bellingham-Anacortes area to be inimical or less than satisfactory for the proper development of Pacific oyster eggs and larvae." [4] A resident of the area described the water as "reddish-brown in color, presumably from the waste sulfite liquor from the pulp mill." [5]

A large fraction of the water pollution from a sulfite mill could be alleviated if the spent cooking liquor were not simply dumped into the nearest body of water. Of the four possible sulfite liquors, only one, calcium bisulfite, cannot be evaporated or burned to recover its chemical and heat value economically. Unfortunately, this base is the oldest and cheapest, and the one most commonly used until recently.

In an attempt to abate pollution, several mills have turned waste calcium bisulfite cooking liquor into saleable by-products. It can be used for improving road surfaces, preparing oil drilling muds, making plastics, tannins, and vanillin. The dissolved wood sugars can be isolated and fermented to ethyl alcohol and sold. Unfortunately, the demand for these by-products is very limited. In 1963 less than 10 percent of the available waste sulfite liquor was used for these purposes. [6]

Other mills have tried to evaporate and burn the waste liquor to recover the heat from the dissolved organic matter in it. However, calcium bisulfite tends to cake and scale on the surfaces of the evaporator and clog up its operation. The calcium salts also coat the walls of the furnace and stacks.

The cost of solving the technical problems of calcium recovery far exceeds the cost of conversion, and in the light of new interest shown by citizens and government toward pollution, more than half of the sulfite mills in the country have converted to a soluble base that has other advantages beyond reduced pollution loads.

The magnesium, sodium, and ammonium bisulfite bases permit a wider variety of woods to be pulped; ammonium based liquor even gives a greater yield per ton of wood. The liquors penetrate the wood chips faster; and some or all of the chemicals can be recovered and reused.

The most practical of the three alternative bases is magnesium bisulfite. Only about six mills have converted to it, at a cost of about $22,000 per ton of daily capacity, or about $11 million for a 500-ton mill.[7] With this base, the liquor can be burned to recover the chemicals in a form ready to be used for a new liquor. The savings in recovered chemicals more than offset the greater cost of magnesium over calcium.[8] More important from an environmental point of view, the BOD load can be reduced from 500 pounds per ton of pulp to 50 pounds per ton. Because the magnesium process relies on recovering chemicals for economic feasibility, air pollution may be a serious problem. The digesters, blow system, preparation of sulfur dioxide, and recovery stacks all emit air pollutants, particularly sulfur dioxide.

A second alternative base, sodium bisulfite, permits liquor treatment and recovery but is not considered economically competitive in today's market. The waste liquor can be burned for heat while chemical ash is recovered as a sodium carbonate and sodium sulfite smelt that can be reprocessed for a new liquor.

The third possible base, ammonium bisulfite, is becoming the most popular of the three. A number of mills converted to this base after World War II because, since it penetrates wood quickly, it allows a 25 percent decrease in the length of the pulping cycle. The higher cost of the chemical is thus offset by a substantial increase in production. With the ammonium base, the liquor can be burned to generate heat and recover sulfur dioxide. The ammonia itself, however, converts to nitrogen and water and cannot be used for a new liquor. Experimentation has suggested ways, such as ion exchange, to convert the nitrogen back into usable form. Still, this is yet to be done commercially.

Alkaline Sulfate or Kraft Pulping

Over 60 percent of the pulp in the United States is produced by the alkaline sulfate, or kraft, method. This is the most economical and versatile process. It can be used to pulp either softwoods or hardwoods. Because of the strength of the paper it produces, kraft pulp is used for most of the paper packaging, heavy cardboard, paper bags, and corrugated boxes where whiteness is not a requirement. Bleaching makes it an important source of pulp for fine papers and food packaging as well.

The yield from the kraft process is slightly less than that from sulfite pulping. But this is made up for by flexibility, adaptability, and the speed of the process. Because of the expense of building and operating the complex chemical recovery system, kraft mills must produce much larger daily tonnages of pulp than sulfite and NSSC mills. Thus, while the average U.S. sulfite mill production is about 200 tons per day, kraft mills typically produce between 700 and 1000 tons per day.

The kraft process was developed in 1884, but it was not until the invention of modern chemical recovery equipment around 1930 and the postwar "packaging revolution" that it gained wide popularity. Since then, kraft production has expanded rapidly, increasing tenfold while sulfite and groundwood production has only doubled.

Because the process depends on recovering heat and chemicals for its economic feasibility, the amount of pulping wastes that are discharged into the water is about 1/20 of that from calcium sulfite pulping. The average BOD load from bleached kraft is about 65 pounds per ton of pulp; from unbleached 35 to 40 pounds per ton. This reduction in water pollution is more than compensated for by an increase in air pollution. Kraft mills emit huge amounts of "industrial fallout" in the form of ash, chemical dust, mists, and odorous gases. Fallout emissions from an uncontrolled 500-ton capacity mill can reach 13,000 pounds per day.[9]

A Public Health Service study, In Quest of Clean Air for Berlin, New Hampshire, described the air pollution problems:

"Among the major nuisances resulting from air pollution in Berlin are foul odors, heavy dustfall, lowered visibility, and soiling. Malodors from the kraft pulping operations are generally considered to be obnoxious, have an adverse psychological effect on the population, and detract from the possible development of tourism. Heavy dustfall in areas near the mill and generally throughout the city leaves deposits on cars, porches, laundry, sills, etc., imparting a dingy appearance to the area."[10]

"Particulate fallout damages materials and painted surfaces and contributes to the general dirtiness of the surrounding area. Discoloration of lead base paint by hydrogen sulfide is common around sources of this gas ... other property damage includes increased corrosion of metals, damage to building materials, increased soiling of laundry, home furnishing, and display merchandise and similar effects."[11]

Kraft mill towns are often "shrouded in an acrid fog." According to a resident of Covington, Virginia, "you can't park your car out overnight ... or else by morning it is covered with a thick dirt scum which is hard to get off ... and you can't hang your clothes out either."[12] Before a precipitator was installed in the Westvaco mill in Covington, this dustfall ranged from a low of 8 to a high of 120 tons per square mile each month.[13] Now it has been reduced to a monthly average of 15 tons per square mile.

Hydrogen sulfide and complex organic sulfur gases called mercaptans are also emitted from kraft mills. These produce the smell familiar to anyone who lives near, or even passes within miles of, a kraft mill. Variously described as that of rotten eggs, rotten turnips, or rotten cabbage, this odor is so pervasive, so annoying, and so difficult to eliminate that the Minister of Resources of British Columbia, Canada, has offered a $250,000 award to "anyone, anywhere, who will develop a method or device to eliminate it."[14] It has been successfully reduced by 97 to 99 percent, to a level of one part per million, in a mill in Sweden. "From a strictly chemical viewpoint, the odorous compounds are essentially eliminated."[15] However, since the human nose can detect the smell at one part per billion, the problem is not yet solved.

In the kraft pulping process, wood chips are cooked in an alkaline liquor solution of sodium hydroxide (caustic soda or "lye") and sodium sulfide (the whole called "white liquor") in one of two types of digester. The older, more traditional "batch" digester cooks one 10- to 20-ton load at a time for 2 1/2 to 5 hours at 176 degrees centigrade. As in sulfite pulping, the load of chips and liquor is then "blown" from the digester and reduced to fibers by the decrease in pressure. With the more modern "continuous" digester, the chips and liquor are fed into one end, cooked, washed, cooled, and blown in a continuous stream. The "continuous" digester allows much greater control of the odorous fumes and gases formed during the cook and produces pulp of a more uniform quality.

Large amounts of the obnoxious, odorous gases associated with the kraft process are formed in the digester. Often, these are collected and condensed to recover heat and to remove turpentine and "tall oil" which can be sold as by-products. The noncondensible parts may then be routed to the recovery furnace or lime kiln for incineration. Unfortunately, they are more often vented directly into the air where they carry their characteristic rotten turnip smell for many miles in the windward direction from the mill. Both hydrogen sulfide and mercaptans, the most odorous of the gases produced, are difficult to condense by the usual industrial methods. Even when they are reduced by burning to one part per million, they still "set nostrils aquiver with revulsion."[16]

After digestion, the "spent" cooking liquor, containing dissolved organic material, is removed from the pulp in as concentrated a form as possible. This "black liquor," named for its midnight black color, is sent to the most important part of the plant (the chemical recovery area) while the pulp, "a dark brown mass the color of wet grocery sacks," is screened to remove knots and uncooked wood particles, washed, filtered, and either stored or made into paper.

In the first phase of the recovery cycle, "weak black liquor" (i.e., that containing spent chemicals and dissolved wood materials) is concentrated in multistage evaporators until it reaches a solids concentration of 40 percent to 55 percent. Prior to entering the recovery furnace, it passes through direct contact evaporators for further concentration. Chemical interactions in this step cause malodorous gases to form. These again may either be controlled or released directly into the air.

The liquor is evaporated to about a 65 percent solids level, "salt cake" (sodium sulfate) is added, and the solution is sprayed into the recovery furnace for combustion. The recovery furnace is the most "prominent, expensive, and temperamental" piece of equipment in a kraft mill. It generates enormous energy, and is "run on the ragged edge of an explosion." In the furnace, the inorganic chemicals in the black liquor are reduced to molten smelt, while the organic solids are incinerated to carbon dioxide and water vapor. The combustion and recovery of heat from the organic materials in the liquor is very important economically. The steam produced is piped back for use in the pulping process and in the multiple effect evaporation stage, while the combustion gases are used in direct contact evaporation prior to being vented out the stack.

Actions in the recovery furnace produce over 50 percent of all the air pollutants associated with the kraft process. The Public Health Service study, In Quest of Clean Air for Berlin, New Hampshire, analyzed the emissions from a recovery system stack and found that 330 pounds of hydrogen sulfide, 50 pounds of sulfur dioxide, and 97 pounds of soot and ash were emitted every hour from a mill producing 600 tons of pulp per day.[17]

The final phase of the recovery cycle is treatment of the recovered inorganic chemicals. Several possible air pollutants (corrosive alkaline mists, large amounts of lime dust, and sodium salts) are formed during this process. In addition, impurities in the molten chemical ash are separated out and may be discharged into the sewer. From the recovery furnace, the molten chemicals, in the form of smelt, flow into tanks and are dissolved in water. The resulting solution, "green liquor," is clarified, washed, and treated with "burned lime" (calcium oxide) to convert an inactive chemical, sodium carbonate, into sodium hydroxide and to produce calcium carbonate. The resulting liquid solution, containing sodium hydroxide and sodium sulfide, is rejuvenated "white" cooking liquor ready for the digesters. The calcium carbonate is collected, washed, and burned in a lime kiln for reuse in treating another batch of green liquor.

References for Chapter 2

1. Handbook of Pulp and Paper Technology, edited by Kenneth W. Britt, Reinhold Publishing Corp., New York, 1964, figures 6-8 of p. 158.

2. "The 70s Beckon: Will We Enter Them with a Half-Hearted Anti-Pollution Plan?" by Sheldon Brooks, Pulp and Paper, January, 1970, p. 97.

3. Pollutional Effects of Pulp and Paper Mill Wastes in Puget Sound, U.S. Department of the Interior, Federal Water Pollution Control Administration, Northwest Regional Office, Portland, Oregon; and Washington State Pollution Control Commission, Olympia, Washington, March, 1967, p. 131.

4. Ibid., p. 156.

5. Letter to the CEP, May 18, 1970.

6. The Pulp and Paper Industry and the Northwest, by J. Alfed Hall, Pacific Northwest Forest and Range Experimentation Station, Forest Service, U.S. Department of Agriculture, 1969, p. 22.

7. Ibid., p. 23.

8. Handbook of Pulp and Paper Technology, p. 171.

9. Letter to the CEP from the Humbolt County Air Pollution Control District, Eureka, California, June 19, 1970.

10. In Quest of Clean Air for Berlin, New Hampshire, by Paul A. Kenline for the U.S. Public Health Service, Division of Air Pollution, 1962, p. 35.

11. Ibid., pp. 33-34.

12. From a letter to the CEP.

13. "Pollution Like this Smoke at Covington Clogs Virginia," by Ken Ringle and Jim Peterson, Washington Post, March 20, 1969.

14. Sports Illustrated, June 15, 1970.

15. Ibid.

16. Ibid.

17. In Quest of Clean Air for Berlin, New Hampshire, p. 9.

Chapter 3

THE STATE OF THE ART

State-of-the-art pollution control is what can be expected. It is the highest level of control technologically and economically feasible. State-of-the-art control is a flexible concept that changes as new and more efficient pollution control methods become available.

In the pulp and paper industry today, with state-of-the-art effort and equipment, process water can be cleaned of over 90 percent of its solid waste load (fibers, dirt, bark, and wood chips) and over 85 percent of the oxygen-robbing dissolved solids and chemicals washed from the pulping and bleaching operations. Even the discoloring matter in effluent can be greatly reduced by physical and chemical types of treatment. Before mill effluent is discharged into public waterways, it can be restored to a nontoxic and aerated condition.

With state-of-the-art air control, the atmosphere around a mill can be saved from a continuous barrage of many tons of soot and ash and gases. Over 99 percent of particulate emissions and over 95 percent of gaseous contaminants may be filtered or rinsed out of the gas stream. However, odor from the kraft process is a persistent problem even with such optimal controls.

All that is required for a company to achieve state-of-the-art pollution control is a willingness and means to invest the time and money that is required to engineer and buy efficient equipment tailored to the particular needs of its groundwood, sulfite, NSSC, and kraft mills. Most important, management must be convinced that state-of-the-art control is a necessity and no longer a matter of choice.

In-Plant Controls

The most obvious pollution control technique is proper operation of a mill. This means avoidance of spills and leakage and limitation of production to the optimum capacity of the equipment. Because of complex chemical reactions that take place during the pulping process, overloading (or improper maintenance) of production equipment can lead to disproportionately high levels of pollution. A classic example is a kraft mill recovery furnace operating at 150 percent of capacity, omitting under such conditions 600 percent of the normal amounts of hydrogen sulfide gas. Unfortunately, overloading of equipment for increased production is quite common. As one leading company executive admitted apologetically:

"The traditional road to promotion in a paper company for the mill manager has been to push his mill's production up further than any other mill manager, mill capacity notwithstanding."

Greater water recyling can also significantly reduce pollution while increasing production and saving valuable fibers, chemicals, and heat. For example, newer sulfite and kraft mills have been built with continuous digesters especially adapted to reuse wash water, condenser water, and other formerly wasted liquids. Because pulp chemicals and water are run in a never-ending stream through the equipment, less water is used, fewer fibers are lost, and heat is conserved.

Installation of efficient spent liquor treatment is one in-plant change that should be mandatory, as keeping waste liquor out of mill effluent automatically eliminates the discharge of many thousands of pounds of BOD. Processes to evaporate and burn spent liquor for chemical recovery, by-product manufacture, or simply clean disposal are available for all chemical pulping methods. In the case of calcium base sulfite mills where liquor treatment often proves too difficult, it seems there is no longer any choice but to convert the mill to another liquor base, or if the cost of conversion is too great, to close the pulping operation down.

On-Site Effluent Treatment

While careful in-plant controls can have a marked effect on reducing a mill's pollution, they are simply not adequate for removing the large amounts of settleable and dissolved solids from effluent. For this, primary and secondary treatment are necessary.

Primary sedimentation, the first step, can remove 75 to 95 percent of the settleable solids while also obtaining a 25 percent BOD reduction. It is a strictly physical process that involves allowing mill effluent to remain still in a settling pond or a large cement clarifying tank until the solid wastes fall to the bottom. The clear water overflows the top. Many clarifying systems are equipped with a skimming device for removal of wood fibers floating on the surface so they can be fed back into the pulp-making process.

Clarifiers are relatively standardized equipment, costing between $300,000 and $1,000,000 plus varying amounts for collection systems and sludge disposal equipment. It is obviously important for efficient primary treatment that the clarifiers or settling ponds be large enough to retain all effluent until all the heavy solid waste matter has settled out.

Partial solid waste removal at most mills is provided by screens and "save-alls." These pieces of equipment are actually installed to reclaim valuable "lost" fibers from the effluent to feed back into the pulpmaking process. While they do help keep substantial amounts of solid waste from being discharged, no mill equipped solely with screens and save-alls can be said to provide full primary treatment.

The accomplishment of primary treatment creates one major subsidiary problem, disposal of the mass of solid wastes that accumulates on the bottom of clarifiers. This "sludge," which may have a consistency similar to Jell-O, can be dehydrated and burned for fuel or simply used for landfill. Unfortunately, as landfill sites become more scarce, the problem of sludge disposal becomes more acute, and more mills will need to incinerate or find new uses for sludge.

Secondary effluent treatment removes 80 to 95 percent of the dissolved

oxygen-demanding organic and inorganic materials in effluent. It is a biological
process that depends on bacteria in the water to consume wastes and a high
level of oxygen to support this bacterial life. Secondary treatment is actually
an attempt to simulate, though at a higher rate, the natural oxidation process
that takes place in streams, lakes and rivers.

Equipment for achieving this biological breakdown is more varied and
more expensive than that for primary treatment. The least expensive and
most common system is a series of holding ponds or naturalization stabilization lagoons. All that is required for such systems is extensive acreage on
which to locate ponds large enough to retain effluent until natural processes
decompose the dissolved wastes. This takes many months, but it saves the
expense of installing special equipment.

Some mills must have holding pond systems. If the river into which their
effluent is discharged has a slow flow or a high temperature, the daily discharge may not be absorbed and dispersed. In order to avoid damaging aquatic
plant and animal life, the effluent must be released very gradually. In extreme
cases, effluent has to be held entirely until certain times of year when there
is a high enough stream flow to permit discharge. Union Camp's mill at
Franklin, Virginia, has one of the largest holding pond systems in the industry.
It covers 1,600 acres, has a capacity of 12 billion gallons, and is capable of re-
taining the 38 million gallons of daily effluent for 7 to 8 months. Over 90 percent BOD reduction is achieved before the effluent is released into the Blackwater River.

In aerated ponds or lagoons, biological oxidation is accelerated. Aerators
force oxygen into the effluent to fortify the waste-consuming bacteria. In 10 to
20 days, 80 to 95 percent BOD reduction is possible. A lagoon that is aerated
can treat up to 20 times as much effluent as one of similar size without aeration. The first surface-aerated system in the pulp and paper industry was developed by Riegel Paper Company and installed at its Riegelwood, North
Carolina, kraft mill in 1964. This achievement won the company a State Wildlife Federation award for water conservation.

The most elaborate and expensive form of secondary treatment, activated
sludge, is used by many municipal sewage plants. This system is particularly
useful in crowded urban locations because it takes up the least space of any
secondary system and purifies water in the shortest time. A biological "sludge,"
rich in bacteria and oxygen, is whipped into the waste water by giant eggbeater-
like aerators. The bacteria quickly attack and break down the waste so that,
within a few hours, effluent can be discharged. Because of the elaborate equipment used and the need for constant supervision and regulation, both capital
and operating costs of activated sludge systems are very high, reaching double
the cost of other secondary treatment methods. Capital costs run between
$3,000,000 and $4,000,000, while annual operation may be over $600,000.

Activated sludge is not a new system. It was first used in the pulp and
paper industry in 1955 at Westvaco's Covington, Virginia, kraft mill. A major
reason the company installed an activated-sludge plant was the seriousness of
the water pollution problem at this location. The Jackson River, into which
the huge (1,353 ton per day) kraft and NSSC mill discharges its 25 million gal-

lons of daily effluent has such low flow during summer months that any substantial waste load does extensive damage, even with the very low BOD load
Westvaco discharges into the Jackson (8 to 10 pounds per ton of pulp).

Tertiary effluent treatment is a catchall phrase for anything beyond secondary, such as the use of chemical methods and such physical techniques as
microscreening and reverse osmosis. The process depends on the nature of
the wastes. In municipal sewage, tertiary treatment removes nitrates and
phosphates; in pulp and paper effluent it removes coloring matter and organic
salts. Interstate Paper Corporation's kraft mill in Riceboro, Georgia (a 500
ton per day unbleached kraft and linerboard mill), installed the industry's first
full-scale color-removal system using lime. The system was built by Rust
Engineering Corp. (a division of Litton Industries) and removes from 85 to 95
percent of the coloring matter from the effluent. A second system was completed in 1970 at Georgia-Pacific's Crossett, Arkansas, kraft mill, and a
third will go into operation at Continental Can's Hodge, Louisiana, mill in
March, 1971.

At International Paper's Springhill, Louisiana, mill a "massive lime" system (using about ten times as much lime as the other systems) is in the pilot
stage. This "massive lime" technique was developed and patented by the National Council for Air and Stream Improvement, the technical and research
organization for the pulp and paper industry.

All industry color-removal projects to date have been partially funded by
Federal Water Quality Administration grants — with the exception of the Georgia-
Pacific project, which was paid for entirely by the company.

Out-of-Plant Water Treatment

Although mill effluent is usually treated in company-owned facilities, there
are alternative treatment methods. In joint municipal-industrial treatment,
effluent may be piped to a municipal sewage plant where it is treated along
with the town's domestic waste. The mill usually pays a significant share of
the capital and operating costs of such installations, either directly or indirectly, as a very large (usually the largest) economic source in the area.
Almost all jointly run plants provide both primary and secondary treatment.
A few are now planning to provide tertiary treatment as well. One example
is the plant in St. Paul, Minnesota, to which the Hoerner-Waldorf NSSC pulp
mill and other local industries contribute effluent.

A second form of external water treatment is direct stream aeration.
This simply entails beating air into the stream around the point of effluent
discharge. This system may be useful as a supplement to, but should never
be considered a replacement for, good basic secondary treatment. Crown
Zellerbach aerates parts of the Pearl River near its Bogalusa, Louisiana,
mill during the three summer months when river flow is very slow and temperature is quite high. The system is serving as a stopgap measure until the
new aerated lagoons for the mill's effluent, now under construction, are completed.

Two methods of effluent disposal often included on lists of treatment are spray irrigation and outfall pipes with diffusers. Both techniques are almost always totally unacceptable, although they are often condoned by state control agencies. Spray irrigation means that untreated or partially treated effluent is just piped to land near a mill and discharged; a simple way, from a corporation's point of view, of avoiding the cost of installing adequate water treatment facilities. While the water and its load of organic wastes may have some soil-fertilizing value, the chemicals and concentrations of salts in the effluent are more likely to degrade soil quality.

With outfall pipes, untreated effluent is carried in another direction for disposal: either out into the ocean to areas of rapid flow or tide, or down into the bottom of large rivers. Holes, called diffusers, spaced along the end of the pipe allow the effluent to disperse over a relatively wide area instead pouring out in one concentrated stream. The usual justification for pipelines rests on the outdated assumption that oceans and large rivers still have an almost infinite assimilative capacity for man's wastes, that if these wastes are diluted so that they are hardly visible and no immediately recognizable harm is done, the pollution must not be destructive. Only last year, Westvaco's new mill at Wickliffe, Kentucky, went into operation with a state-approved outfall pipe that sends its effluent, only primarily treated, into the Mississippi. Washington, Oregon, and California have also been guilty of approving pipes into the Pacific. Many pulp mills on the West coast, including the two large Georgia-Pacific Kraft mills in Samoa, California, and Toledo, Oregon, have spent several million dollars to install this inadequate piece of equipment.

Air Pollution Control

Every mill in the country has to cope with the soot and ash particles emitted from oil, bark, and coal-fired boilers. Mills that burn spent liquor for chemical recovery have the added problem of controlling chemically coated fly ash from the recovery furnaces. Particulate control technology is, however, very advanced, and, as mentioned earlier, 99 percent particulate removal is no longer too much to expect.

For particulate control, the oldest and most widely used devices are mechanical or cyclone dust collectors. They set up a continuous air current that spins particles against the walls inside the pipes leading to the mill stacks. The bits of soot and ash, after striking the walls, fall back down into a hopper at the bottom of the stack, from wich they are periodically removed. Dust collectors cost approximately $100,000 and can achieve 80 to 90 percent removal of large particulate matter.

Their inadequacy in coping with the fine soot and ash emitted from coal boilers was attested to by one company pollution control director:

"Mechanical collectors just get the golf balls. We used to consider them sufficient, but with the new stricter pollution control standards, we would probably have to think of them as antiques."

Mechanical collectors are limited in another way. They do not control gas emissions, and thus are inadequate if high-sulfur oil or coal (over 1.5 percent sulfur content) is burned. Nonetheless almost all mills continue to rely completely on these devices (if they have installed any device at all) to control power boiler emissions. Only rarely, as at the Luke, Maryland, Westvaco mill, have more efficient scrubbers or electrostatic precipitators been used.

Some mills with coal-fired boilers have found the emissions such an annoying problem that instead of installing the expensive scrubbers or precipitators that may be required, they have decided to switch to the use of natural gas, which presents no pollution problem at all. The only condition for doing this is, of course, that a local supply of natural gas exist.

Scrubbers can remove about 95 percent of the particulate matter as well as over 90 percent of the sulfurous gases from the exhaust passing through a mill's stacks. A stream of water, sometimes containing chlorine or other chemicals, is sprayed through the exhaust, "scrubbing" soot and ash and absorbing large volumes of gaseous compounds. Scrubbers may be used on all coal boilers. They may also be used alone, or in conjunction with precipitators on kraft recovery furnace stacks and on stacks venting from the digestor and chemical recovery areas of NSSC, sulfite, and kraft mills. The "venturi-type" scrubber, a particularly efficient model, is the best control currently available for reducing the heavy chemically coated dust emissions from kraft lime kilns. The capital cost of this piece of equipment is usually about $250,000.

One annoying side effect resulting from the use of scrubbing units is an increase in water vapor emissions. Devices called demisters or demister pads, made of fine mesh, or porous materials, may be placed in the neck of the scrubbers to trap large water droplets and help reduce this vapor problem.

Electrostatic precipitators, which have been in use in the industry for over 50 years, are considered one of the most efficient particulate collection devices available today. Improved over the years, they can now remove up to 99.5 percent of particulate emissions. A precipitator consists of a boxlike structure containing positively and negatively charged rods that create an electric field. It is placed at the entrance to a mill stack, and as the gas stream passes through it each soot particle receives a charge and adheres to a rod with the opposite charge. From time to time, the rods are "rapped," and the clinging particulate matter falls to the bottom of the precipitator. At the present time, 99.5 percent efficient electrostatic precipitators should be accepted as necessary for controlling kraft recovery furnace emissions, yet many mills already equipped with older 80 to 90 percent models are only very reluctantly making the necessary modifications to upgrade their control by rebuilding the precipitator or adding a supplemental precipitator or scrubber. The capital cost of a new 99.5 percent precipitator is about $500,000.

The air emissions from groundwood mills can all be easily controlled by mechanical collectors plus perhaps a scrubber, if coal is burned. Sulfite and NSSC mills with liquor recovery should have a number of highly efficient scrubbers or electrostatic precipitators. Only in kraft mills are these pieces of equipment not enough, for there is that additional odor problem to contend with.

Kraft Mill Odor Control

The odorous gases released from the digester and blow pit during pulping and from the evaporators during chemical recovery operations are dealt with in different ways.

Gases formed during pulping are collected and piped to the lime kiln for incineration. A particularly efficient piece of equipment for gas collection, developed by Weyerhaeuser in the early 1950s, is called a "vaporsphere." All relief gases are gathered in this large hollow sphere (the size of a one-story building) and are channeled on for burning.

The most widely accepted method of controlling the odor from the Kraft recovery operation is black liquor oxidation. This treatment is aimed not at capturing and burning odorous gases as just described, but at preventing their formation. The process involves combining oxygen with the dissolved sulfur-containing compounds in the spent liquor. The resulting chemical reaction transforms them into nonodorous gas, instead of allowing them to separate out from the liquor as foul-smelling compounds, mainly dimethyl mercaptan ("skunk smell"). For greatest efficiency, the forced combination of oxygen and the sulfurous compounds should take place at the earliest possible point in the recovery process.

Before the waste liquor even enters the first phase of the recovery cycle (the evaporators), while it is still in its unconcentrated or "weak" condition, it is piped into a black liquor oxidation tower. The tower is packed with a series of metal plates or porous carbon blocks, and as the liquor flows through, streams of air are injected into it. Virtually 100 percent of odorous gas formation can be prevented at this stage by such "weak black liquor oxidation."

Several years ago, some of the mills using this treatment found that, even with 100 percent effective weak black liquor oxidation, a serious odor problem remained. Studies indicated that the odorous gases were reforming as the liquor underwent final evaporation to its concentrated or "heavy" state. In order to achieve more complete odor control, mills then began injecting air into the concentrated liquor also, i.e., performing "heavy black liquor oxidation." By combining these two steps, a kraft mill can greatly reduce the odor emitted.

While the system for black liquor oxidation was developed almost 20 years ago (it was first installed by Weyerhaueser at the company's Springfield, Oregon, mill in 1953) and while the system's cost is not overwhelming (about $500,000) over half the kraft mills in this country have not been equipped with it. J. O. Julson, former director of air and water resources at Weyerhaeuser, stated in an April 1969 article in the industry trade magazine Pulp and Paper that oxidation is a payout process, due to the reduction of sulfur losses. The sulfur chemicals are retained instead of escaping as gases.

One excuse offered by companies operating kraft mills in the South is that the highly resinous woods from which southern pulp is produced create serious foaming problems. When air is injected into the liquor the resin bubbles, clogging machinery.

However, E. R. Hendrikson, one of the foremost air pollution control authorities, suggested in his study of atmospheric emissions in the industry [1] that the possibility of foaming need not prohibit use of any black liquor oxidation. It is simply a major consideration in the selection of a system. He stated that both weak and heavy liquor oxidation can be used effectively on nonfoaming liquor while only concentrated liquor oxidation is practicable where foaming is a significant problem.

A new technique for one-stage oxidation, using molecular oxygen instead of air, was developed by Owens-Illinois and is now in operation at the company's Orange, Texas, mill. The system is expected to provide odor control efficiency comparable to that of the best weak and heavy oxidation systems now in use and of the newer non-direct-contact evaporator systems.

Non-direct contact evaporation systems attack the problem of odorous gas formation at a different point in the chemical recovery cycle, i.e., during final evaporation. In this phase hot combustion gases are usually forced through the liquor to evaporate it. Physical and chemical reactions between the combustion gases and the black liquor release foul-smelling hydrogen sulfide and mercaptans. By eliminating any contact between the liquor and the combustion gases, formation of the odorous compounds is prevented.

Two systems have been developed which keep the gases and liquor separated. In the one engineered by Babcock and Wilcox, called "non-contact," the combustion gases are isolated in pipes and only their heat is transferred to the liquor. In the Combustion Engineering system, called "air cascade evaporation," a stream of hot air, instead of hot combustion gases, is forced through the liquor. Neither system allows the direct contact that causes the formation of odorous gases.

Non-direct contact evaporation systems have long been used abroad. By 1968, they were installed in almost every kraft mill in Sweden. Only in the past two years have American corporations begun to use them. The new American Can mill in Halsey, Oregon, which went into operation only late in 1969, has a Babcock and Wilcox system that has reduced the mill's odor emissions to a bare minimum. Weyerhaeuser's new mill at New Bern, North Carolina, has a Combustion Engineering system.

Installation of these systems in a new mill is relatively simple. However, converting an old kraft mill is extremely expensive as purchase of a new recovery furnace is usually necessary. Crown Zellerbach's installation of a Combustion Engineering system at its Port Townsend, Washington, kraft mill involved a total expenditure of over $13 million.

With the most efficiently operated black liquor oxidation system or with non-direct-contact evaporation, the final odorous gas emissions from the Kraft process may be reduced to one part per million in the air. However, since odorous gases are still detectable and distressing at concentrations of one part per billion, no final solution to this form of pollution is now technologically possible.

Cost of Pollution Control Equipment

Providing the just described state-of-the-art air and water pollution control equipment represents about 10 percent of the capital cost of a new mill. Thus, $4 million of the $40 million cost of the Halsey mill and $6 million of the $50 million cost of the New Bern mill are attributed to pollution control.

Tables 3.1 and 3.2 present the major pollution control methods and equipment needed for a state-of-the-art mill, their uses, and their approximate capital costs. This is not an exhaustive list of all possible equipment and mill modifications, but it includes the major categories of equipment.

Table 3.1
Water Pollution Control Methods

| | | Cost (dollars) | |
Method	Efficiency	@ 20 mil/gal	@ 50 mil/gal
Primary (settling ponds, clarifiers)	75-95% solid waste removal	300,000	750,000
Secondary (holding ponds, aerated lagoons, activated sludge)	80-95% BOD reduction	700,000- 2 million	750,000- 2 million
Tertiary (color removal)	80-95% removal of coloring matter	Systems too new; accurate cost figures not available	

Table 3.2
Air Pollution Control Methods

Method	Where Installed	Efficiency	Cost (dollars)
Particulate control			
Mechanical collectors	Low-sulfur oil, bark boilers	80-90% soot, ash removal	100,000
Electrostatic precipitators	Kraft recovery furnaces	99.5% soot, ash removal	500,000
Scrubbers (particulate, gas)	High-sulfur oil, coal boilers; NSSC, sulfite, kraft recovery	95% gas removal	250,000

Table 3.2
Air Pollution Control Methods (continued)

Method	Where Installed	Efficiency	Cost (dollars)
	furnaces; kraft lime kilns		
Odor control			
Black liquor oxidation	Kraft mills		500,000
Non-direct contact evaporation	Kraft mills		Systems vary too widely to generalize

For older mills, the capital cost of the equipment may be only the tip of the iceberg. Land may need to be acquired; recovery and/or disposal systems may need extensive rebuilding in order to accommodate the equipment. The costs of treatment vary widely depending not only on the age and size of mill but also on its location, type of operation, etc. Where a mill has been built with no thought of controlling waste disposal, it may take considerable effort even to find, much less plug up, all the discharge points. The industry's rule of thumb is that modifications to an older mill cost two to three times as much as installations in a new one.

But these costs are only a partial reflection of the debt that such mills owe to their environmant. For decades, air and water have been cost-free raw materials for some in the industry, even though in pulp production water is second in production importance only to wood. It is only fair for some sort of bill to be presented by downstream users, downwind dwellers, and paid by companies, their stockholders, and consumers of paper products. As Stuart Udall has prophesied:

"The notion that an industry has some kind of divine right to dump whatever kinds of waste it wants in the nearest lake or stream because the industry provides jobs will give way to the principle that pollution control is a normal part of the cost of doing business." [2]

References for Chapter 3

1. Control of Atmospheric Emissions in the Wood Pulping Industry, Vol. 2, U.S. Department of Health, Education, and Welfare, March 15, 1970.

2. The Pulp and Paper Industry and the Northwest, by J. Alfred Hall, Pacific Northwest Forest and Range Experimentation Station, Forest Service, U.S. Department of Agriculture, 1969.

Chapter 4

POLLUTION CONTROL AND THE LAW

Environmental legislation does not reflect available technology, either for standards or for enforcement. The state of the art is in most respects impressive, but is not the basis for either federal or state legislation standards. Rather than setting standards in terms of what is technologically possible, most laws consider only what is "economically feasible." Such feasibility is in most instances determined by the industries themselves. The line between industry advice or consultation and lobbying is fine indeed, and apparently often crossed at both state and federal levels. Such apparent conflict of interest is present in enforcement procedures as well.

Instead of drafting legislation that incorporates all known or possible protective standards and treatment devices or processes, or anticipates problems that, were they to arise, could be solved with the technical means available, legislatures usually employ a negative criterion. They usually respond, in minimal terms, to a generally known, publicized, or complained-of problem.

Two recently constructed pulp mills provide graphic examples of the results of this legislative philosophy. In Louisiana, because the air pollution control agency did not require it, the new Boise Cascade kraft mill was built without installation of any of the well-known and readily available odor control systems. Kentucky, meanwhile, only recently authorized a pipeline and waste diffuser, but without secondary effluent treatment of any kind, for the new Westvaco mill at Wyckliffe, which discharges its effluent into the Mississippi.

Enforcement practices, even of technologically anemic laws, reflect industry representation on regulatory commissions and a judiciary sympathetic to economic rather than environmental concerns. In order to evaluate fairly a company's "pollution control report card," that company must be considered in the context of existing laws, regulations, and enforcement philosophies.

The Laws

The first major national resource conservation legislation was written into federal law 71 years ago. The 1899 Refuse Act (Rivers and Harbors Section) empowered the federal government to bring suit against anyone who dumps any materials into the navigable waters of the United States without authorization by the Corps of Engineers. This law was virtually forgotten for many years, but it has recently become important once more in such widely publicized actions as those against industries discharging mercury into Lake Superior and Puget Sound in mid-1970. Unfortunately, much of the Act's potential power as a rapid and efficient means of abating pollution has been undermined. The Department of Justice has instructed local U.S. Attorneys to avoid its use except in cases of massive "incidents" of specific localized pollution.[1]

Most pollution control legislation was formulated independently by individual states up until 1965. However, the resultant diversity in criteria, standards, and approaches made apparent the need for some standardization of pollution control laws under federal auspices. The federal Water Quality Act of 1965 and the Air Quality Act of 1967 established a framework for national air and water pollution control and the mechanism for abating interstate pollution. The acts also set deadlines for state submission of proposals for control legislation. For interstate waters, the proposals must describe desired use, quality, and an implementation program for attaining the various qualities.

Once approved by the Federal Water Quality Administration, state laws become, in effect, federal regulations also. If states fail to submit legislation by the deadlines, the federal government has the authority to move in and dictate standards. Use of this power under the Water Quality Act has been twice threatened (against Virginia and Iowa) but never invoked. In fact, some states have been given extensions of over three years to submit their water pollution control proposals.

The federal government set a June, 1967, deadline for state submission of proposed water quality legislation. While the states were attempting to set their respective standards, the federal government was determining through research what standards were necessary for preservation of various types of aquatic life. The Department of the interior did not publish the results of this research—a book called Water Quality Criteria (the "Green Book"*)— until April, 1968, ten months after the deadline. Thus, the states did not have the benefit of these data in formulating their regulations, and many states which had made the effort to submit laws by the deadline had to make revisions.

Even though there has been considerable delay on the federal level and disinterest or intransigence on the part of various state legislatures, all of the states have now submitted proposed receiving-water-quality laws.

A major incentive for state participation in the federal program has been the possibility of obtaining financial benefits. States that file acceptable enforcement programs are eligible for federal construction, administrative, and planning grants to cover up to 55 percent of the cost of their state and municipal water quality attainment programs.** A more recent and significant spur to action is the National Environmental Protection Act of 1970, a bill which specifically denies the use of federal funds for any project which has not been planned with explicit provisions for environmental protection.

----

*Further discussion of the "Green Book" appears in Appendix 1.
**Unfortunately, Congress has appropriated only about half of the funds authorized in the 1965 Act. As a result, many planned municipal or industrial-municipal systems, still awaiting the federal check, are far behind schedule.

Even with these incentives, only 18 states have fully approved standards that include the recommended antidegradation clause to guarantee preservation of, at least, the present water quality. Another 9 states have fully approved standards without antidegradation clauses, while 23 are still attempting to gain federal approval.

The response of the states to the provisions of the Air Quality Act of 1967 has been similarly lacking in results. While most states have filed letters of intent to establish air quality criteria for defined "air quality regions," the regions are not yet statewide, and the programs are not very far along.

Standards

The present State pollution control laws look very similar because they are all cast in the mold set forth by the Federal Water Quality and Air Quality Acts. However, there is great variation in the stringency of enforcement and in deadlines.

The Federal Water Quality Administration has described the water quality necessary to permit various water uses (public water supply, pleasure boating, swimming, shellfish harvesting, preservation of various species of fish, and navigation and industrial uses). The use classification of a waterway thus determines what treatment need be given to the wastes discharged into it. It is perhaps reflective of industry's role in the process of setting pollution control standards that many states have classified those rivers serving large industries as suitable for navigation only, a classification that requires virtually no pollution control by dischargers. The Federal Water Quality Administration has repeatedly urged that such waters be reclassified for higher, more appropriate, uses.

Only in a very few states (Washington, Oregon, Georgia, Texas, and California) have portions of the law been set with a real commitment assuring that emissions are limited to the minimum levels attainable. The Georgia Water Quality Control Board required that the Interstate Paper Company install a color removal system at a mill it was building even though the technology for such a system was not yet fully established. Similarly, Washington and Oregon set kraft mill odor and particulate emission standards which major paper companies said could not be met. Having failed in the effort to block the standards, these companies are somehow managing to comply now.

California has the oldest, strictest, and most comprehensive air pollution control program in the nation; both federal and state air quality standards have been guided by its research and experience. Typical air pollution control regulations indicate maximum permissible concentrations in the air of sulfur dioxide and other sulfur compounds, fluorides, carbon monoxide, nitrogen dioxide, visibility-reducing particles (i.e., soot), and oxidant (ozone).

The minimal enthusiasm for setting effective pollution control standards in Virginia and Maine presents a distinct contrast. Virginia only developed a federally approvable pollution abatement program in 1970 and the state air pollution control agency seems unique in its policy of concealing from the public the pollution control status of corporations. (This control board was the only one of the 27 contacted by the CEP for this study that refused to release information on the compliance status of the pulp mills in the state.)

In Maine, where 52 percent of the land is owned by paper companies, the legislature set the latest secondary water treatment deadline of any state (1976). Preliminary plans for compliance are due in 1972, the year many other states are requiring that secondary treatment facilities be constructed and in operation. Maine's legislature has twice refused to pass a statute limiting the permissible temperature of industrial effluent, and has, further, given classifications to many state waterways which tend to solidify the status quo of industrial dumping, rather than defining water uses which would require significant pollution control by industry.

Compliance

Most states attempt to control industrial pollution under a discharge permit system, which requires written state approval of the amount and composition of effluent. If such an approval is received, the company is "in compliance." The conditions of these permits which determine their effectiveness in abating pollution vary greatly. Some permits are contracts requiring considerable abatement activity and the installation of specific pieces of control equipment by fixed deadlines. Others are nonexpiring agreements that the present level of treatment is acceptable. Such permits in California, however, are continually subject to revision by that state.

Unfortunately, some permits are virtual licenses for industries to continue to pollute. The Hammermill Paper Company, at its Erie, Pennsylvania, mill, held a permit requiring no specific abatement activity. It was renewed by the state of Pennsylvania ten times over a twenty-year period, and during these twenty years the company, technically "in compliance" with the law, daily discharged millions of gallons of essentially untreated wastes.

Many states have "grandfather clauses" under which permits are issued. These clauses exempt older, and usually dirtier, facilities from the state's general pollution control requirements. For example, Alabama industries that predate 1949 have seven years rather than five to provide secondary water treatment. Similar clauses appear in most states' air pollution control laws as well.

In some states, "compliance" is meaningless because there is no applicable law. An industry located outside specified "air quality control regions" or in a state that has no standard dealing with the particular kinds of pollution it emits may have heavy particulate or gaseous emissions and still be technically in compliance. This is the case for most of the kraft mills in this study, since most states have not yet specified odorous gas standards. The same general exceptions can apply to water pollution. Most states have set up standards applicable only to specific bodies of water. Thus, any plant situated on a river or lake not covered by a specific designation may be in compliance while having no effluent treatment facilities whatever.

Enforcement

Even though there are obvious flaws in much of the legislation designed to
protect the environment, strict enforcement of laws now on the books would
go a long way toward stopping the deterioration of the air and water. Most
enforcement today is handled on the state level, by the state water quality and
air quality boards. Although the composition of the boards varies widely, 35
of the 50 state boards now include important representatives of the major industrial polluters. For example, each of the six "industry" seats on the Water
Improvement Commission of Alabama is occupied by an executive of a company now involved in pollution abatement proceedings. The rationale for corporate representation on such boards is that the engineering expertise needed
to establish standards rests with industry. While this is true to a certain extent, one may question the likelihood that a representative of a particular industry will press for the same stringency of enforcement as would a professional environmentalist or a concerned citizen. As one federal official put it,
"if you were trying a case against the XYZ paper clip company, would you
want an official of the company on the jury?"[2] And long before one gets to
court, industry advisory boards have veto power over questionnaire and study
proposals of the federal air and water pollution control agencies.[3]

The principal mode of federal antipollution law enforcement has been in-
tercession in proved instances of interstate pollution. In both the federal air
and water quality laws, there are provisions for conferences, six-month waiting periods, recommendations, extra sessions, and reconvention if progress
is not considered satisfactory. At the end of the line, if all else fails, the
federal agency may ask the U.S. Department of Justice to bring an action to
enforce the conference's abatement recommendations. This has happened
twice; in 1961 for water pollution and in 1968 for air pollution. Although extreme cases of environmental degradation have been documented at these conferences, only one case of water pollution has been brought to court. To date
there have been 50 federal water pollution abatement conferences; of these,
15 directly involved pulp and paper mills. Twenty-eight of the 50 conferences
reconvened when satisfactory improvement did not occur after several years.

Sanctions

Enforcement efforts to date have been less than effective because sanctions
are simply not being applied. Agencies may issue an abatement order, but
the most potent penalty, under both state and federal law, is ordering the immediate closing of a facility that continues to pollute after having been instructed to clean up. This power has not been used in one significant case to
date.

Laws now pending before Congress would give governments the authority
to impose fines of up to $10,000 per day for pollution of air or water. How-
ever, historically, the courts seldom impose fines even close to the maximum permitted under existing law; and such maxima are substantially less
than the $10,000 now being proposed. (Recent cases such as the $1 million

fine levied in 1970 against Chevron Oil Company for the Santa Barbara oil
spill may signal a change in judicial thinking.)

It appears that the most effective and widely used weapon at the disposal
of pollution control agencies remains adverse publicity. This can embarrass
a corporation and foment strong public outcries against its pollution. The
fact remains, however, that there is no law behind a newspaper headline, and
any action generated by publicity may be cosmetic rather than fundamental.

The private citizen has had very little power to act against polluters beyond bringing a civil action damage suit. The problem with private litigation
at present is that the citizen must prove that he has suffered substantial loss
or damage in order to gain "standing," i.e., the right to sue. This is often
hard to do. One state, Michigan, has addressed itself to this problem by
giving any citizen the same right to sue polluters as the state attorney general.
Several states are contemplating similar laws.

The Refuse Act of 1899, referred to earlier, also contains a provision
allowing citizens to bring suit and stipulates that a part of any fines levied
be paid to the initiator of the action.

Despite this patchwork of legislation coupled with a growing public de-
sire for environmental protection, many companies protest that compliance
is impossible because laws are changing too rapidly. They complain that
states have not made clear what companies are required to do. This expresses the hope that state-of-the-art pollution control will not be required.
In fact, no company has ever lost a cent by going ahead and installing the
best available pollution control equipment. For example, in the early 1950s,
Weyerhaeuser installed the best available kraft mill odor control systems
at several mills. Today, upgrading them to the state-of-the-art level is costing less than $200,000 each. If the company had waited until today, entirely
new systems would cost at least ten times as much. Similarly, companies
that provided water pollution control at mills as they were built may have
only to deepen existing ponds or add aerators. Such steps have a similar relationship to present-day capital costs for totally new systems. And, of
course, the older systems have provided environmental protection to citizenry and waters through the years.

References for Chapter 4

1. New York Times, June 30, 1970, p. 33.

2. New York Times, December 6, 1970, p. 62.

3. "Business in Government," by Mark J. Green, New Republic, November
14, 1970, pp. 14-15.

Chapter 5

THE STUDY: BACKGROUND AND MAJOR RESULTS

Aims of the Study

The study was undertaken to:

ascertain what the 24 leading producers of pulp and paper in the United States had actually accomplished in controlling pollution at their pulp mills;

evaluate the degree of committment individual company managements have made to pollution abatement; and

estimate what each company may have to spend to provide air and water treatment approaching state-of-the-art control at each of its mills.

No attempt was made to compare mills directly for, as Mr. J. L. McClintock of Weyerhaeuser explained in an interview with the Council, "comparing mills is like comparing apples and oranges. No two are identical and the local environmental problems and priorities are different."

Scope of the Study

In this study, we have investigated the 130 pulp mills operated by the 24 largest pulp and paper companies (in terms of net sales) in the United States. All of the mills involved produce "virgin pulp" by the groundwood, NSSC, sulfite, or kraft process. At some locations, recycled wastepaper is also used as a raw material. Mills where production is limited to the pulping of recycled paper or the refinement of pulp manufactured at another site were not included, nor were any other nonpulping facilities such as chemical plants and lumber mills.

The study has focused on an analysis of "virgin" pulping facilities because their operations, using millions of gallons of water and tons of cordwood daily, as well as many chemicals, are the most polluting of all paper production processes. As such, pulp mills should be the prime target of company efforts and expenditures to control pollution. Performance in this area should thus provide a rough measure of the seriousness of each company's overall abatement activities.

The 130 locations studied produce approximately 87,300 tons of pulp each day, equal to about 68 percent of the total daily output of the industry in 1969. In achieving this production, at least 2.7 billion gallons of water are used and discharged daily. The actual total is considerably higher; International Paper, which operates 24 mills, refused to release any water use figures.

The mills are scattered across the country, with concentrations in the Northwest in Washington and Oregon, in the Midwest in Wisconsin, in the

Northeast in New York and Maine, and across the "Southern Pine Belt" from northern Florida to eastern Texas. (Table 5.1 gives a breakdown of mills by state and city.)

For every company, the research focused on the kind of pollution problems existing at each location, the steps taken and expenditures made to correct the problems, as well as the specific control equipment installed and its effectiveness. An evaluation was made of the problems yet to be solved, the company's plans for solving them, and the probable cost. The legal status of each mill in relation to the pertinent state air and water pollution regulations was also investigated.

Methodology

The Council's staff, over a period of eight months, corresponded with and interviewed the executive in each of the companies charged with managing its air and water pollution control effort. In some cases, managers of individual mills were also interviewed. To understand the range of pollution problems, researchers also visited a number of mill sites, including U.S. Plywood-Champion's kraft mill in Canton, North Carolina (where there is a very serious air pollution problem), the Hoerner Waldorf kraft mill in Missoula, Montana (the subject of a suit by the Environmental Defense Fund and now being rebuilt to "state-of-the-art" standards), and a Hoerner Waldorf NSSC mill in St. Paul, Minnesota (small and old, but well controlled).

Correspondence and interviews were carried on with representatives of air and water pollution regulatory agencies in each relevant state and with environmental groups or citizens in or near pulp mill towns. All federal and state pollution abatement hearings from the past 5 years involving paper and pulp companies were also reviewed. A full bibliography appears at the end of this study.

The 24 companies varied widely in the degree to which they were willing to provide information about their mills. Among the most cooperative were: Georgia Pacific, Kimberly-Clark, Westvaco, St. Regis, Owens-Illinois, Weyerhaeuser, and U. S. Plywood-Champion Papers. Not all of these companies were able to turn in an impressive list of accomplishments: Westvaco has been plagued by a series of old and very polluting mills, St. Regis's pollution abatement effort has lagged far behind other major companies, and Georgia-Pacific has wide gaps in its control record. Still, these companies did indicate a recognition of the public's right to information, and all have indicated plans for a more aggressive control effort.

Some companies were notable for their lack of cooperation. International Paper and Crown Zellerbach, the two largest companies in the industry (representing 30 percent of the capacity of the entire group of 24 companies) were the least cooperative. Crown Zellerbach consistently refused to release information until two weeks after the initial publication of Paper Profits. While International Paper finally did give out limited information on the equipment at its mills, its management continued to protest that the public does not need to know specifics as long as "we indicate the job is being done." Potlatch

Forests and Union Camp were also highly uncooperative. Detailed information
on most of their mills was not difficult to locate, however, as serious pollution problems had already brought extensive criticism and study of these last
two companies.

Presentation of the Data

In Part II of this study, the introduction for each company contains basic information on the major products and consumer brands, financial data, company officers and directors, and a listing of company plants, timberholdings,
and annual pulp production. For many companies, pulp and paper production
comprises most of their annual sales. Others, like Owens-Illinois, American
Can, Boise Cascade, and U.S. Plywood-Champion Papers are much more
widely diversified and, while each is a large producer of pulp and paper, these
products do not dominate total company sales.

The company pollution overview reviews and summarizes the findings
about all the virgin pulp mills run by each company, summarizes the company's pollution control record, and lists the total CEP estimate of what the
company may have to spend for pollution control. All of these estimates are
rough figures based on the capital cost of major pieces of pollution control
equipment. The estimates do not take into account the in-plant changes that
may often be required (e.g. the installation of new sewer systems to transport effluent to clarifiers, or the buildings that must be constructed to house
new kraft recovery units). The figures given should be taken as a general in-
dicator of the costs of cleaning up to an approximately "state-of-the-art" (as
defined in Chapter 3) level of pollution control.

In the mill-by-mill analyses that follow each company overview, the history (including date of construction and/or acquisition), type, and capacity of
the plants are indicated. Each mill is rated in chart form on the basis of the
efficacy of its pollution control system in six different categories: for water —
primary, secondary, and tertiary treatment facilities; for air —control of
particulate emissions on both power and production equipment, and control
of gases and odor.

Opposite each category that applies to a given mill, the specific type of
control equipment installed and its rated or actual efficiency is listed. This
equipment information came mainly from company managements but was supplemented, where possible, by government findings. Overall evaluation of
the ability of this equipment to meet "state-of-the-art" criteria is indicated
by a "$\sqrt{}$" or an "$\times$." Those items with checks are considered adequate,
whereas those rated $\times$ are not. A "—" indicates that no equipment is needed;
and an "$\times$-partial" indicates mills that have no secondary treatment but do
not discharge spent pulping liquor.

Information not included has been indicated in two ways. In some cases,
because of the extremely uncooperative attitude of certain managements discussed earlier, we have been forced to note that data were "not released by
the company." In all other cases, where companies provided enough basic in-

formation for a good evaluation and omitted only details relating to specific emissions levels, an "N.A." (not available) is inserted.

Summary of Findings

The 130 mills surveyed by CEP produce 88,904 tons of pulp daily using at least 2.7 billion gallons of water. The bulk of this production, 76 percent, is the kraft pulp output from 78 large mills. The obnoxious odor that has the most immediate impact on residents of or near kraft mill towns has been adequately controlled at only 32 of these sites. The other 46 have completely failed to provide adequate odor control. Most have never installed black liquor oxidation systems though they have been available for mills which pulp low resin woods for almost two decades. The remaining few have systems which are overloaded to the point of being completely ineffectual.

In terms of production, only 25,540 of the 67,516 tons of kraft pulp are produced with adequate odor control. Four companies control the odor from 100 percent of their production, while eight have no control for any of theirs.

In a broader view of air pollution control, 42,777 tons or less than half of the daily pulp output of the 24 companies is produced with adequate particulate control; while only 44 percent (39,281 tons) is produced without emitting high concentrations of gas and odor.* Only Fiberboard produces 100 percent of its pulp with adequate control of particulate emissions. In contrast, six companies (Diamond, Hammermill, Kimberly-Clark, Marcor, Potlatch and Riegel) produce all of their pulp with no controls at all; and 8 others control the emissions from less than 50 percent of their production.

The breakdown is similar for gas and odor control. Only three companies adequately control the emissions from all production, three control no emissions, and 12 control fewer than 50 percent.

The water treatment findings are no more encouraging. Fully two-thirds of the total 2.7 billion gallons used daily still fail to satisfy the standards for primary and secondary treatment called for by the federal government five years ago. Almost no water treatment is given to 895 million gallons of effluent (33 percent of the total) discharged every day. Another 930 million gallons (34 percent) are treated only for removal of heavy solid wastes. Full secondary treatment is provided for 868 million gallons (31 percent); and a mere 50 million gallons (2 percent), coming from one Georgia-Pacific mill in Crossett, Arkansas, is given tertiary treatment.

Only two of the 24 companies surveyed in this study have records of sustained interest and effort toward achieving excellent pollution control at their mills; these are Owens-Illinois and Weyerhaeuser. All four Owens-Illinois mills have both adequate air and water pollution controls. The extensive lagoon system for effluent treatment at the Company's Valdosta, Georgia, mill was designed and built long before such facilities were being required.

----

*Included in this total are 7,646 tons produced by pulping methods such as groundwood that create no gas and odor problem.

Weyerhaeuser is one of the five largest pulp and paper companies. It operates 11 pulp mills; the 9 mills built since World War II were constructed with state-of-the-art pollution controls (some of which now need upgrading). In addition, the company has innovated many techniques for control, such as the first kraft mill odor control system, simply because it recognized problems and wanted to solve them.

On the other end of the spectrum are companies such as St. Regis, Potlatch, and Diamond International, with records indicating no concern for environmental protection. St. Regis operates 9 pulp mills, of which 7 have neither primary nor secondary water treatment. As a result, 174.3 million gallons of essentially untreated effluent are discharged daily into public waterways. Three of the Company's four large kraft mills have been totally neglected from the point of view of odor control.

Potlatch has serious air pollution problems at both of its two mills and only recently installed primary treatment at one. The extensive pollution caused by its large Lewiston, Idaho, kraft mill has made Potlatch the subject of federal air and water pollution abatement conferences.

Diamond International's four mills are all inadequate in both their air and water pollution control facilities. Only one Diamond mill has even primary water treatment. While the Company indicates "plans" to clean up, there is no intention to do so before state authorities require it. Typical of this attitude, the company opened a brand new tissue mill in Maine this year with no water treatment at all. This mill will be operated with no treatment for the next four years. Why? Because Maine's control requirements do not go into effect until then.

In general, the records of all other companies indicate perhaps one or two mills at which a real effort has been made to clean up and the rest in fair to poor condition. Most, like Diamond International, (with Mead and Hammermill high on the list) have taken a "we will wait until we are told what do do by the state" approach.

To reduce air and water pollution at all 130 mill sites over the next few years, bringing the mills up to state-of-the-art levels would cost the 24 companies, according to the CEP's estimate, a total of almost three-quarters of a billion dollars.

Table 5.1 and Figures 5.1-5.10 provide summary information both on the factual data uncovered by this study and its major conclusions.

Table 5.1
The Locations of the 131 Pulp Mills (Arranged by State and City)

State	Number of Mill Locations	Breakdown City	Company
Alabama	7	Brewton	Marcor
		Butler	American Can
		Coosa Pines	Kimberly-Clark
		Cortland	U.S. Plywood-Champion
		Mobile	International Paper
		Mobile	Scott
		Montgomery	Union Camp
		Selma	Hammermill
Alaska	1	Ketchican	Georgia-Pacific (part owners)
Arkansas	5	Ashdown	Great Northern Nekoosa
		Camden	International
		Crossett	Georgia-Pacific
		Pine Bluff	International Paper
		Pine Bluff	Weyerhaeuser
California	5	Anderson	Kimberly-Clark
		Antioch	Fibreboard
		Fairhaven	Crown Zellerbach
		Red Bluff	Diamond International
		Samoa	Georgia-Pacific
Florida	4	Fernandina Beach	Marcor
		Jacksonville	St. Regis
		Panama City	International Paper
		Pensacola	St. Regis
Georgia	6	Augusta	Continental Can
		Brunswick	Scott; Mead
		Cedar Springs	Great Northern Nekoosa
		Port Wentworth	Continental Can
		Savannah	Union Camp
		Valdosta	Owens-Illinois
Idaho	1	Lewiston	Potlatch
Kentucky	1	Wickliffe	Westvaco

Table 5.1
The Locations of the 131 Pulp Mills (Arranged by State and City) (continued)

| State | Number of Mill Locations | Breakdown | |
		City	Company
Louisiana	7	Bastrop	International Paper
		Bogalusa	Crown Zellerbach
		Hodge	Continental Can
		Port Hudson	Georgia-Pacific
		St. Francisville	Crown Zellerbach
		Springhill	International Paper
		De Ridder	Boise Cascade
Maine	8	Bucksport	St. Regis
		E. Millinocket	Great Northern Nekoosa
		Jay	International Paper
		Millinocket	Great Northern Nekoosa
		Old Town	Diamond International
		Westbrook	Scott
		Winslow	Scott
		Woodland	Georgia-Pacific
Maryland	1	Luke	Westvaco
Michigan	4	Escanaba	Mead
		Muskegon	Scott
		Menominee	Scott
		Ontonagon	Hoerner-Waldorf
Minnesota	4	Cloquet	Potlatch
		International Falls	Boise Cascade
		St. Paul	Hoerner Waldorf
		Sartell	St. Regis
Mississippi	4	Monticello	St. Regis
		Moss Point	International Paper
		Natchez	International Paper
		Vicksburg	International Paper
Montana	1	Missoula	Hoerner Waldorf
New York	10	Corinth	International Paper
		Deferiet	St. Regis
		Lyons Falls	Georgia-Pacific

Table 5.1
The Locations of the 131 Pulp Mills (Arranged by State and City) (continued)

State	Number of Mill Locations	Breakdown City	Company
		Mechanicville	Westvaco
		N. Tonawanda	International Paper
		Ogdensburg	Diamond International
		Niagara Falls	Kimberly-Clark
		Plattsburgh	Georgia-Pacific
		Plattsburgh	Diamond International
		Ticonderoga	International Paper
North Carolina	5	Canton	U.S. Plywood-Champion
		New Bern	Weyerhaeuser
		Plymouth	Weyerhaeuser
		Sylva	Mead
		Riegelwood	Riegel
Ohio	2	Chillicothe	Mead
		Circleville	Marcor
Oklahoma	2	Craig	Weyerhaeuser
		Valiant*	Weyerhaeuser
Oregon	9	Gardiner	International Paper
		Halsey	American Can
		Lebanon	Crown Zellerbach
		Salem	Boise Cascade
		St. Helens	Boise Cascade
		Springfield	Weyerhaeuser
		Toledo	Georgia-Pacific
		Wauna	Crown Zellerbach
		West Linn	Crown Zellerbach
Pennsylvania	3	Erie	Hammermill
		Lockhaven	Hammermill
		Tyrone	Westvaco
South Carolina	2	Charleston	Westvaco
		Georgetown	International Paper
Texas	2	Orange	Owens-Illinois
		Pasadena	U.S. Plywood-Champion

Table 5.1
The Locations of the 131 Pulp Mills (Arranged by State and City) (continued)

| State | Number of Mill Locations | Breakdown | |
		City	Company
Tennessee	2	Harriman	Mead
		Kingsport	Mead
Virginia	5	Big Island	Owens-Illinois
		Covington	Westvaco
		Franklin	Union Camp
		Hopewell	Continental Can
		Lynchburg	Mead
Washington	13	Anacortes	Scott
		Bellingham	Georgia-Pacific
		Camas	Crown Zellerbach
		Cosmopolis	Weyerhaeuser
		Everett	Weyerhaeuser
		Everett	Weyerhaeuser
		Everett	Scott
		Longview	Weyerhaeuser
		Port Angeles	Crown Zellerbach
		Port Townsend	Crown Zellerbach
		Steilacoom	Boise Cascade
		Tacoma	St. Regis
		Wallula	Boise Cascade
Wisconsin	17	Marinette	Scott
		Oconto Falls	Scott
		Appleton	Consolidated
		Biron	Consolidated
		Whiting	Consolidated
		Wisconsin Rapids	Consolidated (2)
		Cornell	St. Regis
		Rhinelander	St. Regis
		Tomahawk	Owens-Illinois
		Kaukauna	Hammermill
		Green Bay	American Can
		Rothschild	American Can
		Nekoosa	Great Northern Nekoosa
		Port Edwards	Great Northern Nekoosa
		Kimberly	Kimberly-Clark
		Niagara	Kimberly-Clark

*Under construction, therefore, not included in totals.

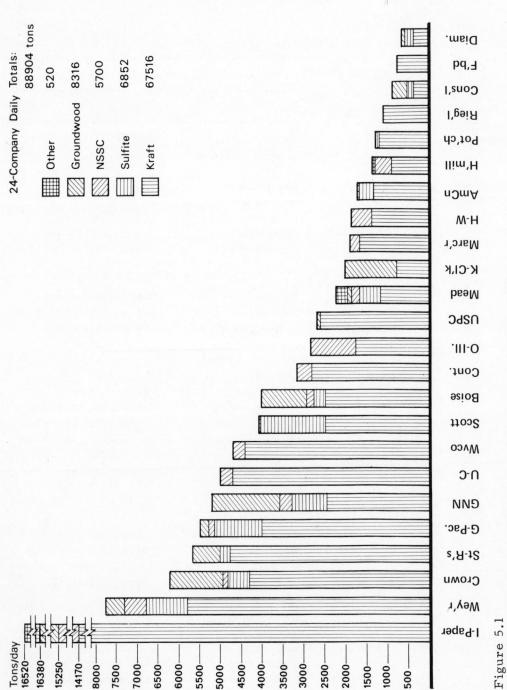

Figure 5.1
Daily Pulp Production (Tons) with Breakdown by Type of Pulp

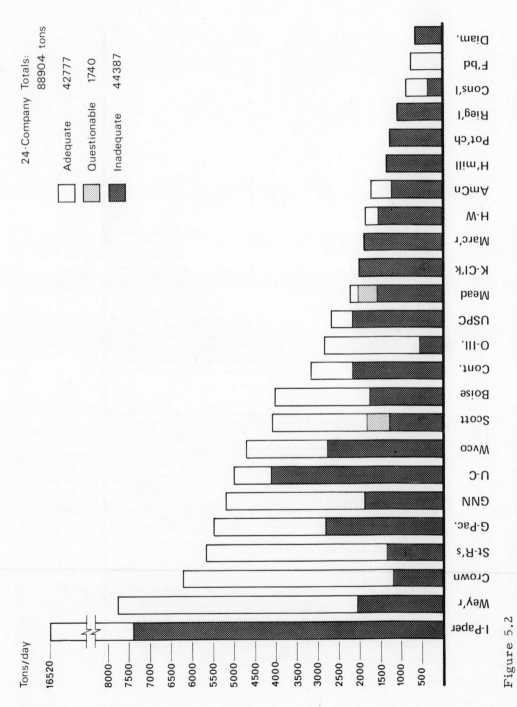

Figure 5.2
Daily Pulp Production with Breakdown by Quality of Particulate Control Per Ton Produced

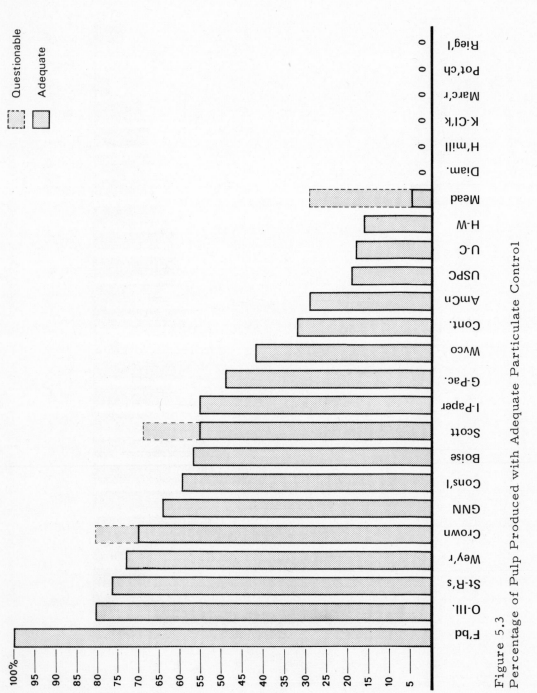

Figure 5.3
Percentage of Pulp Produced with Adequate Particulate Control

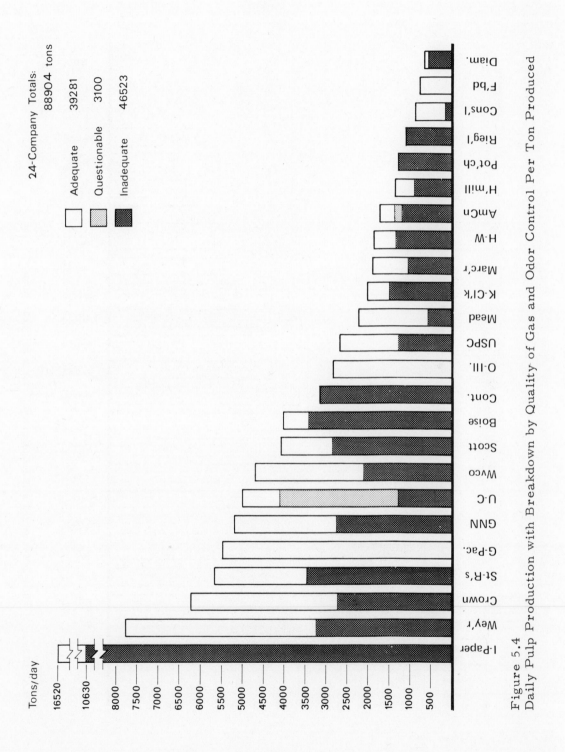

Figure 5.4
Daily Pulp Production with Breakdown by Quality of Gas and Odor Control Per Ton Produced

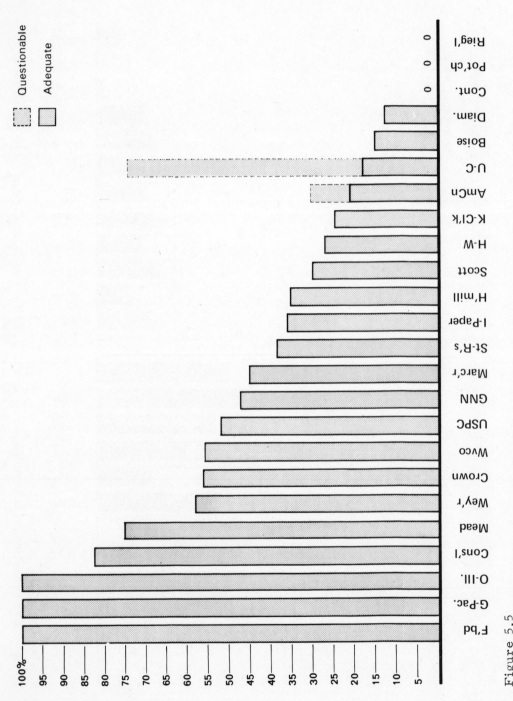

Figure 5.5
Percentage of Pulp Produced with Adequate Gas and Odor Control

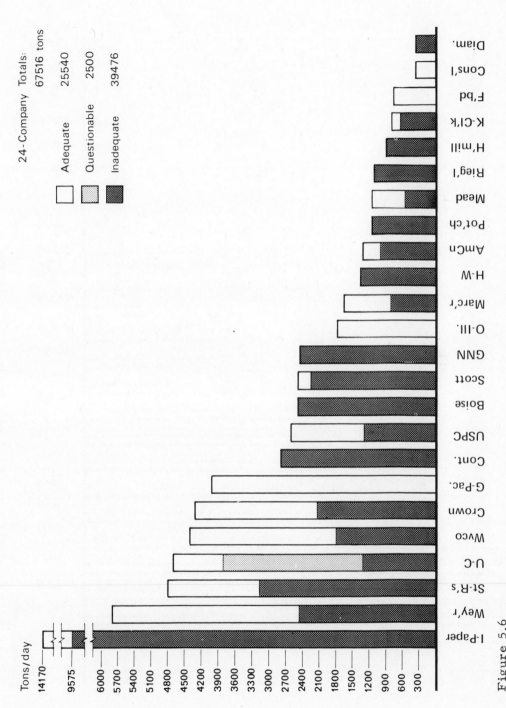

Figure 5.6
Daily Kraft Pulp Production with Breakdown by Quality of Odor Control Per Ton Produced

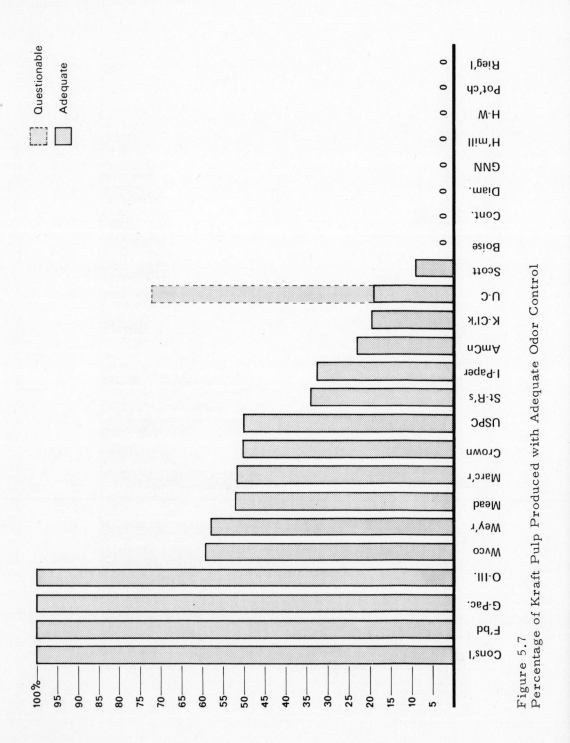

Figure 5.7
Percentage of Kraft Pulp Produced with Adequate Odor Control

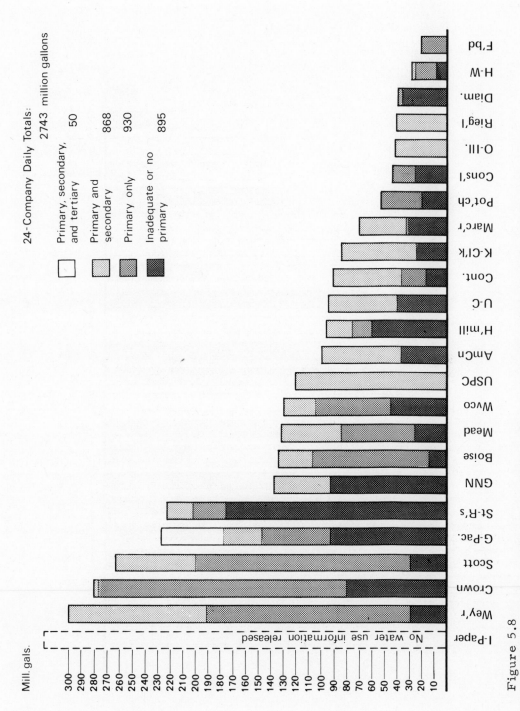

Figure 5.8
Daily Water Use (Millions of Gallons) with Breakdown by Degree of Treatment

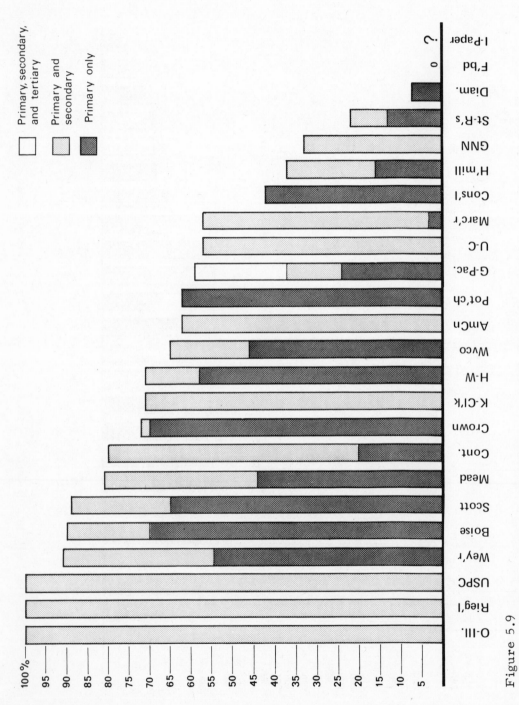

Figure 5.9
Percentage of Water Receiving Treatment Before Discharge

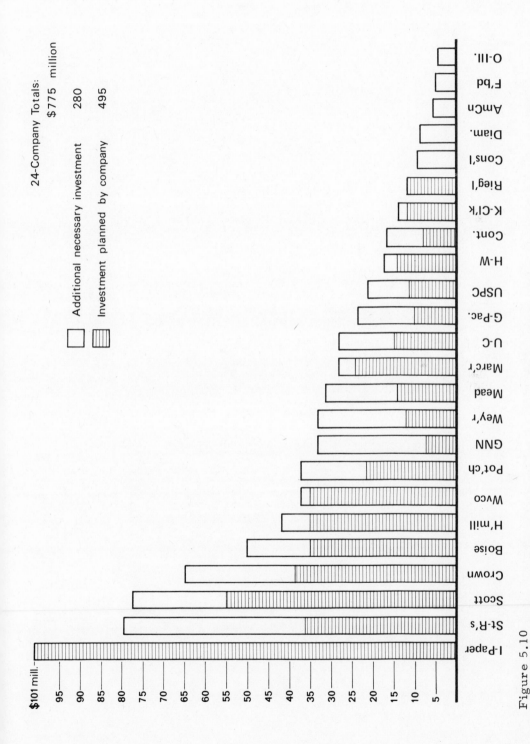

Figure 5.10
CEP Estimate of Total Capital Investment Needed for Each Company to Achieve State-of-the-Art Air and Water Pollution Control at All Mills

Part II

THE TWENTY-FOUR PULP AND PAPER COMPANIES

Chapter 6

AMERICAN CAN COMPANY
American Lane, Greenwich, Connecticut 06831

Major Products:
Metal cans, paperboard products, squeeze tubes, household paper goods, chemicals, patterns.

Major Consumer Brands:
DIXIE cups, AURORA, GALA, NORTHERN toilet and facial tissues, towels, napkins, WAXTEX waxpaper, BUTTERICK and VOGUE patterns.

Financial Data ($ Millions)	1969	1968	1967
Net Sales	1,723.7	1,633.0	1,521.8
Net Income	64.6	77.6	60.5
Capital Expenditures	115.0	164.3	141.2

Pollution Control Expenditures:
N.A.

Annual Meeting:
The annual meeting is held during April in New York.

Officers:
Chairman/President: William F. May, Chappaqua, New York (Director: Bankers Trust Company, Johns Manville Corp.; Trustee: University of Rochester, Brooklyn Polytechnic Institute, Committee for Economic Development)

Outside Directors:
Jervis J. Babb
William E. Buchanan: President, Appleton Wire Works Corp.
Richard R. Hough: Vice President, American Telephone and Telegraph Co.
Donald B. Kepp: Partner-Pitney, Hardin, Kepp, Attorneys
William H. Moore: Chairman of the Board, Bankers Trust Co.
Mundy I. Peale
W. C. Stock: Formerly Chairman of the Board, American Can Co.
C. L. Van Schaick: Formerly Executive Vice President, American Can Co.
J. C. Wilson: Chairman of the Board, Xerox Corp.
E. M. deWindt: President, Eaton, Yale & Towne

Breakdown of Total Company Production (Dollars):
Packaging (beverages, food, drugs, cosmetics,   1,164,107
household and others)

Consumer and service industries (consumer   366,640
products, service products and all others)

Venture Businesses (chemicals, printing,   192,982
and all others)

Timberholdings:
575,000 acres owned, leased, or managed in the United States; of this, owns
79,000 acres. In addition, retains the cutting rights to 2.5 million acres. The
total 1969 value of American Can's holdings (at cost): $30,932,000.

Annual Pulp Production (Estimated on a Basis of 355 Production Days):
578,650 tons (1969).

Company Pollution Overview

Whenever American Can's new Halsey, Oregon, kraft mill is mentioned, the
company's management swells with pride. Halsey, which began operation in
the Fall of 1969, is considered the cleanest mill in the country. Its tight in-
plant controls and primary and secondary water treatment have resulted in a
BOD load of under two pounds per ton of pulp; its odor control is excellent,
and the measured sulfur and particulate emissions are extremely low. American Can has had a stream of pulp and paper company technical managers
visit the mill to study the system.
    Besides the Halsey mill, the company has three other pulping operations;
a large kraft mill at Butler, Alabama, and two old small sulfite mills at
Green Bay and Rothschild, Wisconsin. The four mills produce 1,720 tons of
pulp a day and discharge 92 to 94 million gallons of waste water.
    While Halsey is certainly an outstanding facility, the Butler, Alabama,
kraft mill is in poor condition, with no odor or particulate control. American
Can's two old sulfite operations in Wisconsin are especially sore points. The
company's technical director, Elgin Sallee, spent several hours reveling in
details concerning his Halsey mill controls, but later attempts by the Council
to schedule follow-up meetings to discuss the Wisconsin mills found him
"busy." He never got free again!
    Neither Green Bay nor Rothschild has even primary water treatment, al-
though liquor recovery systems have considerably reduced the pollution loads.
The Green Bay mill is now under an order to abate pollution by the end of
1972. In Green Bay a federal demonstration project, completed in 1969, recommended a joint industrial-municipal treatment plant (using activated sludge).
American Can agreed to join. However, if this plan is accepted the plant will
take years longer to construct than the end of 1972. In this case, American
Can's mill will be given a permit to continue discharging untreated effluent
for a period of years.
    At Rothschild, no definite water improvement plans have been made.
American Can just seems to be stalling committment on the large expendi-

tures that will be needed to reduce this mill's pollution to acceptable levels.

American Can has shown very good progress and initiative in keeping its newest kraft mills clean. However, there is a clear tendency to rest on its laurels and to avoid facing the "difficult-to-cope-with" sulfite water pollution problems at the Wisconsin mills, and air pollution at Butler.

CEP estimates that a $4.75 to $5.25 million capital investment would result in adequate control at all four mills.

Table 6.1
American Can Company: Pulp Production, Water Use, and Pollution Control

Mill Location	Pulp Prod. (TPD)*	Other Prod. (TPD)*	Water Use (MGPD)**	Water Pollution Control			Air Pollution Control	
				Pri.	Sec.	Tert.	Part.	Gas, Odor
Butler, Ala.	1,000	1,250	43-53	✓	✓	✗	✗	✗
Halsey, Ore.	300	200	12-14	✓	✓	✗	✓	✓
Green Bay, Wisc.	220	400	19	✗	✗	✗	✗	—
Rothschild, Wisc.	200	330	18	✗	✗	✗	✓	✗
Total	1,720	2,180	92-104					

*Tons Per Day.
**Millions of Gallons Per Day.

Naheola Mill

Location:
Butler, Alabama

Date Built:
1955

Process:
1,000 TPD (Tons Per Day) bleached and unbleached kraft

Other Production:
350 TPD coated board
900 TPD tissue, towel, napkin and board

Total Water Use:
43-53 MGPD

Source:
Tombigbee River

Discharge:
Tombigbee River

Table 6.2
Naheola Mill: Pollution Control Record

Treatment	Overall Evaluation	Equipment
Water		
Primary*	✓	250-foot diameter clarifier with pressure pump; two 15-acre settling lagoons
Secondary*	✓	40-acre aeration lagoon
Tertiary*	✗	None
Solid waste removal	—	115-acre sludge lagoon
Air		
Particulate (fuel)	✗	Fuel: gas, bark, oil, coal. Dust collectors
Particulate (production)	✗	Recovery furnace, no precipitator. Lime kiln scrubber
Gas and odor	✗	None
Plant Emissions		
To water:		
BOD	N.A.	
Suspended solids	N.A.	
To air:		
TRS**	N.A.	

Table 6.2
Naheola Mill: Pollution Control Record (continued)

Treatment	Overall Evaluation	Equipment
Particulates	N.A.	

*In this and all the following Pollution Control Records, "primary" treatment means the removal of settleable solids; "secondary," removal of dissolved solids; and "tertiary," removal of coloring matter.
**Total Reduced Sulfur. See the glossary in Appendix 1.

Pollution Control Expenditures:
N.A.

Legal Status:
(Water) The mill operates under a permit allowing a 4 ppm (parts per million) dissolved oxygen limit downstream of the mill.
(Air) The mill exceeds state regulations.

Future Plans:
The company's technical director, Elgin Sallee, said "our water treatment handles all 43 million gallons of effluent daily, but sometimes when stream conditions are bad, (two other pulp mills also contribute their effluent upstream) the state requires a shutdown for a couple of days. We need to add more aeration and chemical (tertiary) treatment: also new precipitators and scrubbers."

Profile:
The Naheola mill has good water treatment but needs extensive improvement of its air pollution controls. American Can has made no definite plans to add either particulate or odor control equipment.

CEP Estimate of Minimum Necessary Pollution Control Expenditures Not in Company Budget: $1.25-$1.75 million total, of which $500,000-$1 million is for an odor control system; $250,000 is for improvements to power boiler particulate control; and $500,000 is for a recovery furnace precipitator.

Halsey Mill

Location:
Halsey, Oregon

Date Built:
1969

Process:
300 TPD bleached kraft

Other Production:
200 TPD tissue grades

Total Water Use:
12-14 MGPD

Source:
Willamette River

Discharge:
Willamette River

Table 6.3
Halsey Mill: Pollution Control Record

Treatment	Overall Evaluation	Equipment
Water		
Primary	✓	200-foot diameter clarifier
Secondary	✓	Two 30-acre aeration ponds. Each retains 108 million gallons for about 10 days; Six 40 hp aerators in each pond; effluent then goes to 2.5-million gallon settling pond
Tertiary	X	None
Solid waste removal		Centrifuge; sludge is dewatered and used for landfill
Air		
Particulate (fuel)	✓	Fuel: natural gas, oil. No control needed
Particulate (production)	✓	Recovery furnace: dual chamber precipitator, 99.5% efficient. Lime kiln: venturi scrubber, over 99% efficient. 300-foot stack allows wide dispersal
Gas and odor	✓	Babcock & Wilcox low-odor

Table 6.3
Halsey Mill: Pollution Control Record (continued)

Treatment	Overall Evaluation	Equipment
		system; gases at all steps vented to furnaces or kiln for thermal destruction; non-direct contact evaporator

Plant Emissions

To water:

BOD	414 lb/day (1.3 lb/ton of pulp)*
Suspended solids	N.A.

To air:

TRS	57.7 lb/day (0.19 lb/ton of pulp)*
Particulates	1,000 lb/day (3 lb/ton of pulp)*

*Based on a production rate of 300 TPD.

Pollution Control Expenditures:
$4 million (10 percent of the total capital cost of the mill).

Legal Status:
The mill meets all Oregon state regulations and American Can expects it will have no trouble meeting the 1975 standards.

Future Plans:
None.

Profile:
Halsey Mill, which went into operation in September 1969, has the finest pollution control equipment now available and is considered the cleanest mill in the United States; 10 percent of the capital cost of the mill went into insuring that it would be as clean and as odor free as a kraft mill can be. All odor has not disappeared as this is not yet technologically possible. However, the annoying problem is only periodic and localized. Not only are sulfur and particulate emissions levels very low but the mill effluent's BOD load in the first months of operation was incredibly low (Table 6.4).

The mill also feeds all residual sawdust and chips back into the digesters. The collected amount, about 250,000 tons a year of scrap materials, is efficiently used instead of being added to the pollution load.

A significant reason for the care taken in planning this mill's equipment was the fact that Oregon has the most stringent pollution control laws in the country, and public attitudes in the state have been set against allowing the establishment of any industry in scenic and recreational areas. American Can, anticipating strong opposition, not only did a superior job on its equipment, but also located the mill three miles from the bank of the Willamette River (its water source) so as not to intrude visually on the proposed river edge "green belt" which the State may develop.

CEP Estimate of Minimum Necessary Pollution Control Expenditures Not in Company Budget: None.

Green Bay Mill

Location:
Green Bay, Wisconsin

Date Built:
1910

Process:
60 TPD groundwood
160 TPD bleached sulfite (calcium base)

Other Production:
400 TPD roll toilet paper, towels, facial tissues and napkins

Total Water Use:
19 MGPD

Source:
Fox River

Discharge:
Fox River

Table 6.4
Green Bay Mill: Pollution Control Record

Treatment	Overall Evaluation	Equipment
Water		
Primary	X	Settling pond (inadequate)
Secondary	X (partial)	None, but 90% of spent sulfite liquor is not discharged. (It is condensed, evaporated, and spray-dried; recovered chemicals used in by-products, unused liquor trucked to Rothschild mill and burned)
Tertiary	X	None
Solid waste removal		See above
Air		
Particulate (fuel)	X	Fuel: gas, bark, oil. Information on equipment N.A.
Particulate (production)	—	None needed
Gas and odor	—	Company information N.A.
Plant Emissions		
To water:		
BOD	287 lb/ton of pulp (1967)	
Suspended solids	N.A.	
To air:	N.A.	

Pollution Control Expenditures:
N.A.

Legal Status:
(Water) Under abatement order. By January, 1973, BOD must be reduced to

35 pounds per ton of sulfite pulp (or a maximum of 5850 pounds per day) and suspended solids to under 20 pounds per ton of pulp and paper (or a maximum of 13,350 pounds per day based on the production capacity of the mill as determined by the state).
(Air) N.A.

Future Plans:
A research project conducted from 1958 to 1969 under a federal demonstration grant investigated the technological and economic feasibility of jointly treating the effluent from the Green Bay Metropolitan Sewerage District (including a population of 110,000 plus meat packing, vegetable canning, and other local plants) and the weak effluents from the pulping operations of four local mills: American Can, Charmin Paper Products, Fort Howard Paper and Green Bay Packaging.

After four different activated sludge systems had been tried at pilot plants, it was concluded that a full-scale treatment facility was feasible at an estimated construction cost of $22,178,000 with subsequent annual operating costs of $2,370,000. American Can and Charmin have agreed to participate in the projected plan, while Fort Howard and Green Bay Packaging have rejected it.

Profile:
The Green Bay mill is old and dirty and has totally inadequate air and water pollution controls. The one small settling pond doesn't come close to providing adequate primary treatment, and even after a liquor recovery system was installed, keeping the spent liquor out of mill effluent, the state in 1969 measured a BOD load of 287 pounds per ton of pulp.

American Can has been ordered by the Wisconsin Control Board to cut its BOD load to 35 pounds per ton of pulp by January 1, 1973. If the recommended joint treatment system is selected and approved, the industrial-municipal plant cannot be built and operable by that date. More likely it will not be ready until 1974 or 1976. In this case, American Can's permit to discharge untreated effluent will undoubtedly be extended for the interim years.

Such an extremely large investment for an old small mill would seem unlikely and, not surprisingly, the company gave no indication that a specific program had been outlined or a timetable set.

CEP Estimate of Minimum Necessary Pollution Control Expenditures Not in Company Budget: $2 million total, of which $500,000 is for primary water treatment; $1 million is for secondary treatment; and $500,000 is for improvements to power boiler particulate control.

Rothschild Mill

Location:
Rothschild, Wisconsin

Date Built:
1909

Process:
200 TPD bleached sulfite (ammonia base)

Other Production:
330 TPD special papers for packaging and wrapping, printing papers

Total Water Use:
18 MGPD

Source:
Wisconsin River

Discharge:
Wisconsin River

Table 6.5
Rothschild Mill: Pollution Control Record

Treatment	Overall Evaluation	Equipment
Water		
Primary	X	None
Secondary	X (partial)	Spent liquor treatment system evaporates, condenses, and spray-dries liquor for by-product manufacture, so little waste liquor is discharged
Tertiary	X	None
Solid waste removal		See above
Air		
Particulate (fuel)	✓	Fuel: coal, bark, red liquor. Single precipitator and fly ash arrester on boilers.
Particulate (production)	—	None needed
Gas and odor	X	N.A.
Plant Emissions	N.A.	

Pollution Control Expenditures:
$1.8 million, for a liquor treatment system.

Legal Status:
(Water) Standards have been set by the state control board for the Wisconsin River. The Rothschild mill is not in compliance; however, no abatement order has been issued.
(Air) In compliance.

Future Plans:
None.

Profile:
At the Rothschild mill, American Can completed a $9 million expansion last year providing a new daily machine capacity of approximately 300 tons of high quality uncoated and coated papers. An additional $1.8 million was invested at the pulp mill for a full liquor recovery system, with condenser, evaporator and spray drier. Good particulate and gas controls were also added to the coal boilers.

With almost $2 million already invested in liquor recovery, American Can is unlikely to make any further basic and expensive process changes. The company is equally unlikely to think about closing down the pulping operation. Still, there is a massive water pollution problem here.

Despite this fact, the company has announced no plans for primary or secondary effluent treatment. The state control board has not gotten around to filing an abatement order against the mill, so it appears that the company will take no action until such pressure is applied.

CEP Estimate of Minimum Necessary Pollution Control Expenditures Not in Company Budget: $1.5 million total, of which $500,000 is for primary water treatment, and $1 million is for secondary water treatment (costs are low because of extreme by-product manufacture).

Chapter 7

BOISE CASCADE CORPORATION
Boise, Idaho 83701

Major Products:
Building materials, paper and paper products, homes.

Major Consumer Brands:
DIVCO WAYNE travel trailers, PRINCESS cruisers, KINGSBERRY homes.

Financial Data: ($ Millions)	1969	1968	1967
Net Sales	1,726.0	1,533.1	---
Net Income	84.0	61.3	---
Capital Expenditures	95.5	91.1	---

Pollution Control Expenditures:
Total for 1957-1975 estimated at $65 million. For 1967-1969, $12 million.

Annual Meeting:
The annual meeting takes place in April, in Boise, Idaho.

Officers:
President: R. V. Hansberger, Boise, Idaho (Director: Pennsylvania Mutual
Life Insurance Co.; Pacific Western Pipeline Co.; Western Pacific Railroad;
ABC; First Security Corp.; Idaho Power Co.; VSI Corp.)

Outside Directors:
Eugene R. Black: Director and Financial Consultant, New York
James D. Bronson: Investor and Orchardist, Yakima, Washington
James E. Bryson: Farmer, Newberg, Oregon
Frederick L. Deming: General Partner, Lazard Frères Co., New York
Robert Faegre: Former President, Minnesota and Ontario Paper Co.,
Minneapolis
G. D. Flotron: Investor, St. Louis
Gilbert H. Osgood: Limited Partner, Blunt, Ellis, and Simmons, Chicago
John S. Pillsbury, Jr.: Chairman, Northwestern National Life Insurance Co.
Theodore H. Smyth: Investor, Santa Barbara, California
Hall Templeton: Investor, Portland, Oregon
E. R. Titcomb: President, Rodman Industries, Inc., St. Paul
Joseph L. Weiner: Attorney, New York
Leo D. Welch: Director and Consultant, New York

Breakdown of Company Production (as Restated after Ebasco Industries, Inc.,
Merger):

	1969		1968	
	$ Millions	Percent of Sales	$ Millions	Percent of Sales
Building materials	362	21	347	23
Construction, including housing	550	32	491	32
Recreation	209	12	80	5
Paper products	453	26	417	27
Investment, utility, and other	152	9	198	13

Plants (domestic):
Pulp and paper mills	8
Container plants	18
Envelope	5
Bag and specialty paper	4
Composite can	18
Plywood	9
Mobile homes and recreational vehicles	15
Factory built homes	6
Sawmills	14

Timberholdings:
The company owns or controls 6,284,000 acres in the United States, Canada,
the Philippines, and Colombia. The 1969 value of timberlands owned (cost
less depletion) was $77,280,000. In addition, 443,000 acres in Louisiana are
jointly owned with the Southern Natural Gas Company.

Annual Pulp Production (Estimated on a Basis of 355 Production Days):
1,425,325 tons (1969).

Company Pollution Overview

Boise Cascade is a billion-dollar corporation which ranks fifty-fifth in Fortune's list of the 500 largest American corporations. It also has the second
highest annual sales of the U.S. pulp and paper companies.[1] In the area of
pollution control, the company has the singular distinction of having opened
the only brand new kraft mill in the United States (DeRidder, Louisiana,
where pulping began in late 1969) with no odor control!
     This most recent action is clearly in harmony with Boise Cascade's long
history of "no effort" and "no initiative" in controlling air and water pollution at its mills. The company operates a total of nine domestic pulp mills
at six locations: one in Louisiana, one in Minnesota, and two each in Oregon

and Washington. Four are kraft mills, three are groundwood, one is sulfite and one is NSSC. Daily pulp production is about 4,000 tons, of which 2/3 is kraft pulp.

None of the six locations provides adequate pollution control and five are under state order to do so. Only two of the four kraft mills (DeRidder, Louisiana and St. Helens, Oregon) have adequate particulate control, and none has adequate odor control. Of the company's nine mills, only the sulfite mill in Salem, Oregon, has complete control of both particulate and gaseous emissions.

The treatment of the company's daily 133 million gallons of effluent is no better. Two of the four West Coast mills have no treatment at all. Four locations have primary treatment (DeRidder, Louisiana, Salem and St. Helens, Oregon, and International Falls, Minnesota) but only DeRidder has complete secondary treatment.

Since the establishment of a "pollution task force" in 1967, Boise Cascade has taken some action. It shut down three sulfite pulping operations because they were not worth the investment required for a cleanup. In June, 1970, the company president, R. V. Hansberger, issued a seven-point statement in which he said that it is Boise's policy to conduct its business so as to achieve:

"1. reduction and prevention of air and water pollution to the full limits of existing technology...
2. compliance in letter and spirit with all applicable regulations designed to protect the environment."

Clearly, such a policy of installing equipment "to the full limits of existing technology" would be a totally new direction for the company and it certainly wasn't in effect six months before (when odor control-less DeRidder was opened).

Many plans have been announced, allegedly in line with the new policy. However, the plans do not provide for controls "to the full limits of existing technology" either, but only to the full limits of state orders. In keeping with Mr. Hansberger's policy statement, all six mill locations will have "secondary treatment or the equivalent" in 1972. Mill by mill, this means that DeRidder has it now; Salem will change pulping processes and build an aeration lagoon; St. Helens will share a municipal treatment system; Wallula will probably have a spray irrigation system; Steilacoom hopes that primary treatment will accomplish the 90 percent BOD reduction required by "the letter of the law"; and International Falls is only required to provide 50 percent BOD reduction and only aims to do so.

Another interesting "pollution control" venture was undertaken at the old Vancouver, Washington sulfite mill. The pulping operation was shut down and a pipe was laid to carry pulp from the St. Helens mill to the paper making plant still operating at Vancouver. The company sought to declare the pipe as a tax deductible pollution control expenditure, but the Internal Revenue Service denied it. Nonetheless, closing the sulfite operation was more effective in re-

ducing water pollution at Vancouver than any treatment system, and far
cheaper. Unfortunately, St. Helens does not have any water treatment either.
Despite the IRS decision, Boise Cascade is considering this method of pollution control at International Falls; the pipe will transport pulp there from
Fort Francis, just across the border in Canada.

Although the company produces a substantial amount of paper products,
the major part of its business is linked to construction. It is heavily involved
in land development and second home sites in the West, and has been the subject of strong criticism for its practices, particularly in the Lake Tahoe area.
Not only has Boise Cascade been criticized for not preserving the trees and
ecological balance in its developments, but also for providing only septic
tanks for sewage treatment rather than a community sanitary system. In many
cases the septic tanks have contaminated the groundwater and contributed to
the pollution of one of the most beautiful lakes in the world. Apparently, the
company is not terribly upset about this; it now wants to build a development
in New York State without a community-wide sewage system.

The Council found it quite difficult to obtain information from Boise
Cascade. We spoke for over a month with three nontechnical members of the
company's vast public relations department before learning of the existence
of a "pollution control task force" and its director, Mr. Joe Kolberg. After
another month of attempts to reach him on the telephone, he turned out to be
reasonably informative but extremely slow in providing information. Of
course, his attitude that he has far more important things to do—deal with
state agencies, fight pollution—than gather data for the Council is entirely
reasonable. If only his activities had borne more fruit!

Boise Cascade's tentative pollution control expenditures for 1970-1975
total $35.1 million, and include the costs of new recovery furnaces where
necessary. The Council estimates that an additional $7 million to $15 million
will be needed (for a total of $42 million to $50 million) to adequately clean
up all mills.

Table 7.1
Boise Cascade Corporation: Pulp Production, Water Use, and Pollution
Control

Mill Location	Pulp Prod. (TPD)	Other Prod. (TPD)	Water Use (MGPD)	Water Pollution Control			Air Pollution Control	
				Pri.	Sec.	Tert.	Part.	Gas Odor
DeRidder, La.	1,200	1,050	27	✓	✓	✗	✓	✗
International Falls, Minn.	750	380	35	✓	✗	✗	✗	✗

Table 7.1
Boise Cascade Corporation: Pulp Production, Water Use, and Pollution
Control (continued)

Mill Location	Pulp Prod. (TPD)	Other Prod. (TPD)	Water Use (MGPD)	Water Pollution Control			Air Pollution Control	
				Pri.	Sec.	Tert.	Part.	Gas Odor
St. Helen's Ore.	825	540	40	✓	✗	✗	✓	✗
Salem, Ore.	250	230	18	✓	✗	✗	✓	✓
Steilacoom, Wash.	350	350	6	✗	✗	✗	✗	—
Wallula, Wash.	640	610	7	✗	✗	✗	✗	✗
Total	4,015	3,160	133					

Boise Southern Mill (a joint venture of Boise Cascade and Southern National Gas Company

Location:
DeRidder, Louisiana

Date Built:
1969

Process:
900 TPD kraft
300 TPD groundwood

Other Production:
650 TPD board and paper; 400 TPD newsprint

Total Water Use:
27 MGPD

Source:
Wells

Discharge:
Sabine River

Table 7.2
Boise Southern Mill: Pollution Control Record

Treatment	Overall Evaluation	Equipment
Water		
Primary	✓	5 million gallon clarifier
Secondary	✓	Aerated lagoon, 20-day holding pond
Tertiary	✗	None
Solid waste removal	—	Sludge disposal on site in 160 acre-foot sludge pond
Air		
Particulate (fuel)	✓	Fuel: gas, bark. Mechanical dust collector, 99% efficient
Particulate (production)	✓	Recovery furnace: 99% efficient precipitator. Lime kiln: 99% efficient scrubber
Gas and odor	✗	No odor control system
Plant Emissions		
To water:		
BOD	4,500 lb/day (3.8 lb/ton of pulp)	
Suspended solids	N.A.	
To air:	N.A.	

Pollution Control Expenditures:
$5 million of the $100 million capital cost of the mill was for pollution control. Of this, $3 million was spent for water treatment and $2 million for air pollution control.

Legal Status:
(Water) In compliance; monitoring of the Sabine River is required and shows
summer minimum dissolved oxygen levels downstream of the mill as 7 parts
per million.
(Air) In compliance.

Future Plans:
Tertiary water treatment will be added within five years.

Profile:
Boise Cascade is the largest single landholder in the state of Louisiana, controlling 443,000 acres. Despite the fact that it spent $5 million on pollution
control at the brand-new Boise Southern mill it neglected to spend $500,000
extra for an odor control system. The state air pollution agency agreed with
the company's argument that since the mill is five miles from the town of
DeRidder, odor control was not needed. There is no other new mill in the
country where such an omission has been tolerated. When asked whether in
fact the mill could be considered far enough from the town for odor to be no
problem, Boise Cascade's Manager of Environmental Control commented that
yes, "equipment operating over capacity can be smelled 35 miles away, depending on the wind." But, he added, this mill "cannot, since it is new and
operating at the capacity of the recovery unit." He also added that the typical
recovery unit, "although not this one," will operate at 150 percent of rated
capacity. On this basis, the state of Louisiana took a great deal on faith in
not requiring odor control; mill managers are expected to maximize output
by overloading, and it is hard to believe that this mill will be the sole exception in the industry.

Without an odor control system, no kraft mill can be considered to be
providing acceptable air pollution control, or to be an acceptable neighbor.
It is amazing that a corporation would spend so much money on a mill, and
so much on pollution control, without installing a system to curtail the most
offensive kraft mill pollutant. However, excepting this, the rest of the mill
is up to date, including particulate control, and Boise Cascade is very proud
of it.

Water pollution control at the mill has been very responsibly dealt with
at the urging of two state water pollution control agencies. The Sabine is a
border river, so the strict Texas requirements had to be met also.

Because the stream flow is low at the mill location, monitoring is essential to guard against discharges which would endanger fish. The BOD figures indicate that the mill is providing about 90 percent BOD reduction; very
good secondary treatment by any standard. Also, Boise Cascade has hired a
consultant to test the stream quality and fish life; he reported no change in
water quality or fish life around the mill. The reported oxygen content of
seven parts per million supports this.

The final remaining problem (discoloration) is not solely the mill's fault.
Upstream swamps impart color to the Sabine before it reaches the mill, but

a color removal system will be necessary to prevent extremely discolored waters below the mill.

CEP Estimate of Minimum Necessary Pollution Control Expenditures Not in Company Budget: $2.5-$3 million total, of which $2 million is for tertiary water treatment, and $500,000-$1 million is for an odor control system.

International Falls Mill

Location:
International Falls, Minnesota

Date Built:
1914

Process:
300 TPD bleached and unbleached kraft
150 TPD groundwood
300 TPD groundwood for Insulite Division

Other Production:
1,900,000 square feet of structural insulating board
380 TPD bleached and unbleached groundwood papers and kraft packaging papers

Total Water Use:
35 MGPD

Source:
Rainey River

Discharge:
Rainey River

Table 7.3
International Falls Mill: Pollution Control Record

Treatment	Overall Evaluation	Equipment
Water		
Primary	✓	200-foot diameter clarifier (3.5 million gallon capacity) installed 1970

Table 7.3
International Falls Mill: Pollution Control Record (continued)

Treatment	Overall Evaluation	Equipment
Secondary	X	None
Tertiary	X	None
Solid Waste removal		See above
Air		
Particulate (fuel)	X	Fuel: coal. Multiple cyclone collectors, "65-80% efficient"
Particulate (production)	X	Recovery furnace precipitator. Rated efficiency, 99% actual efficiency, 85% (inadequate)
Gas and odor	X	None
Plant Emissions	N.A.	

Pollution Control Expenditures:
$2.2 million for the clarifier.

Legal Status:
(Water) The mill has a permit for construction in order to be in compliance.
It has until November, 1972, to install and have in operation a secondary
treatment facility if it is not otherwise successful in achieving a 50 percent
reduction in the BOD of its effluent to 50,000 pounds per day.
(Air) The company has filed a compliance program with the state; the timetable has not yet been approved. A state abatement action is being prepared
citing excessive smoke, particulate matter, odor, and noncomplying incineration equipment.

Future Plans:
The company program submitted to the state indicates a $6 million expendi-
ture to: incinerate sludge; change to gas for power (12/70); install a lime
kiln and scrubber (11/71); change to a dry-barking process (11/71); install an
odor control system, and incinerate noncondensible gases (7/72). The dates
indicated are those submitted to the state but not yet approved.

Profile:
This is an extremely old mill, and it shows its age.

Pulp-fiber effluent forms a gelatinous slime on the Rainey River outlet of
Boise Cascade's International Falls mill. Floating globs of this effluent have
fouled fishlines as far as 70 miles downriver, a resident said. Photo by Mike
Zerby, Minneapolis Tribune; © The Minneapolis Star and Tribune Company.

After many decades of virtually no pollution control efforts beyond closing the sulfite mill on the site in 1967, the situation is rapidly changing under state pressure. A clarifier has just been installed, and excessive flyash emissions from the coal boiler are being eliminated by conversion to gas for power. However, these improvements are "a drop in the bucket" compared to the problems that remain. There is no question that the present recovery system is operating far over capacity, thus maximizing odorous emissions to the surrounding area. Over-capacity operation also has reduced the efficiency of the precipitator, and the complete absence of a lime kiln means that many gases are being emitted which are usually incinerated at kraft mills.

The company's environmental director, Joe Kolberg, explained that the recovery furnace is not really "overloaded," since it still converts the kraft chemicals very nicely. It is just that at the high rate of production, very poor performance can be expected on particulate and odor control. "After all," he said, "you can get chemical conversion up to 200 percent of rated capacity." He neglected to mention that at such levels of operation, odorous emissions will increase about 600 percent over those at capacity.[2]

Water pollution has been just as severe as air pollution. Until this fall, the 35 million gallons of mill effluent received no treatment at all. Mr. Kolberg said that the state daily BOD estimate of 94,000 lbs. sounded "about right" although he said that it might be high because it includes the effluent from the board mill. The BOD of the discharge has been somewhat reduced by the recent installation of a clarifier, but secondary treatment will most certainly be required despite the state's seeming sympathy. The other mills in Minnesota are being required to provide 80 percent BOD reductions; Boise Cascade must reduce its discharge by only 50 percent.

The company has been slow to submit an air pollution compliance schedule to the state agencies because of indecision about the future of this mill. Environmental considerations would have required that it shut down long ago. Now it will be economically feasible to install all necessary pollution control equipment (including a new kraft recovery furnace) only if the mill can be expanded simultaneously. But the wood supply in the area is not sufficient to support a larger mill. If the state requires pollution control measures beyond those which Boise Cascade has proposed, the company will probably cease pulping at International Falls and pipe pulp from Fort Francis, just across the Canadian border, where an entire new kraft recovery system with pollution controls is being built. However, even if kraft pulp is not produced at International Falls, there will still be a water pollution problem from the Insulite Division's mill discharges.

CEP Estimate of Minimum Necessary Pollution Control Expenditures Not in Company Budget: $2.5-$9.5 million total, of which $2 million is for secondary water treatment; $500,000 is for a recovery furnace precipitator; and $7 million is for a new recovery furnace. (This furnace may be required, because of the age of the present facility, for adequate odor control.) Or, $2 million total, for a pipeline to bring pulp in from St. Francis if the pulping operation is stopped here.

St. Helens Mill

Location:
St. Helens, Oregon

Date Built:
1926

Process:
825 TPD kraft (of which 700 TPD is bleached)

Other Production:
540 TPD printing papers; kraft wrapping, envelope, gumming, waxing, bag, butchers, and other papers

Total Water Use:
40 MGPD

Source:
Multonomah Channel

Discharge:
Multonomah Channel

Table 7.4
St. Helens Mill: Pollution Control Record

Treatment	Overall Evaluation	Equipment
Water		
Primary	✓	3.5 million gallon clarifier
Secondary	✗	None
Tertiary	✗	None
Solid waste removal	—	Landfill on mill site and in St. Helens
Air		
Particulate (fuel)	✓	Fuel: gas, interruptible oil. No controls needed
Particulate (production)	✓	Two recovery furnace precipitiators, 92%, 99.5% ef-

Table 7.4
St. Helens Mill: Pollution Control Record (continued)

Treatment	Overall Evaluation	Equipment
		ficient. Two lime kiln scrubbers, 90%, 99% efficient
Gas and odor	X	Weak black liquor oxidation (inadequate)

Plant Emissions

To water:

BOD	52,000 lb/day (65 lb/ton of pulp)
Suspended solids	N.A.

To air:

TRS	2,500–3,000 lb/day (about 3.4 lb/ton of pulp)
Particulates	7,625 lb/day (9.4 lb/ton of pulp)

Pollution Control Expenditures:
$4.5 million.

Legal Status:
(Water) The mill is required to install secondary treatment by July, 1971 to limit BOD discharge to 12,000 pounds per day. Present permit limits daily BOD to 55,000 pounds except at times of fish passage when it may be no more than 25,000 pounds.
(Air) On a compliance schedule to meet kraft mill emission regulations.

Future Plans:
A joint secondary treatment system (aerated lagoon) will be installed by January, 1971 at any estimated cost to the company of $2.6 million. The company is also planning to spend $8.5 million for heavy black liquor oxidation, incineration of noncondensible gases, a new lime kiln scrubber, new recovery furnace and precipitator, and new smelt tank vent scrubbers.

Profile:
This mill, which supplies pulp to Boise Cascade's paper mill in Vancouver, Washington, by means of a pipeline, has good particulate control, only fair gas and odor control, and inadequate water pollution control. This is somewhat ironic in light of the fact that the company's sulfite pulp mill in Vancouver was closed down in 1968 because of its water pollution and St. Helens is the replacement source of pulp.

Also, both the mill and town put raw sewage and effluent into a lagoon "only a few hundred yards from the Columbia River."[3]

As a result of state pressure for better water treatment a joint industrial municipal secondary treatment plant, financed by a $2 million city bond issue, is now under construction. Boise Cascade will bear the major part of the bond retirement cost and 90 percent of the operating costs, based on estimates of its daily BOD contribution. When completed, the plant is expected to remove 85 percent of the BOD from the combined effluent.

Similarly, the mill's air pollution improvements have been planned in response to state regulations. By 1975, TRS (total reduced sulfur) emissions must be reduced by 50 percent and particulate emissions by over 50 percent. At present, despite the weak black liquor oxidation, strong mill odors are carried over the city of St. Helens by southern and southeastern winds.

This mill is an excellent example of why stringent state pollution control laws and enforcement are so necessary.

CEP Estimate of Minimum Necessary Pollution Control Expenditures Not in Company Budget: None.

Salem Mill

Location:
Salem, Oregon

Date Built:
1924

Process:
250 TPD bleached calcium-base sulfite

Other Production:
230 TPD white and colored bond writing, envelope, ledger, offset, mimeograph, and bleached sulfite specialty papers

Total Water Use:
18 MGPD

Source:
Mill Creek

Discharge:
Willamette River

Table 7.5
Salem Mill: Pollution Control Record

Treatment	Overall Evaluation	Equipment
Water		
Primary	✓	1.5 million gallon clarifier
Secondary	X (partial)	Pond system for spent liquor; liquor discharged during river's high flow (November-February)
Tertiary	X	None
Solid waste removal	—	Clarified sludge dumped on company-owned Minto Island
Air		
Particulate (fuel)	✓	Fuel: gas, interruptible oil. No controls needed.
Particulate (production)	—	None needed
Gas and odor	✓	Showers on blow stacks for sulfur dioxide control
Plant Emissions		
To water:		
BOD	Summer, 9,800 lb/day (39 lb/ton of pulp) Winter, 150,000 lb/day (600 lb/ton of pulp)	
Suspended solids	N.A.	
To air:	N.A.	

Pollution Control Expenditures:
$800,000 of which $560,000 was for the clarifier.

Legal Status:
(Water) Secondary treatment required by 1972; the permit will limit summer
BOD to 8,000 pounds per day and winter BOD to 10,000 pounds per day. The
present permit limits daily BOD between June and November to 10,000 pounds,
and 135,000 pounds in the winter.
(Air) No state sulfite mill standards yet.

Future Plans:
About $7 million will be spent to improve the pollution control at this mill by
1972. Equipment will include a 150 million gallon aeration lagoon; spent sulfite liquor evaporation and burning system; sulfur dioxide scrubber; and a
complete blow-gas scrubber system.

Profile:
Although some steps have been taken to solve the air pollution problems at
this mill, water treatment has been another story. The mill has been utilizing its holding pond system as an alternative to real water treatment with the
state's sanction. The system is not complex. The effluent is simply stored
for nine months to allow it to slowly and naturally break down. Unfortunately,
nature takes her time about degrading calcium sulfite effluent and during
November and December when effluent is discharged the BOD load is enorm-
ously heavy: 150,000 pounds per day.

When the planned liquor treatment system and lagoon are in, the mill will
be providing legally acceptable treatment, and better than a 90 percent BOD
reduction. However, even at that, 40 pounds of BOD will be discharged per
ton of pulp.

CEP Estimate of Minimum Necessary Pollution Control Expenditures Not
in Company Budget: $2 million total, for secondary treatment to provide
further BOD reduction beyond that required by the state.

Steilacoom Mill

Location:
Steilacoom, Washington

Date Built:
1926; acquired in 1969 from W. Tacoma Newsprint Co.

Process:
350 TPD groundwood

Other Production:
350 TPD newsprint

Total Water Use:
6 MGPD

Table 7.6
Steilacoom Mill: Pollution Control Record

Treatment	Overall Evaluation	Equipment
Water		
Primary	X	None
Secondary	X	None
Tertiary	X	None
Solid waste removal	—	"98% efficient savealls"
Air		
Particulate (fuel)	X	Fuel: gas, hog fuel. No control equipment
Particulate (production)	—	None needed
Gas and odor	—	None needed
Plant Emissions		
To water:		
Suspended solids	7,000 lb/day (20 lb/ton of pulp)	

Pollution Control Expenditures:
$200,000 of which $180,000 was for savealls (1952-1959).

Legal Status:
(Water) State requires primary treatment and outfall pipe.
(Air) In compliance.

Future Plans:
A $1.25 million program is planned for completion by December, 1971. It
will include a clarifier ($300,000), a 3/5-mile long outfall with diffuser
($600,000), a new incinerator ($100,000) and a cyclone scrubber for the bark
boiler ($250,000).

Profile:

This mill, only recently acquired by Boise Cascade, illustrates the relatively small pollution created by an unbleached groundwood pulp mill. The problem is primarily one of solids because there is little dissolved organic matter. Although the previous owners made no substantial pollution control investments, cleanup will be simple and inexpensive. After installation of the clarifier, the company said, mill effluent will have a minimal amount of solids and a BOD content of only about "one part per million," compared to the four parts per million discharged from the first-rate secondary treatment system at the new DeRidder kraft mill.

However, it is extremely hard to find any water with a BOD of 1 ppm and CEP feels that to get an adequate BOD reduction, secondary treatment will be needed here.

CEP Estimate of Minimum Necessary Pollution Control Expenditures Not in Company Budget: $500,000 total, for secondary water treatment (low because mill's water use is relatively small).

Wallula Mill

Location:
Wallula, Washington

Date Built:
1956

Process:
450 TPD kraft
190 TPD NSSC

Other Production:
190 TPD NSSC board
420 TPD linerboard

Total Water Use:
7 MGPD

Source:
Wells, Columbia River

Discharge:
Columbia River

Table 7.7
Wallula Mill: Pollution Control Record

Treatment	Overall Evaluation	Equipment
Water		
Primary	X	None
Secondary	X (partial)	None. However, spent NSSC liquor is not discharged; it is cross-recovered for use in the kraft mill
Tertiary	X	None
Other	—	5-6 million gallon spill pond. Spray irrigation project
Air		
Particulate (fuel)	✓	Fuel: gas, interruptible oil. No controls needed
Particulate (production)	X	Recovery furnace, 80-85% efficient venturi scrubbers (inadequate). Lime kiln: 90% efficient scrubber
Gas and odor	X	None
Plant Emissions	N.A.	

Pollution Control Expenditures:
$300,000 since 1967.

Legal Status:
(Water) Mill is operating under permit; secondary water treatment required by 1972.
(Air) On compliance schedule, does not yet meet required TRS or particulate limits for kraft mills.

Future Plans:
$1 million spray irrigation system; NSSC recovery system and odor control

system by July, 1972 or an entire new kraft recovery unit by January, 1975. The company has budgeted $8.75 million for the necessary air pollution control measures.

Profile:

Although this mill is comparatively young, water treatment and odor control are virtually nonexistent. The company took steps to control particulate emissions some time ago, but the equipment needs improvement because equipment which once may have been the best available is often no longer adequate.

Boise Cascade is uncertain how it will achieve compliance with state water regulations. It is presently conducting a pilot spray irrigation project which Washington State University joined this year. The system "would turn the arid desert-type land in this area into productive acreage" by means of waste water piped from the mill to 1200 acres of farmland which Boise Cascade holds on 20-year lease. If the system is found ecologically safe, it will be enlarged to full scale at a projected cost of about $1 million, substantially less than the cost of primary and secondary water treatment. However, if spray irrigation proves unfeasible and Boise Cascade has to build a conventional treatment system, the mill will not meet the 1972 state deadline.

The air pollution control improvements will depend on whether the mill is expanded because, as a company spokesman explained, it will not be economically feasible to install the best pollution control equipment on a middle-aged, middle-sized mill. He explained that the present rate of return on capital at a mill might be 20 percent; with the installation of "nonproductive" pollution control equipment, this might fall to 10 percent; but if the capacity were increased, the rate of return might go back up to 15 percent.

Boise Cascade may have ignored Wallula's odor problems because an animal feed lot next door to the mill takes first prize in odor production. However, with the 1975 state gas and particulate emission standards on the way, the company is planning to install a new NSSC recovery system and either a new, expanded kraft recovery system or addition to existent kraft recovery equipment of black liquor oxidation and a new scrubber for the lime kiln. Complete rebuilding of both recovery operations seems to be the preferred approach and would cost Boise Cascade a total of $8.75 million.

CEP Estimate of Minimum Necessary Pollution Control Expenditures Not in Company Budget: None.

References for Chapter 7

1. "Top 50," Chem. 26, June, 1970, p. 32.

2. Control of Atmospheric Emissions in the Wood Pulping Industry, Vol. I, by E. R. Hendrickson. Gainsville, Florida: Environmental Engineering, Inc.; and Greenville, South Carolina: J. E. Sirrine Company, March 15, 1970. Pages 4-11, 4-12.

3. Interview with Leo Nickelson, City Recorder, St. Helens.

Chapter 8

CONSOLIDATED PAPERS INCORPORATED
Wisconsin Rapids, Wisconsin 54494

Major Products:
Paper, pulp, paper products, and laminated plastic; Consolidated is the
leading producer of coated packaging paper.

Major Consumer Brands:
None.

Financial Data ($ Millions)	1969	1968	1967
Net Sales	127.7	117.1	115.5
Net Income	5.3	4.7	8.1
Capital Expenditures	6.7	10.0	33.0

Pollution Control Expenditures:
$17,500,000, 1939-1970; in 1969, $1,460,000.

Annual Meeting:
The annual meeting takes place during April in Wisconsin Rapids.

Officers:
Chairman: Stanton W. Mead (President and Director, Wisconsin Valley Im-
provement Company; Director: Illinois National Bank and Trust Co. First
Wisconsin Bankshares Corp.; Emeritus Trustee, Lawrence University)
President: George W. Mead II, Wisconsin Rapids, Wisconsin (Wisconsin De-
velopment Authority member, Governor's Commission on Education member,
Trustee of Lawrence University; Director: Soo Line Railroad, TAPPI (Tech-
nical Association of the Pulp and Paper Industry))

Outside Directors:
Eugene Abegg: Chairman of the Board, Illinois National Bank and Trust Co.,
Director: Central Illinois Electric & Gas, Woodward Governor Company,
Rockford Newspapers, Inc., and others
Mrs. Henry P. Baldwin: Director, University of Wisconsin Foundation, Board
of Curators of State Historical Society; Trustee: Northland College, Beloit
College; member of Wisconsin Development Authority
Merritt D. Hill: Director: Ann Arbor Railroad Co., Dundee Cement Co.
Milo B. Hopkins: Former Executive Vice President, Manufacturers Hanover
Trust; Director: Eltra Corp., James Talcott, Inc., Schenley Industries, Inc.,
Copperweld Steel Co., Brockway Glass Co., Wean-United Inc.
D. Richard Mead: Chairman of the Board, D. R. Mead & Co.; Director:
First National Bank and Trust Co. of Eustis, Fla., Bank of Mount Dora, Fla.,
member of advisory board, First National Bank of Miami

Breakdown of Total Company Production (tons):

Enamel papers	362,475
Packaging papers	21,994
Other papers	5,499
Paperboard	2,706
Paperboard products	32,160
Pulp	17,635

Plants:
The company has nine domestic facilities:

Enamel papers and paperboard (and pulp)	1
Enamel papers (and pulp)	2
Coated packaging and carbonless papers	1
Corrugated containers, cartons, tubes and cores	1
Corrugated packaging, paperboard tubes, and laminates	1
Kraft pulp	1
Mitscherlich sulfite pulp and chemical products	1
Consoweld laminated plastic	1

Timberholdings:
The company owns 585,000 acres of land in Wisconsin, Michigan, Minnesota, South Dakota, and Ontario, Canada. The total 1969 value of Consolidated's holdings was $1,066,291.

Annual Pulp Production:
307,075 tons (1969)

Company Pollution Overview

"Water pollution is being opposed by a rather vociferous and righteously in-dignant minority. When you really think about it, there are very few of your neighbors in your community who really want to fish or swim downstream of your mills even if the water were clear."[1]

Consolidated Papers, Inc. is the "world's largest producer of coated papers" for magazines and books. The company operates five pulp mills in Wisconsin: two at Wisconsin Rapids (bleached groundwood and kraft), one at Whiting (groundwood), one at Appleton (calcium sulfite), and one at Biron (groundwood). These mills produce 865 tons of pulp and 1328 tons of paper daily, using 43 million gallons of water. All but the Appleton mill discharge their effluent into the Wisconsin River.

Consolidated does not have serious air pollution problems. Three of the mills have a "clean" fuel source and the kraft mill was built with particulate and odor controls. Excess particulate emissions (fly ash from coal burning at the Biron mill and dust from the by-product manufacture at Appleton) could be reduced by changing fuel and installing scrubbers.

The water pollution situation is a different story. None of the Consoli-

dated mills has adequate water treatment. The three groundwood mill locations, which also produce 1328 tons of coated and enameled paper daily, have
in-plant controls to reduce fiber and coating material losses, but only two
have primary clarifiers and none has secondary treatment. The clarifier at
the Biron groundwood/paper mill settles out 18,000 pounds of solids daily.
Such outputs are an indication that this treatment is "de rigeur" for papermaking effluent.

Until 1968, Consolidated discharged spent liquor from its two calcium
sulfite mills into the Wisconsin and Fox Rivers. However, the company was
not unaware of the problem and actively sought better solutions to the liquor
disposal problem. In 1939 it joined a cooperative research program and between that time and 1968, spent millions of dollars on projects for liquor disposal such as evaporation and burning, an experimental recovery system,
soil filtration, use of the liquor as a road binder or in the manufacture of
yeast, and a pilot reverse osmosis unit. In 1968, the company concluded that
the problem of calcium sulfite liquor treatment was overwhelming; it closed
down its sulfite mill at Wisconsin Rapids and began operating a new kraft mill
on the site. This one step reduced the Wisconsin Rapids mill's BOD discharge
by 70 percent and accounted for half of the total BOD reduction achieved between 1967 and 1970 by all the pulp and paper mills on the Wisconsin River.
This reduction was in spite of the fact that the new kraft mill was built with
only a six million gallon settling poind to treat 12 million gallons of effluent.

Unfortunately, the sulfite mill at Appleton continues to operate with 25
percent of its waste liquor discharged completely untreated.

Despite creditable efforts to treat spent sulfite liquor, Consolidated's
history is one of avoiding the expense of primary and secondary facilities by
refurbishing or augmenting in-plant controls. These are simply not sufficient
for reducing oxygen-demanding dissolved solids. It would be well if George
Mead II, President of Consolidated, would take his own words to heart: "...
the day of laissez-faire is pretty much behind us... one of the most glaring
areas in which we must show a much more responsible attitude is in the
area of pollution control."[2]

CEP estimates that if all the company's mills are to be cleaned up, a
total investment of $9,290,000 will be required. ($290,000 of this has already
been announced).

Table 8.1
Consolidated Papers Incorporated: Pulp Production, Water Use, and Pollution Control

Mill Location	Pulp Prod. (TPD)	Other Prod. (TPD)	Water Use (MGPD)	Water Pollution Control			Air Pollution Control	
				Pri.	Sec.	Tert.	Part.	Gas, Odor
Appleton, Wisc.	150	--	8	X	X	X	X	X
Biron, Wisc.	200	560	10	✓	X	X	X	—
Whiting, Wisc.	100	313	5	X	X	X	✓	—
Wisconsin Rapids, Wisc.	55	455	8	✓	X	X	—	—
Wisconsin Rapids, Wisc. (kraft)	360	--	12	X	X	X	✓	✓
Total	865	1,328	43					

Appleton Mill (Appleton Division)

Location:
Appleton, Wisconsin

Date Built:
(Acquired) 1916

Process:
150 TPD bleached sulfite (calcium base) (company figure)

Other Production:
By-product manufacture of 110-130 TPD chemicals for oil well drilling industry

Total Water Use:
8 MGPD

Source:
Fox River

Discharge:
Fox River

Table 8.2
Appleton Mill: Pollution Control Record

Treatment	Overall Evaluation	Equipment
Water		
Primary	X	None
Secondary	X (partial)	None. However, 75% of the spent liquor is not dis-charged, but is used for by-product manufacture
Tertiary	X	None
Other	—	Industry's first large-scale pilot reverse osmosis unit to clean dilute pulp effluent, partially financed by FWQA grant
Air		
Particulate (fuel)	✓	Fuel: natural gas. No con-trols needed
Particulate (production)	X	Wet "sly" scrubber col-lects dust from by-product manufacture from one drier. Two other driers have no controls
Gas and odor	X	None

Plant Emissions

To water:

BOD   30,880 lb/day
                       (268 lb/ton of pulp)

Table 8.2
Appleton Mill: Pollution Control Record (continued)

Treatment	Overall Evaluation	Equipment
Suspended solids	8,260 lb/day (55 lb/ton of pulp)	
To air:	N.A.	

Pollution Control Expenditures:
To date $2.3 million has been spent, primarily for the spent liquor recovery system.

Legal Status:
(Water) The mill was cited in 1968 as not being in compliance, and is presently under abatement order.
(Air) Wisconsin air standards under review by E.P.A.

Future Plans:
A $40,000 scrubber will be installed on the second dryer this year.

Profile:
At Appleton, Consolidated pioneered in concentrating spent sulfite liquor for chemical recovery. In 1953, a $500,000 evaporator was installed which reduced the daily BOD discharge from 90,000 pounds to 40,000 pounds. Two more evaporators were installed in 1962 and 1967 and about 75 percent of the pulping wastes are now recovered in a saleable by-product form.

Unfortunately, even with evaporation and recovery, the BOD discharge is still 268 pounds per ton of pulp; this is not bad for sulfite mills but too high to meet the standard set for the Fox River. The mill was cited in 1968 and received orders in December, 1969, to achieve a daily BOD load of 4,900 pounds (35 pounds per ton) and a suspended solids load of 2,800 pounds (20 pounds per ton).

According to Mr. Donald Pryor, Manager of Environmental Control, these orders "are very difficult to meet ... We can't afford to spend the money needed to meet the standards and our alternatives at the present time are limited. One consideration is how much waste we can recover as a by-product." This mill is facing the same problems as are many other small, marginally profitable calcium sulfite mills. The technology for cleaning up is limited and expensive; there is, in this case, little land available for large scale primary and secondary treatment; and joint industrial-municipal treatment is uneconomical.

Consolidated President Mead's 1967 prediction has been fulfilled.

"Frankly, the paper industry, for the most part, in the past has been brushing (pollution abatement) under the rug in hopes that it will not have to pay the piper. This day, if it hasn't ended, will end very soon."[3]

CEP Estimate of Minimum Necessary Pollution Control Expenditures Not in Company Budget: $2 million total, for primary treatment and equivalent secondary treatment by using 100 percent of spent sulfite liquor in by-product manufacture.

Biron Mill (Biron Division)

Location:
Biron, Wisconsin

Date Built:
1894; purchased 1911; modernized and expanded since

Process:
200 TPD bleached groundwood (company figure)

Other Production:
560 TPD enamel book paper (company figure)

Total Water Use:
10-11 MGPD

Source:
Wisconsin River

Discharge:
Wisconsin River

Table 8.3
Biron Mill: Pollution Control Record

Treatment	Overall Evaluation	Equipment
Water		
Primary	✓	150-foot diameter clarifier for 85% of pulp and paper wastes installed January, 1970. Three sludge lagoons (100- by

Table 8.3
Biron Mill: Pollution Control Record (continued)

Treatment	Overall Evaluation	Equipment
		400-foot) for further settling of 18,000 lb/day of solids
Secondary	X	None
Tertiary	X	None
Solid waste removal	—	Landfill
Other	—	Savealls on paper machines, screens, incineration of woodroom bark
Air		
Particulate (fuel)	X	Fuel: coal. Scrubber, 93-94% efficient (inadequate)
Particulate (production)	—	None needed
Gas and odor	—	None needed
Plant Emissions		
To water:		
BOD	N.A.	
Suspended solids	10,000 lb/day (50 lb/ton of pulp; company figure)	

Pollution Control Expenditures:
$425,000 for primary treatment for pulp- and paper-making wastes; $50,000 for scrubbers and dust collectors.

Legal Status:
(Water) Review hearing in late 1970 set a combined BOD limit for both Wisconsin Rapids mills and the Biron mill of 19,150 pounds per day, to be achieved by September 1975.

(Air) Wisconsin air standards under review by EPA.

Future Plans:
N.A.

Profile:
This is Consolidated's oldest but not quite worst mill and is located four miles upstream from the Wisconsin Rapids Division mills. It produces 200 tons of bleached groundwood pulp and 560 tons of enamel book paper daily. A clarifier, installed in January, 1970, has reduced the daily discharge of fibers, clay and solids from 13.33 tons to 5 tons although only 85 percent of the effluent is treated: there is no secondary treatment.

   CEP Estimate of Minimum Necessary Pollution Control Expenditures Not in Company Budget: $1.25 million total, of which $1 million is for secondary treatment, and $250,000 is to improve the coal boiler scrubber.

Whiting Mill (Wisconsin River Division)

Location:
Whiting, Wisconsin

Date Built:
1919, purchased 1945

Process:
100 TPD bleached groundwood

Other Production:
93 TPD coated packaging and carbonless paper
220 TPD enamel book

Total Water Use:
5 MGPD

Source:
Wells, Wisconsin River

Discharge:
Wisconsin River

Table 8.4
Whiting Mill: Pollution Control Record

Treatment	Overall Evaluation	Equipment
Water		
Primary	X (partial)	Filter-press screen collects, dewaters 3 TPD of solids (installed, 1968). 430,000 gallon capacity lagoon system retains paper mill effluent for 1 day
Secondary	X	None
Tertiary	X	None
Solid waste removal	—	Incineration of screened solids
Other	—	Savealls. Reused white water (4,400 gallons/day)
Air		
Particulate (fuel)	✓	Fuel: gas. No controls needed
Particulate (production)	—	None needed
Gas and odor	—	None needed

Plant Emissions

To water:

BOD	5,500 lb/day	
Suspended solids	18,920 lb/day	

Pollution Control Expenditures:
The company stated that such figures were "not available. We attempt to achieve minimum losses through process design, etc. Dollar costs would not be meaningful." However, a March, 1970, publication, "Report on Consolidated's Environmental Protection Program" announced that $369,000 was spent by the Wisconsin River Division for savealls and other devices between 1945 and 1969.

Legal Status:
(Water) The mill operates under orders which expire September, 1971.
(Air) Wisconsin air standards under review by EPA.

Future Plans:
A $250,000 settling lagoon system will be installed at the end of 1971 to re-
duce the solids discharged by 80 percent.

Profile:
Overall water treatment at this mill is inadequate. Since in-plant equipment
did not satisfactorily reduce the BOD, and the settleable solids load increased
over the past three years, primary treatment lagoons are now finally planned.

   CEP Estimate of Minimum Necessary Pollution Control Expenditures Not
in Company Budget: $1 million total, for secondary treatment.

Wisconsin Rapids Mill (Wisconsin Rapids Division)

Location:
Wisconsin Rapids, Wisconsin

Date Built:
1900

Process:
55 TPD bleached groundwood

Other Production:
80 TPD linerboard, container, and other stock
375 TPD enameled printing papers

Total Water Use:
8 MGPD

Source:
Wisconsin River

Discharge:
Wisconsin River

Table 8.5
Wisconsin Rapids Mill (Wisconsin Rapids Division): Pollution Control Record

Treatment	Overall Evaluation	Equipment
Water		
Primary	✓	Two clarifiers (200,000 and 75,000 gallon, installed in 1961, 1970). Centrifuge
Secondary	✗	None
Tertiary	✗	None
Solid waste removal	—	Company landfill
Other	—	Savealls on paper machines
Air		
Particulate (fuel)	—	None needed; hydroelectric and steam power generated at kraft mill on site
Particulate (production)	—	None needed
Gas and odor	—	None needed
Plant Emissions	N.A.	

Pollution Control Expenditures:
N.A.

Legal Status:
(Water) Review hearing in late 1970 set a combined BOD limit for both
Wisconsin Rapids mills and the Biron mill of 19,150 pounds per day, to be
achieved by September, 1975.
(Air) In compliance; no emissions.

Future Plans:
Possible third clarifier for paper mill wastes.

Profile:
The main pollution load from this old mill results from the coating and en-
ameling paper process. However, clarifiers and the centrifuge recover 4.5

of every 5 tons of coating solids processed. Bleaching wastes require, and do not receive, secondary treatment.

CEP Estimate of Minimum Necessary Pollution Control Expenditures Not in Company Budget: $2 million total, of which $1 million is for secondary treatment, and $1 million is for a third clarifier.

Wisconsin Rapids Mill (Kraft Division)

Location:
Wisconsin Rapids, Wisconsin

Date Built:
1968 (converted to replace sulfite mill)

Process:
360 TPD bleached kraft

Other Production:
None

Total Water Use:
12 MGPD

Source:
Wisconsin River

Discharge:
Wisconsin River

Table 8.6
Wisconsin Rapids Mill (Kraft Division): Pollution Control Record

Treatment	Overall Evaluation	Equipment
Water		
Primary	X	6 million gallon settling pond with 12-hour retention time (inadequate)
Secondary	X	None
Tertiary	X	None
Solid waste removal	—	Landfill, incineration

Table 8.6
Wisconsin Rapids Mill (Kraft Division): Pollution Control Record (continued)

Treatment	Overall Evaluation	Equipment
Other	—	In-mill system to recycle spilled chemicals and fibers
Air		
Particulate (fuel)	X	Fuel: coal. Cyclone scrubber, rated 90% efficient (inadequate)
Particulate (production)	✓	Recovery furnace: precipitator, 99% efficient. Lime kiln: venturi scrubber
Gas and odor	✓	Digester gas scrubber. Weak black liquor oxidation (Lundberg odor control system). 299-foot recovery stack

Plant Emissions

To water:

BOD	39,360 lb/day (101 lb/ton of pulp; state figure)
Suspended solids	N.A.
To air:	N.A.

Pollution Control Expenditures:
$1.1 million of the $37 million total capital outlay for the mill was for pollution control. Of this, over $650,000 was for particulate and odor control.

Legal Status:
(Water) The review hearing in late 1970 set a combined BOD limit for both Wisconsin Rapids mills and the Biron mill of 19,150 pounds per day, to be met by September, 1975.
(Air) Wisconsin air standards under review by EPA.

Future Plans:
The company is planning to build an aeration lagoon and to improve in-plant

collection and recovery of fibers and chemicals. It is also studying the possibility of non-direct contact evaporators.

Profile:

This mill, "the largest capital improvement in Consolidated's history," was converted between 1966 and 1968 to replace 175 TPD of sulfite production. Water pollution control was the motivation for its conversion. The company had not been able to sufficiently reduce its sulfite pollution and decided to replace the sulfite mill with a kraft mill. The change did effect a 70 percent (53,000 pounds per day) reduction in the BOD discharge from the Wisconsin Rapids pulping operations. However, the mill was converted without any secondary water treatment.

This inadequacy at a new mill is even more difficult to understand since the company did install extensive air pollution control equipment. It, unfortunately, has not worked as well as expected because mechanical difficulties have interrupted the mill's continuous operation, but it is there and it does have the potential for efficiency. The only explanation seems to be that Consolidated was satisfied with the automatic 70 percent BOD reduction achieved by changing processes, and overlooked company President Mead's insight that "In the area of water pollution most of us are still talking about the assimilative capacity of streams. I am afraid that the day is going to come only too soon when the public will not even tolerate this approach..."[4]

CEP Estimate of Minimum Necessary Pollution Control Expenditures Not in Company Budget: $2.75 million total, of which $500,000 is for improved primary treatment, $2 million is for secondary treatment, and $250,000 is for improved coal scrubbers.

References for Chapter 8

1. From a speech at Western Michigan University, April 6, 1967, by George W. Mead II, president of Consolidated.

2. Idem.

3. Idem.

4. Idem.

Chapter 9

CONTINENTAL CAN COMPANY
633 Third Avenue, New York, New York 10017

Major Products:
Cans, paper cartons and bags, paper cups, metal closures, and glassware.

Major Consumer Brands:
BONDWARE paper plates, HAZELWARE glassware, SUPER-SAK bags.

Financial Data: ($ Millions)	1969	1968	1967
Net Sales	1,780.0	1,507.9	1,397.6
Net Income	90.4	83.5	78.1
Capital Expenditures	132.4	85.6	81.6

Pollution Control Expenditures:
$13,500,000, 1965-1969; $40,000,000 (estimated), 1970-1975.

Annual Meeting:
The annual meeting takes place during April in New York City.

Officers:
Chairman/President: Ellison L. Hazard, Darien, Connecticut (Director:
Kennecott Copper Corp., Charter New York Corp.)

Outside Directors:
Paul C. Cabot: Chairman of the Board, State Street Investment Corp.
Stephen M. DuBrul, Jr.: Partner-Lehman Brothers
George P. Edmonds: Honorary Chairman of the Board, Wilmington Trust Co.
Thomas C. Fogarty: Retired Chairman of the Board, Continental Can Co.
Helmer R. Johnson: Attorney, Partner, Wilkie, Farr and Gallagher
George A. Murphy: Chairman of the Board, Irving Trust Co.
Charles E. Saltzman: Partner, Goldman, Sachs and Co.
Lloyd B. Smith: Chairman of the Board and Chief Executive Officer, A. O.
Smith Corp.
Frederick J. Stale: Chairman, Department of Nutrition, Harvard School of
Public Health
Jack I. Straus: Chairman of the Executive Committee, R. H. Macy and Co.
Leslie H. Warner: President, General Telephone and Electronics Corp.

Breakdown of Total Company Production (Percent of Sales):
Metal operations	51.9
Paper operations	20.5

Plastics, flexible packaging, closures, and glassware   10.0
International   17.6

Plants:
Continental Can has 233 plants in the United States, Canada, Europe, and
South America.

Timberholdings:
Continental Can has 1,400,000 acres in seven southern states. The total 1969
value of the company's holdings: $55,480,000.

Annual Pulp Production (Estimated on a Basis of 355 Production Days):
1,116,475 tons

Company Pollution Overview

Continental Can, the world's largest packaging company, is also the nation's
largest collector and recycler of waste paper fibers. As such, it is proof
that recycling on a large scale can be economically as well as ecologically
sound. In 1969, Continental Can was responsible for collecting 1.1 million
tons of waste paper or approximately 10 percent of the total collected in the
nation, which filled 50 percent of the company's raw-material needs.

Continental annually produces about 1.5 million tons of pulp and paperboard from four large southern kraft mills and seven mills (not included in
this study) which use reclaimed fibers from the recycled paper. The four
kraft mills, located in Augusta and Port Wentworth, Georgia, Hodge,
Louisiana, and Hopewell, Virginia, use 88-90 million gallons of water to
produce 3145 tons of pulp daily.

The company has a history of installing pollution control equipment only
in response to outside prodding from citizens or government agencies. Once
prodded, it seems to go to great lengths to install the best that is available.
This has inevitably resulted in a strange unevenness in the total company pollution control record. The Hodge plant, which already has the lowest per ton
rate of discharge of suspended solids of any kraft mill in the world (according
to the company) is adding what the FWQA considers the most advanced water
treatment system in the country. In contrast, Port Wentworth has only primary treatment and the Hopewell mill's 18.2 million gallons of daily effluent
receive only partial primary treatment in a 1.5 million gallon fiber settling
pond.

The overall pattern in air pollution control is similarly spotty. As part
of a $100 million expansion project, the Hodge mill is installing the best avail-
able equipment to control particulate gaseous and odorous emissions. At
Hopewell, two new 99+ percent efficient precipitators have been installed on
the recovery furnaces because residents blamed the mill for the heavy dustfall in the area. At Augusta, the world's first brine scrubber using weak brine
as the scrubbing medium will be installed on one recovery furnace by early

1972. Yet, three mills continue to burn bark with only mechanical dust collectors to control fly ash emissions, and only Hodge will have complete odor control. (Augusta will have partial odor control by early 1972.)

The result of Continental's patchwork problem-solving is that after a $10.6 million investment in water treatment over the past five years, two mills have no secondary systems, but one will soon have tertiary. After a $3 million expenditure for clean air, two towns are no longer covered with dust, but three will continue to smell for a while.

The company still has a long way to go before all its mills reach the standards projected for Hodge. CEP estimates that the required expenditure will be approximately $16.5 million, of which $7.7 million has already been allocated by the company.

Table 9.1
Continental Can Company: Pulp Production, Water Use, and Pollution Control

Mill Location	Pulp Prod. (TPD)	Other Prod. (TPD)	Water Use (MGPD)	Water Pollution Control			Air Pollution Control	
				Pri.	Sec.	Tert.	Part.	Gas, Odor
Augusta, Ga.	700	600	38	✓	✓	✗	✗	✗
Port Wentworth, Ga.	625	625	18	✓	✗	✗	✗	✗
Hodge, La.	820	700	14-16	✓	✓	✗	✗	✗
Hopewell, Va.	1,000	1,000	18.2	✗	✗	✗	✓	✗
Total	3,145	2,925	88.2 -90.2					

Augusta Mill

Location:
Augusta, Georgia

Date Built:
1960

Process:
700 TPD bleached kraft

Other Production:
600 TPD bleached kraft board

Total Water Use:
38 MGPD

Source:
Savannah River

Discharge:
Savannah River

Table 9.2
Augusta Mill: Pollution Control Record

Treatment	Overall Evaluation	Equipment
Water		
Primary	✓	38-45 million gallon clarifier
Secondary	✓	38-45 million gallon combination aerated and stabilization lagoon and holding pond (also handles effluent from Cox Newsprint)
Tertiary	X	None
Solid waste removal	—	Landfill
Air		
Particulate (fuel)	X	Fuel: bark. Mechanical collectors (inadequate)
Particulate (production)	X	Two recovery furnaces: one with 99% efficient precipitator, one with 84% efficient scrubber (inadequate)
Gas and odor	X	None

Table 9.2
Augusta Mill: Pollution Control Record (continued)

Treatment	Overall Evaluation	Equipment
Plant Emissions		
To water:		
BOD	14,400 lb/day (20 lb/ton of pulp)	
Suspended solids	N.A.	
To air:	N.A.	

Pollution Control Expenditures:
N.A.

Legal Status:
(Water) In compliance.
(Air) Not in compliance, but a schedule for compliance due July 1, 1970.

Future Plans:
An estimated $2.5 million will be spent on pollution control at this mill in the next five years. A $500,000 Babcock & Wilcox brine scrubber using brine as a scrubbing medium will be installed on one recovery furnace by early 1972. This is the first such system to be put in operation anywhere in the world, and is expected to achieve 97 percent particulate removal.

The mill is also working to improve the efficiency of its lime kiln scrubber, precipitators, and aerators.

Profile:
This mill provides excellent water treatment for the much-beleaguered Savannah River. Odor should be reduced when planned changes are complete, but particulate and odor problems will remain.

CEP Estimate of Minimum Necessary Pollution Control Expenditures Not in Company Budget: $1.2 million total, of which $500,000 is for black liquor oxidation; $500,000 is for improved bark burner particulate control; and $200,000 is to improve the efficiency of the precipitator on the recovery furnace.

Port Wentworth Mill

Location:
Port Wentworth, Georgia

Date Built:
1948

Process:
625 TPD unbleached kraft

Other Production:
625 TPD wet finish liner

Total Water Use:
18 MGPD (6 MGPD for cooling)

Source:
Savannah River

Discharge:
Savannah River

Table 9.3
Port Wentworth Mill: Pollution Control Record

Treatment	Overall Evaluation	Equipment
Water		
Primary	✓	190-foot diameter clarifier treats 12 million gallons of process water (installed 1968)
Secondary	X	None
Tertiary	X	None
Solid waste removal	—	Sludge dewatered and used for landfill
Air		
Particulate (fuel)	X	Fuel: gas, bark, oil. No equipment

Table 9.3
Port Wentworth Mill: Pollution Control Record (continued)

Treatment	Overall Evaluation	Equipment
Particulate (production)	X	Two recovery furnaces: precipitators, 90% efficient (inadequate). Lime kiln: scrubber, 90% efficient (inadequate)
Gas and odor	X	None

Plant Emissions

To water:

BOD	13,600 lb/day (22 lb/ton of pulp)	
Suspended solids	N.A.	
To air:	N.A.	

Pollution Control Expenditures:
Company information not available.

Legal Status:
(Water) The 1965 Federal Abatement Conference at Savannah required the mill
to reduce its BOD load of 22,134 pounds per day by 25 percent. The installation of the primary treatment clarifier reduced the BOD by 40 percent to
13,600 pounds per day. The state also required a solids reduction of 90 percent. A second abatement hearing in 1969 required secondary water treatment
by December 1972.
(Air) A state air emission survey of the plant found that it exceeds "all allowable emissions standards."

Future Plans:
The mill is now working with the town of Port Wentworth on a regional system to treat municipal and industrial wastes by the December 1972 deadline.

Profile:
This mill was described in the 1965 Proceedings of the Conference on Pollution of the Lower Savannah River as the "uppermost source of industrial pollution in the lower Savannah River." (p. 82). As a result of this conference,
and its recommendations that exterior water treatment facilities be installed,

the mill tightened in-plant control of wastes, consolidated its sewer system, and constructed a primary treatment clarifier for process water only. These steps reduced the BOD to an acceptable level, but achieved only half of the required suspended solids reduction. At the reconvened conference in 1969, Continental Can was instructed to install a secondary treatment system by the end of 1972 in order to satisfy the 90 percent solids removal requirement. A proposed joint treatment facility will achieve this.

Despite adequate water pollution control, the mill will still require considerable modernization and upgrading in order to comply with state air pollution laws.

CEP Estimate of Minimum Necessary Pollution Control Expenditures Not in Company Budget: $4.25 million total, of which $500,000 is for improved power boiler particulate control; $1 million is for two new recovery furnace precipitators; $250,000 is for a new lime kiln scrubber; $500,000 is for black liquor oxidation; and $2 million is for joint water treatment.

Hodge Mill

Location:
Hodge, Louisiana

Date Built:
1928

Process:
620 TPD unbleached kraft
200 TPD NSSC

Other Production:
700 TPD kraft wrapping; converting and specification kraft; extensible paper; bags, butcher's paper; and corrugating medium

Total Water Use:
14-16 MGPD

Source:
Wells

Discharge:
Dugdemona Bayou

Table 9.4
Hodge Mill: Pollution Control Record

Treatment	Overall Evaluation	Equipment
Water		
Primary	✓	200 acres of natural stabilization ponds continuously discharge weak effluent. 500-acre holding basin with 110-day retention capacity discharges effluent only in the winter
Secondary	✓	
Tertiary	✗	Under construction, start-up in March 1971
Solid waste removal	—	Sludge retained in huge ditch
Air		
Particulate (fuel)	✓	Fuel: gas. No equipment needed
Particulate (production)	✗	Three recovery furnaces: one without precipitator, two with 85% and 90% efficient precipitators. NSSC liquor is also burned in recovery furnace and cross-recovered for use in kraft process. Two lime kilns (pre-war vintage): 90% efficient scrubbers (inadequate)
Gas and odor	✗	None
Plant Emissions		
To water:		
BOD	3,280 lb/day (5 lb/ton of pulp; company figure)	
Suspended solids	N.A.	
To air:	N.A.	

Pollution Control Expenditures:
Company information not available.

Legal Status:
(Water) In compliance; operating with permit.
(Air) State hasn't yet surveyed pulp mill emissions.

Future Plans:
As part of a $100 million modernization and expansion project, by 1973 this
mill will be equipped with state-of-the-art pollution control equipment. A
$4.2 million water treatment system including a company-developed process
for color removal using lime, and a 12 million gallon per day capacity multistage clarifier will be completed in February 1971. The sludge, composed of
colored matter, settled solids, and lime, will be dewatered and incinerated in
the lime kiln. A $750,000 Federal Water Quality Administration grant helped
Continental Can develop this color removal process. The entire system will
be the first in the world to provide four levels of water purification and the
company feels that it will revolutionize waste water treatment.

Other equipment to reduce pollution will include new scrubbers on two
new lime kilns and, on a new recovery furnace, Combustion Engineering's
non-direct contact evaporators to reduce odor. These facilities, along with a
new power boiler, a continuous digester, and "the world's largest kraft papermaking machine," will double the mill's production capacity to 500,000 tons of
unbleached kraft paper per year.

Profile:
The elaborate water system at this mill will augment what would ordinarily be
considered adequate summer water treatment. Despite the fact that Hodge has
"the lowest suspended solids discharge of any kraft mill in the world" and the
very low BOD count of 5 pounds per ton of pulp further treatment is necessary
because Dugdemona Bayou is only 3 feet wide and, during several months, its
entire flow is mill effluent.

By 1973, when the mill's $100 million modernization is complete it will
clearly be the best level of state-of-the-art control.

CEP Estimate of Minimum Necessary Pollution Control Expenditures Not
in Company Budget: None.

Hopewell Mill

Location:
Hopewell, Virginia

Date Built:
1928

Process:
850 TPD kraft
150 TPD semichemical

Other Production:
1000 TPD test chipboard for laminating paper, folding boxboard, cylinder
kraft specialities

Total Water Use:
18.2 MGPD

Source:
James River

Discharge:
James River into Bailey's Bay

Table 9.5
Hopewell Mill: Pollution Control Record

Treatment	Overall Evaluation	Equipment
Water		
Primary	X	1.5 million gallon settling pond (inadequate)
Secondary	X	None
Tertiary	X	None
Other	—	Two savealls (installed, 1970)
Air		
Particulate (fuel)	X	Fuel: coal, oil and bark. Mechanical dust collectors (inadequate)
Particulate (production)	✓	Two recovery furnaces: one with a 99.5% efficient precipitator, a second precipitator installed in 1970. Lime kiln: scrubber, 90% efficient
Gas and odor	X	None

Table 9.5
Hopewell Mill: Pollution Control Record (continued)

Treatment	Overall Evaluation	Equipment
Plant Emissions		
To water:		
BOD	41,000 lb/day (41 lb/ton of pulp)	
Suspended solids	42,000 lb/day (42 lb/ton of pulp)	
To air:	N.A.	

Pollution Control Expenditures:
$1,500,000 ($750,000 for each of the new precipitators).

Legal Status:
(Water) Not in compliance. The mill is required by the state to upgrade its water treatment to primary and secondary by July, 1972.
(Air) No information available from state agency.

Future Plans:
This mill discharges 18.2 million gallons of effluent per day. 1.5 million gallons of this is treated in the settling pond; the rest is directly discharged through the main mill sewer into the James River. Continental Can feels that the July 1, 1972 state deadline for installation of secondary treatment is difficult to meet; is has to date presented a preliminary and "confidential" proposal and has hired a consulting firm to study the possibility of building a central treatment plant for all the industrial and municipal wastes in the area including those from several chemical companies. The Virginia agency does not think that such a facility would be adequate without pretreatment of the wastes at the plant sites.

Profile:
This mill has been discharging untreated waste water into the almost hopelessly polluted James River for over 40 years; yet it feels that the state's 1972 "cease and desist" requirement is unreasonable. Aside from installation of two savealls and an obviously inadequate settling pond with a capacity of only 8 percent of the daily effluent, no steps have been taken, and the company's preliminary abatement proposal is held confidential.

    In 1967 the state surveyed the daily mill discharge at 39,000 pounds of

BOD and 51,000 pounds of suspended solids. In 1970, after installation of the savealls and settling pond, the suspended solids in the effluent had dropped to 42,000 pounds per day, but the BOD had risen 2,000 pounds to 41,000 per day. These figures indicate the urgency of installing an adequate water treatment system.

CEP Estimate of Minimum Necessary Pollution Control Expenditures Not in Company Budget: $3.3 million total, of which $300,000 is for primary water treatment; $2 million is for secondary treatment; $500,000 is for black liquor oxidation; $250,000 is for improvement or replacement of the lime kiln scrubber; and $250,000 is for improved power boiler particulate control.

Chapter 10

CROWN ZELLERBACH CORPORATION
One Bush Street, San Francisco, California 94119

Major Products:
Publication papers, packaging, industrial and business papers, lumber and plywood.

Major Consumer Brands:
ZEE napkins, paper towels and toilet tissues, CHIFFON household paper goods.

Financial Data: ($ Millions)	1969	1968	1967
Net Sales	919.3	865.5	788.8
Net Income	54.0	59.5	49.4
Capital Expenditures	82.7	62.4	76.7

Pollution Control Expenditures:
$78,000,000 from 1968-1975 (company estimate)

Annual Meeting:
The annual meeting takes place during April in San Francisco.

Officers:
President and Chief Executive Officer: C. R. Dahl

Outside Directors:
A. H. Dean: Sr. Partner, Sullivan & Cromwell, New York
W. B. Dunckel: Director, Bankers Trust New York Corporation
J. R. Fluor: Chairman, The Fluor Corp.
M. C. Mumford: Chairman, Lever Bros.
C. M. Pigott: President, Pacific Car & Foundry
Reed O. Hunt: Retired Chairman of Executive Committee, Crown Zellerbach

Breakdown of Total Company Production (Dollars and Percent of Sales, 1969):

Packaging	$240,513,000	26 percent
Wood products	144,866,000	16
Paper products		
Publication papers	112,069,000	12
Tissue	90,541,000	10
Industrial and business	113,730,000	12

Breakdown of Total Company Production (Dollars and Percent of Sales, 1969): (continued)

Distribution $ 183,223,000   20 percent

Miscellaneous   34,340,000   4

Timberholdings:
The company own three million acres in the U.S. The total 1969 value of Crown Zellerbach's holdings (cost basis): $ 102,300,000.

Annual Pulp Production:
2,245,375 tons (1969)

Company Pollution Overview

Besides being the world's second largest paper producer, Crown Zellerbach was the second most uncooperative company in this study. Crown Zellerbach operates 14 pulp mills at nine locations in the United States, seven in the Pacific states and two in Louisiana. At four locations (Camas, St. Francisville, Bogalusa, and Wauna) there is more than one type of pulp mill. One of the mills (St. Francisville) is owned jointly with Time, Inc. Total daily pulp production is about 6300 tons, two-thirds of which is kraft pulp. The mills use 280 million gallons of water per day.

Crown Zellerbach has gone beyond state requirements in providing air pollution control equipment, but has waited for prodding to install water treatment facilities. All the mills have equipment for monitoring air and water emissions, but the company refused to release any results. Also, as construction dates were available for only two mills, it was very difficult to determine where age contributed to pollution problems. Faced with these handicaps, the CEP was able to evaluate air pollution control systems only if there was external evidence, such as emissions data or state action, indicating adequacy or inadequacy.

On this basis, four of the six kraft mills have adequate odor and particulate control: Fairhaven, California; St. Francisville, Louisiana; Wauna, Oregon; and Port Townsend, Washington. In fact, Port Townsend has the best air pollution control of any old kraft mill in the United States and is an outstanding example of what can be done to upgrade an old mill to meet and even be the model for the strictest kraft mill air pollution standards in the country. Renovation of this 1928 mill cost about $ 14 million, and it now complies with sulfur and particulate emissions regulations which other kraft mills in the state will not meet for three or four more years.

However, the company's largest kraft mill, at Bogalusa, as well as the one at Camas, Washington, has inadequate odor and particulate controls. These facilities are on established time schedules for improvements.

From the point of view of water pollution control, Crown Zellerbach has an abysmal record. There is no treatment at all at Fairhaven and St. Francisville and Port Angeles. The company's large sulfite mill at Camas uses the calcium sulfite process and for years discharged "spent" pulping chemicals untreated into the Columbia River. However, the small Lebanon sulfite mill pulps with an ammonia base and uses some spent liquor for by-product manufacture and incinerates the rest. Two mills (Bogalusa and Lebanon, Oregon) have partial aerated treatment projects; these provide the equivalent of primary treatment but do not operate year-round.

Crown Zellerbach's history in water pollution control is not good; but the company is beginning to take some corrective steps. The Camas sulfite mill is being converted to a recoverable base. St. Francisville will have primary and secondary treatment by 1973; and two mills (Wauna and Bogalusa) will have secondary treatment four years from now (1974). Bogalusa has twice been cited at federal water pollution abatement conferences in 1963 and 1968.

Crown Zellerbach's inadequacy is an indictment not only of the company, but of shortsighted state authorities who persist in the fanciful notion that large bodies of water are able to absorb infinite waste loads. Seven of Crown Zellerbach's mill locations are on extremely large rivers or bays where primary treatment and outfall pipes have been certified as adequate treatment. They are not!

Crown Zellerbach was initially cooperative with the Council but refused to respond to several months of additional inquiries following up a primary telephone interview. This refusal was so complete that we were not even able to learn the reasons for it. The company was at least consistent; it was the only one of the 24 companies to refuse to talk about any of its mills, including the better ones. The last word heard from Dr. Herman Amberg, Environmental Control Director, in June '70, was: "Your material has been sent to the main office." All attempts to recover it were fruitless until two weeks after the initial publication (and press coverage) of Paper Profits. At that time, CEP was delighted to receive a 13-page letter, dated January 13, which answered nearly all the outstanding questions. Also included in the correspondence was the suggestion that "perhaps the time has come when pollution control data should be collected and made available at one national source ... This could be an appropriate agency of the Federal Government or a private research institution positioned to provide this service to the public."

CEP's response was: "There is such an organization and it is we."

Crown Zellerbach has indicated a planned $38.3 million investment in pollution control at its pulp mills. CEP estimates that the company would have to spend an additional $26.1 million (bringing the total to $64.4 million) to provide adequate air pollution control and secondary water treatment at all locations.

Table 10.1
Crown Zellerbach Corporation: Pulp Production, Water Use, and Pollution
Control

Mill Location	Pulp Prod. (TPD)	Other Prod. (TPD)	Water Use (MGPD)	Water Pollution Control			Air Pollution Control	
				Pri.	Sec.	Tert.	Part.	Gas, Odor
Fairhaven, Calif.	500	--	30	X	X	X	✓	✓
Bogalusa, La.	1,500	1,750	30	✓	X	X	✓	X
St. Francisville, La.	720	770	35	X	X	X	✓	✓
Lebanon, Ore.	105	100	4	✓	✓	X	?	✓
Wauna, Ore.	1,200	760	50	✓	X	X	✓	✓
West Linn, Ore.	340	500	15	✓	X	X	?	—
Camas, Wash.	1,225	1,050	90	✓	X	X	X	X
Port Angeles, Wash.	195	400	14	X	X	X	?	—
Port Townsend, Wash.	420	400	12	✓	X	X	✓	✓
Total	6,205	5,730	280					

Fairhaven Mill

Location:
Fairhaven, California (owned jointly with Simpson Timber Co.)

Date Built:
1968

Process:
500 TPD bleached kraft

Total Water Use:
30 MGPD

Source:
Mad River

Discharge:
Pacific Ocean

Table 10.2
Fairhaven Mill: Pollution Control Record

Treatment	Overall Evaluation	Equipment
Water		
Primary	X	None
Secondary	X	None
Tertiary	X	None
Other	—	Outfall pipe, diffuser
Air		
Particulate (fuel)	?	Fuel: hog. Multiple cyclone collector
Particulate (production)	✓	Recovery furnace: precipitator and secondary wet scrubber
Gas and odor	✓	Black liquor oxidation; system for odorous gas collection and incineration

Plant Emissions

To water:

BOD   36,100 lb/day
(72 lb/ton of pulp)

Suspended solids   29,600 lb/day
(59 lb/ton of pulp)

To air:

Table 10.2
Fairhaven Mill: Pollution Control Record (continued)

Treatment	Overall Evaluation	Equipment
TRS	643 lb/day (1.3 lb/ton of pulp; company data. The total TRS from all sources is given by the county as 805 lb/day, or 1.6 lb/ton of pulp)	
Particulates	3,000 lb/day	

Pollution Control Expenditures:
$2.7 million for three pipe lines; $1.2 million for air pollution control.

Legal Status:
(Water) Has permit from regional water board.
(Air) In compliance with Humboldt County requirements.

Future Plans:
Company information not released.

Profile:
Although the company refused to release efficiency data on pollution control equipment at this mill, it is obviously high, and control is excellent. The fact that county kraft mill regulations are the strictest in the country may be partially responsible, although much of the equipment was installed before these regulations were adopted. The county permit limits the mill's TRS emissions to 1300 pounds daily; they are well under this amount. In addition, a 310 ft. high stack provides dispersion of the gases. Recovery furnace TRS emissions are very low: 1/2 pound per ton of pulp.

The water pollution control picture, as everywhere on the West Coast, is not good. The state finds discharge of untreated waste to the ocean to be an acceptable water treatment system.

CEP Estimate of Minimum Necessary Pollution Control Expenditure Not on Company Budget: $6.75 million total, of which $3 million is for primary water treatment; $3.5 million is for secondary treatment (activated sludge); and $250,000 is for improved power boiler particulate control.

Bogalusa Mill

Location:
Bogalusa, Louisiana

Date Built:
1918

Process:
1,350 TPD bleached and unbleached kraft
150 TPD NSSC

Other Production:
1,350 TPD container liner, chipboard, corrugating, paper bag, wrapping,
specialty papers; kraft boxboard; laminate and asphalt waterproof papers;
400 TPD kraft wrapping and bag; solvents and chemicals

Total Water Use:
30 MGPD

Source:
Bogue Lusa and wells

Discharge:
Pearl River

Table 10.3
Bogalusa Mill: Pollution Control Record

Treatment	Overall Evaluation	Equipment
Water		
Primary	√	300-foot clarifier (installed, 1964)
Secondary	X	Spent NSSC pulping liquor is cross-recovered for kraft pulping
Tertiary	X	None
Other	—	Direct river aeration, 90 days/year (installed, 1968); outfall, diffuser

Table 10.3
Bogalusa Mill: Pollution Control Record (continued)

Treatment	Overall Evaluation	Equipment
Air		
Particulate (fuel)	✓	Fuel: natural gas. No controls needed
Particulate (production)	✓	Two recovery furnaces: one with 95% efficient precipitator, one with 99%. Lime kiln: venturi scrubber
Gas and odor	✕	Black liquor oxidation
Plant Emissions		
To water:		
BOD	54,200 lb/day (36 lb/ton of pulp)	
Suspended solids	26,000 lb/day (19 lb/ton of pulp)	
To air:		
TRS	24,351 lb/day (16 lb/ton of pulp)	
Particulates	16,030 lb/day (10 lb/ton of pulp)	

Pollution Control Expenditures:
$1.2 million was spent on pollution control, of which $120,000 was for the
aeration system. Operating costs are $110,000 annually.

Legal Status:
(Water) Not in compliance, secondary treatment required. Two Federal water
pollution abatement conferences cited this mill. The first in October 1963
said that only primary water treatment was required to attain the state permit
allowance of 62,000 pounds of BOD. The second session of the conference in
November, 1968, decided that the mill should provide secondary water treatment.
(Air) State program has not investigated pulp mills yet.

Future Plans:
$4 million secondary water treatment system by 1972 and $3.7 million odor control system for digester and evaporator gases to be completed in 1974.

Profile:
Planned equipment will vastly improve the condition of this mill. At present, residents report that "the whole town stinks from the mill." By the end of 1971, when additional odor control is installed, this problem should be greatly reduced. At present, the mill outpollutes the town of Bogalusa in Pearl River discharges, even though the town has no sewage treatment facilities. Perhaps by the end of 1972, when Crown Zellerbach has water treatment, the citizens of Bogalusa will see more precisely why they need the sewage plant bond issue which they have twice voted down.

It is nice that the mill has adopted the stop-gap measure of stream aeration while the secondary treatment system is being planned and built. However, in the winter when the aeration facility is not in operation, the BOD discharge must be quite high.

There is little that distinguishes the Bogalusa mill from any other large southern facility; air pollution control has been neglected for a lack of state prodding and the water pollution, which has received federal attention, has not been alleviated by secondary treatment any earlier than might otherwise have been expected.

CEP Estimate of Minimum Necessary Pollution Control Expenditures Not in Company Budget: $1 million total, for additional particulate controls.

St. Francisville Mill

Location:
St. Francisville, Louisiana (paper mill jointly owned with Time, Inc.)

Date Built:
1959

Process:
500 TPD bleached and unbleached kraft
220 TPD bleached groundwood

Other Production:
770 TPD coated paper and printing papers

Total Water Use:
35 MGPD

Source:
Mississippi River

Discharge:
Mississippi River

Table 10.4
St. Francisville Mill: Pollution Control Record

Treatment	Overall Evaluation	Equipment
Water		
Primary	X	None
Secondary	X	None
Tertiary	X	None
Other	—	Outfall pipe
Air		
Particulate (fuel)	✓	Fuel: natural gas. No controls needed
Particulate (production)	✓	Recovery furnace: 98% efficient electrostatic precipitator. Wet scrubber on smelt dissolver vent
Gas and odor	✓	Black liquor oxidation; digester gases incinerated

Plant Emissions

To water:

BOD   69,000 lb/day
                       (138 lb/ton of pulp)

Suspended solids   240,000 lb/day
                       (480 lb/ton of pulp)

To air:

TRS   3,000 lb/day
                       (6 lb/ton of pulp)

Particulates   12,000 lb/day
                       (24 lb/ton of pulp)

Pollution Control Expenditure:
$950,000 for air pollution control.

Legal Status:
(Water) Secondary treatment required. Mill has discharge permit.
(Air) State has not yet investigated pulp mills.

Future Plans:
Primary water treatment (1972), secondary water treatment (1973).

Profile:
While particulate and odor control systems are adequate at St. Francisville, this mill site has as little water pollution control as any U.S. mill and is a blot on both Crown Zellerbach and Time, Inc. Large quantities of bleaching and pulping wastes are discharged into the Mississippi without any treatment. The BOD is inordinately high and there has been no change in the water situation since a fish kill was caused by mill discharges in October 1966.

Crown Zellerbach has taken a very leisurely attitude toward the provision of water treatment systems, in another situation where state authorities and company management have been content to continuously pollute a body of water because it is large. It it necessary to mention how large the Great Lakes are? And how polluted? No body of water has infinite capacity to assimilate wastes: Any form of initiative by either the state or company would have resulted in the entire water treatment system being completed by 1972, the planned completion date of only the primary system.

Additional pollution control funds might be spent to speed up the planned water control installations. The BOD and suspended solids measurements indicate how badly effluent treatment is needed. The planned water controls should be adequate.

CEP Estimate of Minimum Necessary Pollution Control Expenditure Not on Company Budget: $3.8 million total, of which $800,000 is for primary water treatment; and $3 million is for secondary water treatment.

Lebanon Mill

Location:
Lebanon, Oregon

Date Built:
1889

Process:
105 TPD unbleached ammonia sulfite

Other Production:
100 TPD unbleached sulfite wrapping, butchers paper; sulfite corrugating
medium; chemical by-products

Total Water Use:
4 MGPD

Source:
South Santiam River

Discharge:
Mark Slough to South Santiam River

Table 10.5
Lebanon Mill: Pollution Control Record

Treatment	Overall Evaluation	Equipment
Water		
Primary	✓	Screens and small lagoon
Secondary	✓	Spent liquor not discharged, but evaporated and dried for by-product manufacture; FWQA demonstration project (aerated lagoon) treats the remainder
Tertiary	X	None
Air		
Particulate (fuel)	?	Fuel: oil and hog fuel. Company information N.A.
Particulate (production)	—	None needed
Gas and odor	✓	Sulfur dioxide from digester controlled with a scrubber

Plant Emissions

To water:

Table 10.5
Lebanon Mill: Pollution Control Record (continued)

Treatment	Overall Evaluation	Equipment
BOD	4,500 lb/day (43 lb/ton of pulp)	
Suspended solids	2,945 lb/day (28 lb/ton of pulp)	
To air:	N.A.	

Pollution Control Expenditures:
The company contributed $350,000 and the government contributed $505,000 to the aeration project for equipment and research.

Legal Status:
(Water) In compliance; has permit specifying maximum BOD discharge of 10,000 pounds per day when river is at low flow. Future permit specifications set maximum BOD load of 2,000 pounds per day in June-November, and 3,000 pounds per day in winter.
(Air) In compliance; sulfite mill regulations under development.

Future Plans:
N.A.

Profile:
In 1954 Crown Zellerbach installed the first spent sulfite liquor evaporation and drying system in Oregon at this mill. The BOD discharge reflects the fact that spent pulping liquor is not dumped. Compliance with the lower state BOD discharge limit is expected to be achieved by the federally supported aerated lagoon system.

   CEP Estimate of Minimum Necessary Pollution Control Expenditure Not on Company Budget: $250,000 total, for improved power boiler particulate control.

Wauna Mill

Location:
Wauna, Oregon

Date Built:
1969

Process:
750 TPD bleached kraft
450 TPD groundwood

Other Production:
360 TPD newsprint
200 TPD tissue
200 TPD kraft specialties

Total Water Use:
50 MGPD

Source:
Columbia River

Discharge:
Columbia River

Table 10.6
Wauna Mill: Pollution Control Record

Treatment	Overall Evaluation	Equipment
Water		
Primary	✓	265-foot diameter clarifier
Secondary	X	None
Tertiary	X	None
Solid waste removal	—	Sludge dewatering equipment
Other	—	Outfall pipe with diffuser
Air		
Particulate (fuel)	✓	Fuel: gas. No controls needed
Particulate (production)	✓	Recovery furnace: precipitator rated efficiency of 98.6%. Venturi scrubber on lime kiln; demister on smelt vent

Table 10.6
Wauna Mill: Pollution Control Record (continued)

Treatment	Overall Evaluation	Equipment
Gas and odor	✓	Black liquor oxidation; non-condensible gases burned

Plant Emissions

To water:

BOD	90,000 lb/day (75 lb/ton of pulp)
Suspended solids	11,400 lb/day (9.5 lb/ton of pulp)

To air:

TRS	1,395 lb/day (2 lb/ton of kraft pulp; state data)
Particulates	2,920 lb/day (3.8 lb/ton of kraft pulp. Figures are for emission from recovery furnace only. State data)

Pollution Control Expenditures:
The company reports that $3.2 million, or 7.5 percent of mill capital cost, was for pollution control.

Legal Status:
(Water) In compliance; has permit.
(Air) In compliance.

Future Plans:
Secondary water treatment; no date specified.

Profile:
This demonstrates what a nice mill Crown Zellerbach can build when it wants to. Air pollution control is excellent, and was installed before Oregon's tough

air laws went into effect. It would have been nicer if secondary water treatment had also been part of the original plan; most new mills do provide secondary treatment at start-up.

CEP Estimate of Minimum Necessary Pollution Control Expenditures Not on Company Budget: $3.5 million total, for secondary water treatment.

West Linn Mill

Location:
West Linn, Oregon

Date Built:
1889 and 1914

Process:
340 TPD groundwood

Other Production:
500 TPD uncoated book, groundwood specialty and double-coated book papers; bleached kraft specialty papers

Total Water Use:
15 MGPD

Source:
Willamette River

Discharge:
Willamette River

Table 10.7
West Linn Mill: Pollution Control Record

Treatment	Overall Evaluation	Equipment
Water		
Primary	✓	Three concrete sedimentation basins
Secondary	X	None
Tertiary	X	None

Table 10.7
West Linn Mill: Pollution Control Record (continued)

Treatment	Overall Evaluation	Equipment
Air		
Particulate (fuel)	?	Fuel: oil, gas, sawmill waste (hog fuel). Cinder collector and shower in stack
Particulate (production)	—	None needed
Gas and odor	—	None needed
Plant Emissions		
To water:		
BOD	17,000 lb/day (50 lb/ton of pulp)	
Suspended solids	8,426 lb/day (24.7 lb/ton of pulp)	
To air:		
TRS	N.A.	
Particulates	3,600 lb/day	

Pollution Control Expenditures:
$801,340 for primary treatment.

Legal Status:
(Water) State permit requires secondary treatment by July 1972. The present BOD limit is 19,000 pounds per day; after treatment, it will be 4,000 pounds per day, or 12 pounds per ton of pulp.
(Air) In compliance.

Future Plans:
Secondary water treatment by mid-1972.

Profile:
This mill needs better water treatment, having had only the most primitive sort. Moreover, the present BOD allowance is much too high for a ground-

wood mill. But Crown Zellerbach seems to be only complying with the law as necessary.

CEP Estimate of Minimum Necessary Pollution Control Expenditure Not on Company Budget: $2.25 million total, of which $2 million is for secondary water treatment and $250,000 is for improved power boiler particulate control.

Camas Mill

Location:
Camas, Washington

Date Built:
Company information not released

Process:
780 TPD bleached kraft
420 TPD calcium-base sulfite
25 TPD groundwood

Other Production:
1,050 TPD bleached and unbleached sulfite and kraft converting and specialty papers, bags, towels and waxed papers; napkins and tissue

Total Water Use:
90 MGPD

Source:
La Camas Lake, wells, and Columbia River (in the summer)

Discharge:
Columbia River

Table 10.8
Camas Mill: Pollution Control Record

Treatment	Overall Evaluation	Equipment
Water		
Primary	✓	Clarifier installed in 1968
Secondary	✗	None
Tertiary	✗	None

Table 10.8
Camas Mill: Pollution Control Record (continued)

Treatment	Overall Evaluation	Equipment
Solid waste removal	—	FWQA project for primary sludge disposal
Other	—	Outfall and diffuser; holding pond on Lady Island
Air		
Particulate (fuel)	?	Fuel: gas and hog fuel. Information on controls N.A.
Particulate (production)	X	Recovery furnace precipitator: 95% efficient
Gas and odor	X	No odor control system; no system for the control of sulfur dioxide from sulfite mill operations

Plant Emissions

To water:

BOD   460,000 lb/day
                   (383 lb/ton of pulp)

Suspended solids   25,000 lb/day
                   (20 lb/ton of pulp)

To air:

TRS   10,000 lb/day
                   (13 lb/ton of kraft
                   pulp)

Particulates   33,200 lb/day
                   (27 lb/ton of kraft
                   pulp)

Pollution Control Expenditures:
$3.7 million for water treatment facilities; $500,000 of the $848,000 cost of the sludge utilization and disposal project was contributed by Crown Zellerbach, $350,000 by the federal government.

Legal Status:
(Water) In compliance.
(Air) Kraft mill will meet state schedule; does not now. Company required to cease open burning.

Future Plans:
There are three mills on this site and the company says it will spend $30.6 million to clean up the two large ones: $12.7 million for the kraft mill, $15.6 million for the sulfite mill. This includes the costs of conversion to magnesium bisulfite pulping, installation of a sulfur dioxide absorption system (for 96 percent removal), a liquor recovery system in the sulfite mills. At the kraft mill, new recovery furnaces, collection and incineration of noncondensible gases in the lime kiln, an odor control system, and a system for lime kiln particulate control are planned.

There are special, separate plans for 1972 compliance with the open burning prohibition. The state-approved schedule allows 24 months for study, six months for engineering, and six months for construction of necessary facilities. "Crown Zellerbach needs three years because it intends to control its air pollution with methods and devices that have never been developed," a representative explained.[1]

Profile:
The sulfite mill improvements are planned for 1971 completion. The kraft mill program, which has a 1975 state deadline, will be completed by 1974.

As a result of the massive planned air pollution control program, the company expects odorous TRS emission to decline from 10,000 pounds per day (13 pounds per ton of kraft pulp) to 939 pounds per day (1.2 pounds per ton of kraft pulp): an 85 percent reduction. Particulate emissions will also be cut from 33,200 pounds per day to 4400 pounds per day (also 85 percent reduction).

The company seems to have adequately evaluated what changes need to be made at this site, but the cost of the pollution control effort is somewhat inflated. For example, an $8 million kraft recovery furnace is not totally a pollution control device. Only the attendant odor-control system, less than 1/10th of the cost, should be considered as "antipollution." Similarly, conversion to magnesium pulping is a break-even proposition. It allows recovery of the expensive chemicals, a saving in fuel consumption (probably), and usually expansion of pulping production capacity.

Consequently, of the $30.6 million which the company claims will improve environmental quality, only about $15 million will be expended without a calculable economic return.

CEP Estimate of Minimum Necessary Pollution Control Expenditure Not on Company Budget: $3.5 million total, for aeration equipment for holding ponds (to provide secondary treatment).

Port Angeles Mill

Location:
Port Angeles, Washington

Date Built:
1919

Process:
195 TPD groundwood

Other Production:
120 TPD refiner groundwood
400 TPD newsprint and directory paper

Total Water Use:
9.7 MGPD (1965 figure including sulfite mill which was closed in 1968)

Source:
Elwha River

Discharge:
Puget Sound

Table 10.9
Port Angeles Mill: Pollution Control Record

Treatment	Overall Evaluation	Equipment
Water		
Primary	X	N.A.
Secondary	X	None
Tertiary	X	None
Other	—	Outfall and diffuser
Air		
Particulate (fuel)	?	Fuel: hog. Information on controls N.A.
Particulate (production)	—	None needed
Gas and odor	—	None needed

Table 10.9
Port Angeles Mill: Pollution Control Record (continued)

Treatment	Overall Evaluation	Equipment
Plant Emissions		
To water:		
BOD	33,535 lb/day (172 lb/ton of pulp)	
Suspended solids	15,000 lb/day (77 lb/ton of pulp)	

Pollution Control Expenditures:
Company information N.A.

Legal Status:
(Water) In compliance. Has discharge permit. The 1967 Puget Sound abatement conference cited the sulfite mill for effluent which harmed water life; the sulfite mill was closed prior to completion of that Puget Sound study. (Air) In compliance.

Future Plans:
Primary treatment to be installed in 1971.

Profile:
The closing of the sulfite mill no doubt improved water quality in the area. But while groundwood mill effluent is not as polluting as that from chemical pulping, the large amount of paper produced at this site creates a very high BOD discharge. There is no excuse for the lack of any water treatment at a mill with such high amounts of discharged pollutants.

CEP Estimate of Minimum Necessary Pollution Control Expenditure Not on Company Budget: $3 million total, $1 million for primary water treatment (cost is low because mill is small); $2 million for secondary water treatment.

Port Townsend Mill

Location:
Port Townsend, Washington

Date Built:
1928

Process:
420 TPD bleached kraft

Other Production:
400 TPD kraft wrapping paper, bag, liner, saturating, wet strength, bag and
wrapper; colored linerboard, pulp board, box board

Total Water Use:
N.A.

Source:
Little Quilcene and Big Quilcene Rivers

Discharge:
Puget Sound

Table 10.10
Port Townsend Mill: Pollution Control Record

Treatment	Overall Evaluation	Equipment
Water		
Primary	√	Clarifier started up in late 1970
Secondary	X	None
Tertiary	X	None
Other	—	Outfall pipe and diffuser
Air		
Particulate (fuel)	?	Fuel: hog. Information on controls N.A.
Particulate (production)	√	Recovery furnace: 99% efficient precipitator. Scrubber on smelt vent
Gas and odor	√	Non-direct contact evaporation system; digester gases incinerated in hog fuel burner

Table 10.10
Port Townsend Mill: Pollution Control Record (continued)

Treatment	Overall Evaluation	Equipment
Plant Emissions		
To water:		
BOD	15,770 lb/day (37.5 lb/ton of pulp)	
Suspendid solids	7,000 lb/day (17 lb/ton of pulp)	
To air:		
TRS	400 lb/day (1 lb/ton of pulp)	
Particulate	2,300 lb/day (5.5 lb/ton of pulp)	

Pollution Control Expenditures:
About $12.5 million; $2.5 million for water, and about $10 million for air.

Legal Status:
(Water) In compliance.
(Air) Meets new kraft mill standards far ahead of schedule, except for lime kiln which will be in compliance on schedule.

Future Plans:
Lime kiln modifications.

Profile:
This mill is the crowning glory of the Crown Zellerbach empire. It shows that an old kraft mill can meet the strictest air pollution regulations. State officials were so pleased with the air pollution controls at this mill that its emissions were used at the basis for the state standards. This mill is living proof that an old kraft mill can be upgraded and provide satisfactory odor control. Daily TRS emissions were reduced from 7700 pounds per day to 864 pounds per day; and daily particulate emissions were cut from 70,000 pounds to 13,000 pounds, and will be further reduced to 2300 pounds in early 1975 when the kiln modifications are complete. Water pollution however was ignored until the present construction of a primary treatment clarifier.

CEP Estimate of Minimum Necessary Pollution Control Expenditure Not on Company Budget: $2 million total, for secondary water treatment.

Reference for Chapter 10

1. Pulp and Paper, June, 1969, p. 127.

Chapter 11

DIAMOND INTERNATIONAL
733 Third Avenue, New York, New York 10017

Major Products:
Paperboard cartons, pulp products, posters, matches, woodenware.

Major Consumer Brands:
VANITY FAIR, BLUE RIBBON household paper products; BICYCLE, BEE, TALLY-HO, and CONGRESS playing cards.

Financial Data ($ Millions)	1969	1968	1967
Net Sales	498.1	489.3	447.0
Net Income	35.7	35.5	32.6
Capital Expenditures	30.2	22.9	22.5

Pollution Control Expenditures:
N.A.

Annual Meeting:
The annual meeting takes place during April in Wilmington, Delaware.

Officers:
Chairman: William H. Walters, Manhasset, New York (Director: Security National Bank of Long Island; Trustee; Emigrant Savings Bank, New York City)
President: Richard Walters, Manhasset, New York

Outside Directors:
Peter Berkey III: President Peter Berkey Foundation
Joseph C. Brennan: Chairman of the Board and Chief Executive, Emigrant Savings Bank
Dr. Walter C. Langsam: President, University of Cincinnati
John T. Madden: Retired, Former Chairman of the Board and Chief Executive, Emigrant Savings Bank
Bert W. Martin: President, Berada Corp.
Thomas J. McHugh: Retired; Former Chairman of the Board, Atlantic Lumber Co.
William L. Pfieffer: President, Albany Savings Bank; Trustee, Albany Medical College

Breakdown of Total Company Production (Percent of Sales):

Packaging products

Lumber and retail 25

Consumer products 13

Pulp, paper and paperboard 12

Commerical printing, advertising and other 6

Foreign 4

Machinery 3

Plants:
Diamond International has 62 plants located in 52 towns and cities in the United States.

Timberholdings:
The company owns 1,330,000 acres of timberlands, of which 873,000 are in Maine, and the balance is primarily in California and the Pacific Northwest. The total 1969 value of Diamond's holdings (at cost): $15,872,000.

Annual Pulp Production:
223,650 tons (1969) (Estimated on the basis of 355 production days)

Company Pollution Overview

Diamond International has shown almost no interest in trying to prevent pollution at any of its four pulp mills. Only one mill has even primary water treatment and none has adequate air emissions control. Nor has the company shown any initiative in planning future facilities. Treatment is being installed at the slowest possible rate on the longest possible timetable.

This situation would be much worse if Diamond were a major pulp manufacturer. Fortunately, the company operates only three small groundwood mills in Red Bluff, California and Ogdensburg and Plattsburgh, New York, and a larger sulfite, kraft and tissue mill complex in Old Town, Maine. The daily pulp production at California and Maine totals 630 tons using 32.5 million gallons of water; the company would release no production information on the New York mills. Two additional mills in Lockland and Middletown, Ohio, which make paperboard from recycled fiber, are not included in this study.

Diamond International's lack of concern for the environment is most clearly shown at Old Town, Maine, where the company is taking full advantage of the state's extraordinarily long timetable to avoid making any immediate efforts to reduce its water pollution. Thirty million gallons of untreated water are discharged daily into the Penobscot River. This situation would be somewhat more understandable if the mill were old, small, and only marginally profitable. But Old Town is Diamond International's largest plant and the only one to have been expanded in this decade: a kraft mill was added in 1965 and a tissue plant in 1970.

In answer to a question about the future pollution control plans at a

Diamond mill, the company's vice-president for public affairs, Ray Dubrowin, stated: "We are working on a total pollution control system—air, water, everything—but I can't tell you anything about it. Just that it will solve all the problems."

The fact that the word "environment" is not even mentioned in Diamond International's 1969 annual report causes doubt on the likelihood of the company coming up with a complete revolutionary pollution control system. CEP estimates that if Diamond International is to solve all of the pollution problems at all of its mills, a total investment of $8.55 million will be required.

Table 11.1
Diamond International: Pulp Production, Water Use, and Pollution Control

Mill Location	Pulp Prod. (TPD)	Other Prod. (TPD)	Water Use (MGPD)	Water Pollution Control			Air Pollution Control	
				Pri.	Sec.	Tert.	Part.	Gas, Odor
Red Bluff, Calif.	80	--	2.5	√	X	X	X	—
Old Town, Me.	550	100	30	X	X	X	X	X
Ogdensburg, N.Y.	--*	--*	4.4	X	X	X	X	—
Plattsburgh, N.Y.	--*	--*	1.6	X	X	X	X	—
Total	630	100	38.5					

*Company production figures N.A.

Red Bluff Mill

Location:
Red Bluff, California

Date Built:
1957

Process:
80 TPD chemi-mechanical unbleached groundwood

Other Production:
Molded pulp products

Total Water Use:
2.5 MGPD

Source:
Wells

Discharge:
Sacramento River

Table 11.2
Red Bluff Mill: Pollution Control Record

Treatment	Overall Evaluation	Equipment
Water		
Primary	✓	5 million gallon settling pond in which company raises salmon to illustrate that no damage is done to fish
Secondary	X	None
Tertiary	X	None
Solid waste removal	—	Landfill
Other	—	Outfall pipe installed, 1970
Air		
Particulate (fuel)	X	Fuel: sawdust (waste from sawmill on site). No control equipment
Particulate (production)	—	None needed
Gas and odor	—	None needed
Plant Emissions	N.A.	

Pollution Control Expenditures:
$900,000.

Legal Status:
(Water) The mill is in compliance; under permit since January 1963.
(Air) N.A.

Future Plans:
Secondary water treatment.

Profile:
This is the only one of Diamond International's four mills which has any water treatment at all. As such, perhaps it should receive a gold star for uniqueness. The system is not outstanding, but the effluent does meet the requirements set by the Central Valley Water Quality Board. The company further explained that "upgrading of the treatment facility will be based on pilot plant work and studies now in progress. This includes Diamond's work as well as that being done in conjunction with state and federal agencies regulating river conditions in this region." Technology of secondary treatment, of course, is very well established and has long been available for all sorts of pulp mills, but it is less necessary for unbleached groundwood pulp than for any other type of production.

CEP Estimate of Minimum Necessary Pollution Control Expenditures Not in Company Budget: $1.25 million total, of which $250,000 is for power stack particulate control and $1 million is for secondary water treatment.

Old Town Mill (owned by the Penobscot Company, a Diamond Subsidiary)

Location:
Old Town, Maine

Date Built:
1918, 1965 (kraft), 1970 (tissue mill)

Process:
200 TPD bleached calcium sulfite
350 TPD bleached kraft

Other Production:
100 TPD tissue and toweling

Total Water Use:
30 MGPD

Source:
Penobscot River

Discharge:
Penobscot River

Table 11.3
Old Town Mill: Pollution Control Record

Treatment	Overall Evaluation	Equipment
Water		
Primary	X	None
Secondary	X	None
Tertiary	X	None
Solid waste removal	—	Company landfill for barking wastes
Other	—	Savealls; barking waste screen; 80 TPD uncooked chips reclaimed
Air		
Particulate (fuel)	X	Fuel: oil; no equipment
Particulate (production)	X	Recovery furnaces: 90% efficient precipitators. Lime kiln: no control equipment
Gas and odor	X	None
Plant Emissions	N.A.	

Pollution Control Expenditures:
N.A.

Legal Status:
(Water) Tissue mill is operating under permit due to expire in 1976. The company states: "Standards are not applicable until 1976; we will meet them then."
(Air) State regulations not yet established.

Future Plans:
The company has extensive plans for meeting the state pollution control deadline when the time comes and apparently not a moment before. Water treatment facilities will include a primary settling pond and extended aeration to
achieve an expected 95 percent reduction in BOD and solids in the effluent.

For air pollution control, a new recovery furnace with precipitator is
"postulated to reduce current odor by over 95 percent" by burning the gases
from the evaporators. This unit and scrubbers to reduce sulfur dioxide emissions from the fuel boilers are scheduled to begin operation in 1971. A new
lime kiln is also planned.

Profile:
This mill provides a perfect example of why the state of Maine, which allows
four more years than any other state before requiring pollution control, should
be condemned. Maine set a 1972 deadline for filing plans while other states
expect completed facilities by that date. At present the mill's water pollution
control is almost nonexistent, and, by state consent, nothing need be done
about it until 1976. The effluent from 200 tons of bleached calcium sulfite pulp
contains 100,000 to 200,000 pounds of BOD; the 350 tons of bleached kraft probably contribute 17,500 pounds. The effects of this total daily pollutant load of
at least 117,500 pounds of BOD on the already sorely taxed Penobscot River
can be easily imagined.

There is no excuse for this pollution to continue for any longer than the
time required to install adequate water treatment, particularly since the state
allowed two of the three mills on the site to be built in 1965 and 1970 with no
abatement facilities. Diamond will continue to "pollute by permit" for six
years, and the state is as responsible as the company.

CEP Estimate of Minimum Necessary Pollution Control Expenditures Not
in Company Budget: $3.5 million total, of which $500,000 is for primary water
treatment; $2 million is for secondary water treatment; $500,000 is for power
stack particulate control; and $500,000 is for a recovery furnace precipitator.

Ogdensburg Mill

Location:
Ogdensburg, New York

Date Built:
1945

Process:
Groundwood (production figure not available)

Other Production:
Molded pulp products

Total Water Use:
4.4 MGPD

Source:
St. Lawrence River

Discharge:
St. Lawrence River

Table 11.4
Ogdensburg Mill: Pollution Control Record

Treatment	Overall Evaluation	Equipment
Water		
Primary	X	None
Secondary	X	None
Tertiary	X	None
Air		
Particulate (fuel)	X	Fuel: coal. Mechanical dust collectors, 95% efficient (inadequate)
Particulate (production)	—	Not needed
Gas and odor	—	Not needed
Plant Emissions	N.A.	

Pollution Control Expenditures:
N.A.

Legal Status:
(Water) On September 30, 1969, the company filed a preliminary report and schedule for compliance. In November 1969 it was cited by the state for "excess settleable solids, sludge deposits, and discoloration of the river."
(Air) The coal boiler is not in compliance.

Future Plans:
The mill is planning to change its fuel source from coal to oil. Diamond has

announced that it is installing a primary water treatment system, approved by
the New York State Department of Health, on the following schedule:

September 30, 1969 - preliminary report filed
June 30, 1970   - report on 6 months operation of a pilot treatment
                              plant filed
May 1, 1971   - submit construction drawings
October 1, 1971   - begin construction
October 2, 1972   - complete construction

In other words, the mill will have primary water treatment 35 months
after being cited by the state for causing "excess settleable solids, sludge
deposits, and discoloration" of the St. Lawrence River. Primary treatment
facilities usually require at most 18 months from conception to completion
of construction.

Profile:
This mill, although only producing groundwood pulp and molded pulp products,
is a significant local source of water pollution. Exactly how large a source
cannot be determined because Diamond International will not release the mill's
daily production figure to anyone. Fortunately, however, New York State laws
are somewhat stricter than those in Maine, and the company is being required
to have some water treatment by the end of 1972.

CEP Estimate of Minimum Necessary Pollution Control Expenditures Not
in Company Budget: $2.3 million total, of which $300,000 is for primary water
treatment and $2 million is for secondary water treatment.

Plattsburgh Mill

Location:
Plattsburgh, New York

Date Built:
1934

Process:
Groundwood (production figures N.A.)

Other Production:
Molded pulp products

Total Water Use:
1.6 MGPD

Source:
Municipal water system

Discharge:
Lake Champlain

Table 11.5
Plattsburgh Mill: Pollution Control Record

Treatment	Overall Evaluation	Equipment
Water		
Primary	X	None
Secondary	X	None
Tertiary	X	None
Air		
Particulate (fuel)	X	Fuel: coal, bark. Bark boilers covered by 99% efficient mechanical dust collectors
Particulate (production)	—	None needed
Gas and odor	—	None needed
Plant Emissions	N. A.	

Pollution Control Expenditures:
N. A.

Legal Status:
(Water) Not in compliance; not operating under approved state permit. This mill was cited in August, 1966, for excess settleable solids and sludge deposits. No remedial action has been taken yet.
(Air) In May, 1970, New York State reported that abatement action may be required because of complaints about the dust from the coal boilers.

Future Plans:
Beginning in 1972 the mill's entire effluent will be discharged into an enlarged

municipal treatment system which is now under construction. The effluent will receive primary and secondary treatment.

Profile:
Fortunately for Lake Champlain, this is only a small groundwood mill.

CEP Estimate of Minimum Necessary Pollution Control Expenditures Not in Company Budget: $1.5 million total, of which $500,000 is for improved bark and coal boiler particulate controls and $1 million is for joint industrial municipal water treatment (planned but no cost given).

Chapter 12

FIBREBOARD CORPORATION
55 Francisco Street, San Francisco, California 94133

Major Products:
Paperboard and packaging, lumber, plywood and other wood products. Other products include high temperature insulations, inks and printing plates. A subsidiary is engaged in land development; another in collecting waste paper for recycling.

Major Consumer Brands:
Products are marketed under PABCO label.

Financial Data: ($ Millions)	1969	1968	1967
Net Sales	181.8	184.0	153.4
Net Income	6.9	8.5	3.8
Capital Expenditures	11.0	10.4	9.6

Pollution Control Expenditures:
$7.5 million ($1 million from 1967 to 1970)

Annual Meeting:
The annual meeting is held during April in San Francisco, California.

Officers:
President: George W. Burgess, San Francisco, California (Director: Dymo Industries Inc.)

Outside Directors:
Joseph A. Moore, Jr.: President, Moore Dry Dock Co.
Robert A. Magowan: Chairman of the Executive Committee, Safeway Stores, Inc.
William L. Keady: President, Advalloy, Inc.
Herman Phleger: Attorney-Partner, Brobeck, Phleger and Harrison
Emmett G. Solomon: Chairman of the Board, Crocker Citizens National Bank
Peter E. Haas: President, Levi Strauss & Company
Alexander M. Wilson: Executive Vice President, Utah Construction and Mining Co.

Breakdown of Total Company Production:

Paperboard	508,000 tons per year
Carton	150,000 tons per year
Container	3,200,000 (in thousands of square feet per year)

Plants:
Fibreboard has 19 domestic pulp and paperboard and packaging plants:
Board   5
Carton   7
Container   7

Timberholdings:
Fibreboard owns 165,174 acres in the United States. The total 1969 value of
Fibreboard's holdings: $7,399,439.

Annual Pulp Production:
266,817 tons (1969)

Company Pollution Overview

Fibreboard, ranked the 25th largest U.S. paper and pulp company in a 1969
industrial magazine listing, has operated two pulp mills: one in Antioch,
California, and one in Port Angeles, Washington.

    The Port Angeles mill was in irreparably bad shape. It was an old small
ammonium sulfite and groundwood mill and was cited for pollution at the 1967
Puget Sound Federal Abatement Conference. The mill then had totally inadequate water and air pollution control equipment. All waste sulfite liquor was
discharged into the sound. Despite the federal notice, until October 1970,
Fibreboard made no move to correct the mill's pollution, stalling the inevit-
able decision to close down or begin a major expansion and modernization.
The permit deadline for primary water treatment was September 1970. Fibreboard closed down the mill in December 1970.

    Fibreboard's San Joaquin mill at Antioch, California, where stringent
Bay-area smog regulations are in force, has first-rate air pollution control,
but has been operating without primary or secondary water treatment since
1949. Even today, in the midst of a major expansion, there are no plans to
build the obviously indispensible facilities. The company's director of pollution affairs said only that new water laws are being established and that when
they are "we will take steps to meet the new regulations."

    With this negligent and foot-dragging company attitude, it need simply be
mentioned that luckily, Fibreboard has only one mill left. The Council estimates that cleaning it up would entail an investment of $4.75 million.

Table 12.1
Fibreboard Corporation: Pulp Production, Water Use, and Pollution Control

Mill Location	Pulp Prod. (TPD)	Other Prod. (TPD)	Water Use (MGPD)	Water Pollution Control			Air Pollution Control	
				Pri.	Sec.	Tert.	Part.	Gas, Odor
Antioch, Calif.	750	750	20	X	X	X	✓	✓

San Joaquin Mill

Location:
Antioch, California

Date Built:
1949, expansion in 1967 (and in 1970-1972)

Process:
500 TPD bleached and unbleached kraft
250 TPD "hi yield kraft" (groundwood mill, supplementary production)

Other Production:
250,000 tons per year paperboard manufactured from pulp produced at San Joaquin (750 TPD corrugating medium, kraft linerboard and bleached kraft carton boards)

Water Use:
20 MGPD

Source:
San Joaquin River

Discharge:
San Joaquin River

Table 12.2
San Joaquin Mill: Pollution Control Record

Treatment	Overall Evaluation	Equipment
Water		
Primary	X	None
Secondary	X	None
Tertiary	X	None
Other	—	"Extensive internal control within mill"
Air		
Particulate (fuel)	√	Fuel: 1.4% sulfur oil
Particulate (production)	√	Three recovery furnaces, each has precipitator followed by scrubber; overall efficiency is 99.5%; lime kilns have high pressure drop
Gas and odor	√	Vent gases are incinerated in lime kilns; heavy black liquor oxidation
Plant Emissions	N.A.	

Pollution Control Expenditures:
To date, Fibreboard estimated $7.5 million has been spent for pollution control at this mill, of which about $1 million was spent since 1967.

Legal Status:
(Water) The mill is in compliance with a long-held permit from the regional water board.
(Air) In compliance with the strict bay area air pollution regulations. The control agency commented: "This plant for over a decade has had excellent air pollution control facilities."

Future Plans:
Between 1970 and 1972 the mill will be expanded from 750 to 900 tons per day

of pulp production. Along with this expansion, there are plans for additional air pollution control measures. The mill will use natural gas and supplemental oil for power, thus virtually eliminating the power boiler pollution problem. This will occur "in the very near future." No further water treatment is now planned.

Profile:
At this medium-sized mill, Fibreboard has done only the "necessary" about pollution control. As the air pollution standards were strict, the mill invested heavily in air control equipment. However, the mill has succeeded in operating with only in-plant water control for 21 years. Clearly, Fibreboard will install no more treatment until required. Mr. Walter Simon, Manager for Fibreboard Water Resources and Effluent Control, stated that the mill is meeting state requirements. True, but only because the state has given the mill a permit to discharge minimally treated waste water. He added: "Future requirements for discharge into the river are uncertain. However, they will undoubtedly be changed in the future and we will take steps to meet new regulations when they are established."

CEP Estimate of Minimum Necessary Pollution Control Expenditures Not on Company Budget: $4.75 million total, of which $750,000 is for primary water treatment and $4 million is for secondary water treatment (activated sludge necessary since little land available.)

Chapter 13

GEORGIA-PACIFIC CORPORATION
900 S.W. Fifth Avenue, Portland, Oregon 97204

Major Products:
Paper, wood, gypsum, and chemical products.

Major Consumer Brands:
CORONET tissue goods, paper towels and napkins, ROYAL OAK charcoal
briquets, and M.D. toilet tissues.

Financial Data ($ Millions)	1969	1968	1967
Net Sales	1,160.2	1,023.9	885.7
Net Income	91.8	76.6	58.6
Capital Expenditures	245.0	149.4	125.2

Pollution Control Expenditures:
$35 million, 1960-1970; $10 million, 1970-1972 (estimated)

Annual Meeting:
The annual meeting is held during April in Portland, Oregon.

Officers:
Chairman/President: Robert B. Pamplin (Trustee: Lewis and Clark College)

Outside Directors:
S. Clark Beise: Chairman of the Executive Committee, Bank of America, NT
and SA
Harvey C. Fruehauf, Jr.: President, The Fruehauf Foundation
James M. Hait: Chairman, FMC Corporation
James F. Miller: President, Blythe and Co., Inc.
Grayson M-P Murphy: Partner, Shearman and Sterling
Stuart T. Saunders: Chairman, Penn Central Co.
John F. Watlington, Jr.: President, Wachovia Bank and Trust Co.

Breakdown of Total Company Production (Percent of Sales, 1969):

Softwood, hardwood, plywood and specialties	37
Pulp and paper	29
Lumber	16
Gypsum	6
Other (including chemicals)	12

Plants:
Georgia-Pacific has 50 domestic plants and six in Canada. The breakdown of
the domestic plant production is

Lumber, softwood, hardwood, and specialty plywood	83
Pulp and paper	15
Gypsum	12
Other (including chemicals)	40

Timberholdings:
Georgia-Pacific is one of the largest private owners of timberlands in the
country, owning 3.5 million acres (equal in size to Rhode Island and Connecticut). In addition, Georgia-Pacific owns one million acres in Brazil and Eastern Canada and has timber concessions consisting of 632,000 acres in the
Philippines and Equador, and cutting rights to 855,000 acres in Alaska. In the
planning stage are harvest rights to 800,000 more acres in Indonesia.

Owned:	4.5 million acres
Cutting rights:	1.5 million acres
Total:	6 million acres

The total 1969 value of Georgia-Pacific's holdings: $265,970,000.

Annual Pulp Production:
2,195,000 tons (1969; includes United States and Canada)

Company Pollution Overview

Georgia-Pacific ranks among the top ten producers of paper and paperboard
in the United States. The company, headquartered in the Northwest, has in
the past few years shown strong initiative in solving many pollution problems
at some of its largest mills, yet has taken minimal or no steps at the others.
Georgia-Pacific produces pulp at 15 locations in the United States and one
in Canada. At six of these locations, pulp is manufactured from reclaimed
fiber only. This production totals 189,000 tons per year and has not been included in this study. Annual pulp production at the other nine locations totals
1.9 million tons. These nine mills are scattered over the United States: five
large kraft mills in Crossett, Arkansas; Woodland, Maine; Port Hudson,
Louisiana; Samoa, California; and Toledo, Oregon; two large sulfite mills;
one in Bellingham, Washington, the other (half owned with Ketchican Pulp
Company) in Ketchican, Alaska; and two small NSSC mills in New York State
(Plattsburgh and Lyons Falls). These nine locations produce a total of 5,470
tons of pulp a day and discharge 227 million gallons of waste water.
    Georgia-Pacific is a rare example of a company which has provided ex-
cellent gas and odor control at all mills including the five kraft mills. At
Samoa, California, the company has particularly good control of odor and
gas emissions. On the other hand, six of the mills could use improved particulate control, especially on the fuel boiler stacks, which have inadequate
dust collectors.

Since 1967, Environmental Controls Director Matthew Gould has concentrated on bringing water treatment at the large kraft mills up to required state standards. Four now meet or exceed these standards, and the fifth, at Port Hudson (acquired in 1969), will have primary treatment by 1972. In line with this effort, both eastern kraft mills (Crossett and Woodland) have fine primary and secondary treatment and Crossett even has tertiary treatment for removing color from the effluent.

The two western kraft mills, in which Georgia-Pacific has recently invested several million dollars, also now meet state standards; but unfortunately, no glory is due either the states or the company. California and Oregon control authorities endorsed Georgia-Pacific's plan to install ocean outfall pipes with diffusers in place of water treatment. Thus, 38 million gallons of mill effluent (solid wastes and chemicals) are simply dumped far out in the ocean each day on the old assumption that in so much water, all will be diluted and will disappear. Mr. Gould's explanation that, even under close scrutiny, no damaging effects on ocean ecology have been noted, is not reassuring. By the time the cumulative direct and indirect effects of these wastes are "noted," corrective action may come too late and too slowly, costing additional millions, which, in the case of these mills, could have been spent on REAL preventive measures now.

Besides Georgia-Pacific's five kraft mills, the other four are highly inadequate from the point of view of pollution control. The sulfite mill in Alaska has no secondary water treatment and inadequate particulate control. The other three mills, the sulfite mill at Bellingham, Washington, and the two New York State NSSC mills, have no water treatment at all. Bellingham recovers two-thirds of its spent liquor (thus achieving an 85 percent BOD reduction), but the two NSSC mills dump their waste water and pulping liquor directly into receiving waters. All three mills were cited for water pollution, two in 1966 and one in 1967. Not until 1970 were programs for joint municipal-industrial treatment approved at Bellingham and Plattsburgh. It will then be six years after the state orders before effluent from these two plants will begin receiving treatment. At the Lyons Falls mill ("the problem mill of the company") applications for state aid have been turned down. The company says if the Federal Economic Development Authority also refuses aid, the pulping operations will probably have to close.

Georgia-Pacific budgeted a total of $ 20 million between 1968 and 1972 for pollution control. About half of this has been spent, and the remaining $ 10 million will be invested in adequate water treatment planned at Port Hudson, Plattsburgh and Bellingham. CEP estimates that if the nine mills are kept in operation, at least another $ 13.6 million will be needed to adequately control both air and water pollution at all locations.

It should be noted that Georgia Pacific's Matthew Gould gave many hours of his time to CEP, openly discussing each of his mills in detail as well as many technical aspects, pros and cons of various pieces of control equipment. He added considerably to the Council's ability to present the facts, and showed his awareness that the public deserves to know what is being done and still needs to be done about pollution control.

Table 13.1
Georgia-Pacific Corporation: Pulp Production, Water Use and Pollution
Control

Mill Location	Pulp Prod. (TPD)	Other Prod. (TPD)	Water Use (MGPD)	Water Pollution Control			Air Pollution Control	
				Pri.	Sec.	Tert.	Part.	Gas, Odor
Ketchican, Alaska	630		42	✓	X	X	X	✓
Crossett, Ark.	1,250	760	50	✓	✓	✓	✓	✓
Samoa, Calif.	550		25	X	X	X	✓	✓
Port Hudson, La.	600		20	X	X	X	X	✓
Woodland, Me.	800	500	30	✓	✓	X	✓	✓
Lyons Falls, N.Y.	70	160	4	X	X	X	X	—
Plattsburgh, N.Y.	70	155	4	X	X	X	✓	—
Toledo, Ore.	1,000	880	13	✓	X	X	X	✓
Bellingham, Wash.	500	535*	39**	X	X	X	X	✓
Total	5,470	2,990	227					

*11,000 GPD of ethyl alcohol are also produced.
**Process water only.

Ketchican Mill

Location:
Ketchican, Alaska (Ketchican Pulp Co.)

Date Built:
1954

Process:
630 TPD magnesium bisulfite pulp (company figure)

Other Production:
None; pulp is produced for market

Total Water Use:
42 MGPD

Source:
Ward Creek (three miles above the mill)

Discharge:
Ward Cove (salt water tidal bay of Pacific Ocean)

Table 13.2
Ketchican Mill: Pollution Control Record

Treatment	Overall Evaluation	Equipment
Water		
Primary	✓	Series of large and small screens remove 85% of settleable solids
Secondary	X	None; however, waste liquor is not discharged but is treated for chemical recovery
Tertiary	X	None
Air		
Particulate (fuel)	X	Fuel: low sulfur oil and bark. Bark boiler has mechanical collector (inefficient)
Particulate (production)	X	Four recovery furnaces with mechanical collectors; secondary wet scrubbing of furnace exhaust as it passes through a series of four towers packed with porous material
Gas and odor	✓	97% sulfur dioxide removal, but inadequate particulate removal; demister on central stack to cut vapor emissions

Table 13.2
Ketchican Mill: Pollution Control Record (continued)

Treatment	Overall Evaluation	Equipment
Plant Emissions		
To water:		
BOD	N.A.	
Suspended solids	8,000 lb/day (12.7 lb/ton of pulp)	
To air:	N.A.	

Pollution Control Expenditures:
Since 1966, the only expenditure for pollution control was $100,000 for two new screens for improved solids removal from effluent.

Legal Status:
(Water and Air) Water and air standards not applied to this mill yet. Hearing scheduled for water study January 1971.

Future Plans: Primary and secondary treatment for sanitary wastes of mill's 500 employees scheduled for completion 1972.

Profile:
This large sulfite mill, tucked away on the coast of Alaska, has operated for 16 years now with relatively little thought given to its pollution problems. "We've known for ten years," said Mr. Roland Stanton, the company's Technical Director, "that we were going to have to do something about our water pollution. The effluent pH is too low. It is 4.5, while the cove level is 5.8. There are too many dissolved solids. When the effluent gets into the cove, we can see it actually floating on the water's surface. It just sits there on top of the water. We don't think it affects lower water ecology, but we haven't tested or measured it yet."[1]

As for air, Mr. Stanton commented: "Particulate emissions haven't been measured yet, but we just don't think they would meet any standards. We have measured sulfur dioxide emissions, though, and they are low—we get 97 percent removation—down to 300 ppm, whereas some of the utilities in the northwest may have about 2000 ppm. We have somewhat of a problem with water vapor which causes a fog and heavy mists, and it settles on the highways."[2]

This honest recognition of Ketchikan's problems has so far led to little concrete action. In the past four years, while $11.5 million was spent for a new recovery furnace and new lumber cutting equipment, the only pollution control expenditure was the $100,000 for the new screens. Mr. Stanton did indicate that more substantial efforts are beginning, such as the treatment planned for the mill's sanitary wastes. And what else: "Well," he added, "next summer we're hiring some students to come and study the quality of our effluent." He concluded philosophically, "Alaska is a little far behind in this area."[3]

CEP Estimate of Minimum Necessary Pollution Control Expenditure Not on Company Budget: $2.75 million total, of which $2 million is for secondary water treatment; $500,000 is for an electrostatic precipitator; and $250,000 is to improve power boiler particulate control.

Crossett Mill

Location:
Crossett, Arkansas

Date Built:
1960

Process:
1,250 TPD bleached and unbleached kraft (company figure; the mill is currently being expanded from 840 to 1,250 TPD)

Other Production:
Three paper mills on site produce 760 tons of paper a day, one mill produces 440 tons of wrapping and butcher papers, asphalting, envelopes, variety bags and crepe papers; the second 210 tons of milk cartons, plates and trays, tags and file folders; and the third 110 tons of bathroom and facial tissues, napkins and towelling

Total Water Use:
50 MGPD (9 MGPD is cooling water only)

Source:
Oachita River, wells

Discharge:
Oachita River

Table 13.3
Crossett Mill: Pollution Control Record

Treatment	Overall Evaluation	Equipment
Water		
Primary	✓	300-foot diameter clarifier
Secondary	✓	10 day retention in 240-acre aerated lagoon
Tertiary	✓	Slaked lime color removal system (company anticipates 80-90% effectiveness)
Air		
Particulate (fuel)	✓	Fuel: gas and bark. Multiple cyclone collectors
Particulate (production)	✓	Three recovery furnaces (including a new 700-ton capacity one) with electrostatic precipitators and scrubbers. Three lime kilns with scrubbers; two scrubbers and one precipitator are 99.5% efficient
Gas and odor	✓	Black liquor oxidation
Plant Emissions	N.A.	

Pollution Control Expenditures:
$5.1 million has just been spent (as part of a $90 million expansion of the mill) for completing installation of almost all needed pollution control equipment (two new 99.5 percent efficient scrubbers, a new 99.5 percent efficient electrostatic precipitator, an aerated lagoon and tertiary water treatment).

Legal Status:
(Water) This mill's water control more than meets state requirements. Georgia-Pacific has just put in tertiary treatment, not required by the state, and paid the entire cost of the program. The Sportsmen's Foundation gave one of its water quality awards to Crossett for its excellent water treatment.

Future Plans:
None.

Profile:
Georgia-Pacific, anticipating the need to put the latest and most effective controls on this large 10-year-old kraft mill, proceeded to equip it so that its control will be excellent, except the particulate control on its fuel boilers, which could be improved.

CEP Estimate of Minimum Necessary Pollution Control Expenditure Not on Company Budget: $500,000 total, to improve power boiler particulate control.

Samoa Mill

Location:
Samoa, California

Date Built:
1965

Process:
550 TPD bleached kraft

Total Water Use:
25 MGPD

Source:
Mad River

Discharge:
Pacific Ocean

Table 13.4
Samoa Mill: Pollution Control Record

Treatment	Overall Evaluation	Equipment
Water		
Primary	X	None
Secondary	X	None
Tertiary	X	None

Table 13.4
Samoa Mill: Pollution Control Record (continued)

Treatment	Overall Evaluation	Equipment
Other	—	One-half mile ocean outfall pipe with diffusers
Air		
Particulate (fuel)	X	Fuel: gas and bark. Cyclone collectors, 90% efficient
Particulate (production)	✓	Two recovery furnaces with electrostatic precipitators followed by scrubbers, 95% efficient
Gas and odor	✓	Heavy black liquor oxidation; fuel gases from cascade evaporators scrubbed before release from 300-foot main stack
Plant Emissions	N.A.	

Pollution Control Expenditures:
No information available.

Legal Status:
(Water) The mill has a permit from the state to discharge its waste water by means of an outfall and diffuser.
(Air) In compliance.

Future Plans:
None.

Profile:
When this large mill was built in 1965, optimal air pollution control was obligatory; first, because the mill's location was only one mile northwest of Eureka, California, with a prevailing wind blowing toward the town, and second, because Humboldt County had (and still has) the strictest sulfur emission standards of any county in the United States. As a result, both precipitators and scrubbers were built on the recovery furnaces, and the weak black liquor oxidation system, when found to be "good but not excellent," was sup-

plemented in 1969 by additional oxidation of heavy black liquor. The current mill sulfur emissions, 1,025 pounds per day, are now well below the county limit of 1,300 pounds.

In the area of mill effluent, similar standards were not set. Georgia-Pacific, with state endorsement, chose to avoid treating its waste water at all. The mill's pulping process water and bleach plant wastes are now discharged through a $3 million outfall pipe into the ocean, a practice authorized by the state, defended by Georgia-Pacific, and believed completely unacceptable by CEP.

CEP Estimate of Minimum Necessary Pollution Control Expenditure Not on Company Budget: $3.25 million total, of which $250,000 is to improve power boiler particulate control; $1 million is for primary water treatment; and $2 million is for secondary water treatment.

Port Hudson Mill

Location:
Port Hudson, Louisiana

Date Built:
1968 (acquired 1969)

Process:
600 TPD bleached kraft (company figure)

Total Water Use:
20 MGPD

Source:
Deep wells

Discharge:
Mississippi River

Table 13.5
Port Hudson Mill: Pollution Control Record

Treatment	Overall Evaluation	Equipment
Water		
Primary	X	None
Secondary	X	None

Table 13.5
Port Hudson Mill: Pollution Control Record (continued)

Treatment	Overall Evaluation	Equipment
Tertiary	X	None
Air		
Particulate (fuel)	✓	Fuel: gas; no equipment needed
Particulate (production)	X	One recovery furnace equipped with an inefficient venturi scrubber
Gas and odor	✓	Furnace exhaust gases scrubbed by waste liquor
Plant Emissions	N.A.	

Pollution Control Expenditures:
None

Legal Status:
(Water) The state has required this mill to install primary water treatment by 1972.
(Air) No state kraft mill regulations set yet. However, Georgia-Pacific environmental controls director conceded that the mill has particulate problems.

Future Plans:
As of September 1970, engineering studies were underway for primary and secondary water treatment. Plans are being made for a new electrostatic precipitator to replace the old recovery furnace scrubber. Georgia-Pacific has budgeted $5 million for cleaning up this mill by 1973.

Profile:
The Port Hudson Mill was built only two years ago, recently enough to expect considerable thought would have gone into pollution control. Yet no water treatment was planned, and air control is inadequate.

Georgia-Pacific only acquired the mill in 1969, and therefore cannot be held responsible for the inadequate controls originally installed. The state of Louisiana can be blamed for allowing the mill to begin production without equipment that should have been a prerequisite to any operation. As a result of this lack of planning, by the time adequate controls are provided, the mill will have been allowed to pollute the Mississippi River and the atmosphere for four years.

CEP Estimate of Minimum Necessary Pollution Control Expenditure Not on Company Budget: $250,000 total, for a lime kiln scrubber.

Woodland Mill

Location:
Woodland, Maine

Date Built:
1910; 1966 converted from calcium sulfite and groundwood to kraft and a new groundwood mill replaced the original one

Process:
600 TPD bleached kraft,
200 TPD groundwood (company figure)

Other Production:
500 TPD newsprint, market pulp, a small quantity of fine paper, and corrugating medium

Total Water Use:
30 MGPD

Source:
St. Croix River (flows between United States and Canada)

Discharge:
St. Croix River

Table 13.6
Woodland Mill: Pollution Control Record

Treatment	Overall Evaluation	Equipment
Water		
Primary	√	190-foot diameter clarifier
Secondary	√	70-foot diameter clarifier with lime treatment used for removal of 80-90% of

Table 13.6
Woodland Mill: Pollution Control Record (continued)

Treatment	Overall Evaluation	Equipment
		the color and an additional 45% of the BOD
Tertiary	X (partial)	As above
Solid waste removal	—	Clarifier and wood room wastes are dewatered and incinerated: 40 tons of solids per day
Air		
Particulate (fuel)	✓	Fuel: bark, clarified sludge, and, as a supplement, oil. Cyclone collectors.
Particulate (production)	✓	Two recovery furnaces, electrostatic precipitators. Lime kiln, gas scrubber
Gas and Odor	✓	Black liquor oxidation
Plant Emissions	N.A.	

Pollution Control Expenditures:
Over $ 5 million is now being spent on water treatment facilities to meet requirement of the United States-Canadian Joint Commission for the St. Croix River.

Legal Status:
(Water) This mill is way ahead of the State of Maine requirements, which call for secondary water treatment in 1976.
(Air) State regulations still undefined, but this mill has generally good odor and particulate control.

Future Plans:
Director of Environmental Controls, Matthew Gould, said, "The company is considering addition of secondary (heavy) black liquor oxidation to supplement the weak black liquor oxidation now used, since this combination has achieved such good results in the Samoa, California mill."

CEP Estimate of Minimum Necessary Pollution Control Expenditure Not on Company Budget: $500,000 total, to improve power boiler particulate control.

Lyons Falls Mill

Location:
Lyons Falls, New York

Date Built:
Ca. 1900

Process:
70 TPD bleached NSSC pulp (company figure)

Other Production:
160 TPD business papers

Total Water Use:
4 MGPD

Source:
Black River

Discharge:
Black River

Table 13.7
Lyons Falls Mill: Pollution Control Record

Treatment	Overall Evaluation	Equipment
Water		
Primary	X	None
Secondary	X	None; waste liquor and bleaching plant effluent discharged into Black River
Tertiary	X	None
Air		
Particulate (fuel)	X	Fuel: oil. Cyclone collector, 90% efficient

Table 13.7
Lyons Falls Mill: Pollution Control Record (continued)

Treatment	Overall Evaluation	Equipment
Particulate (production)	—	No liquor recovery and burning; therefore no particulate emissions
Gas and odor	—	None needed
Plant Emissions	N.A.	

Pollution Control Expenditures:
No information available.

Legal Status:
(Water) The State of New York cited the mill in January 1966 for discharge of excess solids and sluge deposits.
(Air) Report to state under review, boiler under compliance.

Future Plans:
None finalized.

Profile:
For the past four years, since the state abatement order, the Lyons Falls mill has continued to discharge completely untreated waste water, including spent pulping liquor and bleaching chemicals, into the Black River. The company has stalled on providing treatment, since this small old mill could be, at best, only marginally profitable now. According to Georgia-Pacific's Environmental Director, Matthew Gould, an application by the mill and the municipality for state aid for a joint treatment plant was denied, as the state has no provision for funding projects outside urban areas. (Lyons Falls has a population of about 300, of which more than half work for the mill.)

Application for aid from the Economic Development Authority has been made. Mr. Gould noted: "If no help is provided, we will have to close the mill. It has been the problem of our company, but it is the only major employer in the area, and Lewis County already has 6 percent unemployed."

CEP Estimate of Minimum Necessary Pollution Control Expenditure Not on Company Budget: $3.25 million total, of which $250,000 is to improve power boiler particulate control; $1 million is for primary water treatment; and $2 million is for secondary treatment. The same degree of control might be obtained if the company invested $1 million in a joint treatment plant.

Plattsburgh Mill

Location:
Plattsburgh, New York

Date Built:
1965

Process:
70 TPD NSSC pulp (weak hydrosulfite bleaching)

Other Production:
155 TPD of tissue, facial tissue, napkin, toilet, towel

Total Water Use:
4 MGPD (being reduced to 2.75 MGPD by increased recycling)

Source:
Lake Champlain

Discharge:
Lake Champlain

Table 13.8
Plattsburgh Mill: Pollution Control Record

Treatment	Overall Evaluation	Equipment
Water		
Primary	X	None
Secondary	X	None; waste liquor bleaching chemicals discharged into Lake Champlain
Tertiary	X	None
Air		
Particulate (fuel)	√	Fuel: oil, hydroelectric power. Gas scrubbers on boilers
Particulate (production)	—	None needed

Table 13.8
Plattsburgh Mill: Pollution Control Record (continued)

Treatment	Overall Evaluation	Equipment
Gas and odor	—	No liquor burned, so no particulate or gas emissions
Plant Emissions	N.A.	

Pollution Control Expenditures:
No information available.

Legal Status:
(Water) Mill cited by New York State for water pollution (excess solids and sludge deposits and discoloration in Lake Champlain), August, 1966.
(Air) Report to state under review, boiler under compliance.

Future Plans:
The mill, the City of Plattsburgh, and other local industries have signed construction contracts for a joint industrial-municipal treatment plant where all effluents will be pumped and treated according to state standards. Georgia-Pacific's share of the capital cost of the plant will be approximately $900,000, and annual operation will cost the company about $220,000.

    CEP Estimate of Minimum Necessary Pollution Control Expenditure Not on Company Budget: None.

Toledo Mill

Location:
Toledo, Oregon

Date Built:
1960

Process:
1,000 TPD unbleached kraft

Other Production:
880 TPD kraft linerboard

Total Water Use:
13 MGPD

Source:
Olalie Creek

Discharge:
Pacific Ocean

Table 13.9
Toledo Mill: Pollution Control Record

Treatment	Overall Evaluation	Equipment
Water		
Primary	✓	5 million gallon clarifier for waste water from paper production area only; 90% solids removal
Secondary	X	None
Tertiary	X	None
Solid waste removal	—	A new $3 million pipeline with diffusers transports the remaining mill waste water directly into the Pacific
Air		
Particulate (fuel)	X	Fuel: oil. No equipment
Particulate (production)	X	Three recovery furnaces with precipitators and scrubbers—90% efficient; one lime kiln with gas scrubber—90% efficient
Gas and odor	✓	Black liquor oxidation; noncondensible gases burned in lime kiln
Plant Emissions	N.A.	

Pollution Control Expenditures:
$3 million for pipeline.

Legal Status:
(Water) The State of Oregon considers pipeline in compliance.
(Air) Mill in compliance with particulate standards (current emissions 5595 pounds per day, or 5.5 pounds per ton of pulp), but sulfur emissions need further reduction (current emissions 1500 pounds per day i.e., 1.5 pounds per ton).

Future Plans:
Company adding secondary (heavy) black liquor oxidation to the mill this year.

Profile:
This large 1,000 ton per day kraft mill was built with good particulate and gas controls, which now need substantial improvement because they are operating at only about 90 percent efficiency. Between 95 percent and 99 percent efficiency should be expected today.

Where air control was given considerable attention, water treatment was not. The first major expenditure in this area was the $3 million recently invested in a "pipeline with diffuser," through which the mill now daily discharges 13 million gallons of virtually untreated effluent, full of solid waste and chemicals, directly into the Pacific Ocean.

Georgia Pacific's environmental director said: "Our pipeline is the most studied pipe in the United States. However, there have been no detectable effects on the ocean or its ecology. In fact, fishermen fish at the outfall where the warm water attracts the fish." He added, "the effluents are not highly toxic, but are similar to vegetable decay products."

Several studies have been made to ascertain whether foamed Kraft mill effluent from the Toledo mill is toxic to shellfish. One, by Courtwright and Bond [4], commented as follows:

"This research is concerned with a potential pollution problem heretofore overlooked because of the generally accepted attitude that the ocean has an infinite capacity for dilution of wastes ... Previous works ... have indicated that considerable toxic material is associated with the foam fraction of kraft waste. Lignins in kraft wastes apparently precipitate upon contact with salt water ... This precipitate and possibly other material rises to the surface where it is agitated by wind and wave action and foams profusely. It is then carried to the shore where large quantities accumulate. These blankets of potentially toxic foam may be deposited many miles from the outfall sewer. When this occurs, concentrations lethal to some marine animals may accumulate in the tidepools. For example, the shellfish embryos are highly susceptible to low concentrations of kraft mill effluents as evidenced by inhibition of normal development of bay mussel embryos exposed to the effluent."

What happens is that foam concentrates toxic materials of the kraft mill effluent, although it does not concentrate common pollution measures such as BOD or COD. The conclusions of the study were:

"1. Accumulations of foamed kraft mill effluent are four to five times more toxic than the waste prior to oceanic discharge as determined by bioassays with mussel larvae."

"2.Although low temperatures enable the fluffy sculpin to withstand high concentrations of kraft wastes for extended time periods, toxic effects of the waste can adversely affect their survival."

CEP Estimate of Minimum Necessary Pollution Control Expenditure Not on Company Budget: $3.25 million total, of which $250,000 is to improve power boiler particulate control; $1 million is for primary water treatment; and $2 million is for secondary treatment.

Bellingham Mill

Location:
Bellingham, Washington

Date Built:
1938

Process:
500 TPD bleached sulfite-calcium base

Other Production:
200 TPD tissue, napkin, and towelling
45 TPD of paperboard
11,000 gallons per day of ehtyl alcohol
170 TPD of lignin products
120 TPD of chlorine-caustic chemicals

Total Water Use:
39 MGPD (for pulping alone)

Source:
Lake Whatcom

Discharge:
Puget Sound (Bellingham Bay)

Table 13.10
Bellingham Mill: Pollution Control Record

Treatment	Overall Evaluation	Equipment
Water		
Primary	X	None
Secondary	X (partial)	No equipment, but only 1/3 of the spent liquor is discharged. 2/3 is used in by-product manufacture. The resulting 85% BOD reduction is equivalent to secondary treatment
Tertiary	X	None
Air		
Particulate (fuel)	X	Fuel: bark. Mechanical collectors
Particulate (production)	—	No liquor burned, so no controls needed
Gas and odor	✓	All stacks emitting sulfur dioxide are scrubbed with spent caustic soda from the chemical manufacturing operations on the site. The company claims that this "mill has the lowest sulfur dioxide emission in the world"
Plant Emissions	N.A.	

Pollution Control Expenditures:
No information available.

Legal Status:
The Georgia-Pacific mill was cited at the 1967 Federal Water Pollution Control Administration Puget Sound Enforcement Conference for pollution of Bellingham Bay, and causing low dissolved oxygen counts and damage to oyster larva beds. It was required to reduce discharge of waste pulping liquor, provide primary water treatment and construct an outfall pipe with diffusers. The mill was also required to dredge and dispose of the sludge it had deposited

over many years in the harbor and to modify its chip barge unloading procedures to eliminate spillage of chips. The state first set a May, 1968, deadline for increased liquor recovery and a September, 1972, deadline for water
treatment. However, the permit subsequently issued extended the first deadline by five years (to May, 1973) and set no deadline for water treatment or
sludge removal. Bellingham was one of the plants indicted by the federal
government in July, 1970, for mercury pollution. The charge was dropped;
the government had erroneously accused Georgia-Pacific of dumping 40 pounds
of mercury daily into the bay. The actual daily discharge of 10 pounds has
since been reduced to about one-half pound.

Future Plans:
(Water) A joint industrial-municipal treatment plant by January, 1973. The
capital cost to Georgia-Pacific will be about $6.1 million.
(Air) State sulfite standards under development.

Profile:
A Spring 1970 study for the City of Bellingham recommending construction
of a joint treatment plant analyzed area effluents. The city would contribute
organic wastes from its 28,000 residents; eight local waterfront food processing plants would contribute a BOD load equivalent to 145,000 people, and
the Georgia-Pacific mills a BOD load equivalent to a population of one million.

This mill's wastes, including bleach plant chemicals and considerable
waste pulping liquor, have been discharged over the past 32 years directly
into Bellingham Bay, reportedly turning the water near the mill reddish
brown and causing anaerobic conditions in parts of the bay.

The proposed plant giving primary and secondary effluent treatment
would mean a total investment of $11.3 million, with Georgia-Pacific's share,
$6.1 million. The state in September, 1970, accepted the plan with a January,
1973, deadline.

CEP Estimate of Minimum Necessary Pollution Control Expenditure Not
on Company Budget: $250,000 total, to improve power boiler particulate con-
trol.

References for Chapter 13

1. Remark made during an interview by CEP, August, 1970.

2. Idem.

3. Idem.

4. "Potential Toxicity of Kraft Mill Effluent After Oceanic Discharge," by
Robert C. Courtwright and Carl E. Bond, The Progressive Fish Culturist,
October, 1969, p. 207.

Chapter 14

GREAT NORTHERN NEKOOSA CORPORATION
75 Prospect Street, Stamford, Connecticut 06901

Major Products:
Newsprint, containerboard, fine papers, plywood.

Major Consumer Brands:
NEKOOSA Offset, NEKOOSA Opaque papers.

Financial Data (pro-forma, $ Millions)	1969	1968	1967
Net Sales	340.7	296.8	259.4
Net Income	20.3	19.9	17.5
Capital Expenditures	20.0	73.0	46.0

Pollution Control Expenditures:
$10 million (1970 prospectus) plus an additional $6.3 million spent separately by Nekoosa Edwards.

Annual Meeting:
N.A.

Officers:
Chairman/President: Peter S. Paine

Outside Directors:
Hoyt Ammidon: Chairman of the Board, U.S. Trust Co. of New York
Richard G. Croft
Ralph J. Kraut: Chairman and Chief Executive Officer, Giddings & Lewis, Inc.
E. Spencer Miller: President, Maine Central Railroad Co.
Minot K. Milliken: Vice President & Treasurer, Deering Milliken, Inc.
John J. Neely
John A. Puelicher: President, Marshall & Ilsley Bank
Walter D. Sanders: Attorney, Sanders, Mottola & Haugen
Frederick K. Trask, Jr.: Partner, Payson & Trask

Breakdown of Total Company Production (Percent of 1970 Sales):

Paper and paperboard	80
Newsprint	12
Groundwood specialty	13
Coated paper	4
Containerboard	21

Business communication and printing papers	24
Technical papers	6
Butler Paper Company	15
Pulp, by-product, stumpage	4
Plywood	1

Plants:
Great Northern Nekoosa has eight manufacturing plants in five states; Butler Paper Company, a subsidiary, has 28 warehouse locations.

Timberholdings:
Great Northern Nekoosa owns more than 2,200,000 acres of timberland in Maine (the equivalent of 11 percent of the land in the state) out of a total of more than 2,700,000 acres in the U.S. owned or leased by the company or subject to long-term cutting agreements. The total 1970 value of Great Northern Nekoosa's holdings: $17,629,000.

Annual Pulp Production:
1970s, 1,842,450 tons, estimated on 355 days of production per year.

Company Pollution Overview

The Great Northern Nekoosa Corporation was formed in the Spring of 1970 by a merger of Great Northern Paper Company and Nekoosa Edwards Paper Company. Both companies have traditionally manufactured limited lines of paper. Nekoosa Edwards produces about 1,000 tons per day of specialized high quality, high strength, fine printing papers; Great Northern produces about 4,000 tons per day of container board, newsprint, and groundwood specialty papers for use in catalogs and paperback books. The new company operates six mills (two in Maine, two in Wisconsin, and one each in Arkansas and Georgia), which produce 5190 tons of pulp and 4865 tons of paper daily using 138 million gallons of water.

Neither company has a laudatory water pollution control record, but Great Northern puts Nekoosa Edwards to shame by comparison. Each company operates three mills, two old and inadequately controlled, one new and relatively clean. Nekoosa Edward's two old mills at Nekoosa and Port Edwards, Wisconsin, have been discharging essentially untreated bleached kraft and sulfite effluent for decades. At Port Edwards, in fact, the company went to the expense of converting to a recoverable magnesium base pulping process without installing a recovery system. This is both ecologically and economically unsound because the magnesium chemicals are too expensive to pour down the drain.

Great Northern's two old mills have no primary or secondary treatment facilities either, but the company has taken some affirmative action. The Millinocket, Maine sodium sulfite mill was recently converted to the world's largest magnesium sulfite pulping operation. As a result of this $15 million project, the spent sulfite liquor is now recovered and the mill's daily BOD discharge has been reduced by approximately 75 percent. Water pollution at the company's other old mill (East Millinocket, located 7 miles from Millinocket) has also been reduced by a process change, though less spectacularly. The mill switched from chemi-groundwood to straight groundwood pulping and thus removed the major source of pollution. These improvements have had a great effect on the Penobscot River: salmon reappeared after a century-long absence.

All four of Great Northern Nekoosa's old mills sorely need primary and secondary water treatment systems. Great Northern will have primary facilities by 1973 at both Maine mills, but has no plans for installing secondary treatment. Nekoosa plans "to meet state requirements" by eventually installing primary and secondary treatment, but no construction has begun yet.

Water treatment at the company's two new bleached kraft mills also illustrates the somewhat less effective efforts made by Nekoosa. Primary and secondary treatment facilities have been constructed at both mills. However, those at the Great Northern Cedar Springs mill achieve excellent results (BOD discharge is only 8 pounds per ton of pulp), while those at Nekoosa's Ashdown, Arkansas mill, which is one-fourth as large and four years newer than Cedar Springs, do not do one-fourth as well. The BOD discharge there is approximately 41 pounds per ton of pulp, after clarification and lagooning. The State of Arkansas is requiring further secondary treatment facilities there by 1972.

Air pollution control within the new company does not present such a contrast. Neither Ashdown nor Cedar Springs has odor control or formal plans for installing it. Particulate control is inadequate at three of the four old mills ; and there is no odor control at Nekoosa, Wisconsin, the only old mill which emits gases.

Great Northern Nekoosa has plans to spend $6 to $7 million to improve pollution control at its six mills. CEP estimates that an additional $17.7 to $25.7 million would be needed to insure good air and water controls at all locations.

Table 14.1
Great Northern Nekoosa Corporation: Pulp Production, Water Use, and
Pollution Control

Great Northern

Mill Location	Pulp Prod. (TPD)	Other Prod. (TPD)	Water Use (MGPD)	Water Pollution Control			Air Pollution Control	
				Pri.	Sec.	Tert.	Part.	Gas, Odor
Cedar Springs, Ga.	2,000	2,000	26	✓	✓	X	✓	X
E. Millinocket, Me.	920	1,080	20	X	X	X	—	—
Millinocket, Me.	1,300	935	30	X	X	X	X	—
Total	4,220	4,015	76					

Nekoosa

Mill Location	Pulp Prod. (TPD)	Other Prod. (TPD)	Water Use (MGPD)	Pri.	Sec.	Tert.	Part.	Gas, Odor
Ashdown, Ark.	400	200	20	✓	✓	X	✓	X
Nekoosa, Wisc.	340	400	30	X	X	X	X	X
Pt. Edwards, Wis.	230	360	12	X	X	X	X	—
Total	970	960	62					

Ashdown Mill

Location:
Ashdown, Arkansas

Date Built:
1967-1968

Process:
400 TPD bleached kraft

Other Production:
200 TPD business communications paper for writing, printing, publishing, and converting

Total Water Use:
20 MGPD

Source:
Millwood Lake Reservoir

Discharge:
Red River

Table 14.2
Ashdown Mill: Pollution Control Record

Treatment	Overall Evaluation	Equipment
Water		
Primary	✓	Clarifier, 10 million gallon capacity (all high-solids water is piped to clarifier)
Secondary	✓	Natural stabilization lagoon, 3200 acre-foot capacity (additional aeration equipment needed)
Tertiary	X	None
Solid Waste Removal	—	Clarifier sludge is used for landfill
Air		
Particulate (fuel)	✓	Fuel: natural gas. No controls needed
Particulate (production)	✓	One recovery furnace, precipitator over 97% efficient. Lime kiln: scrubber, 97% efficient
Gas and odor	X	None

Table 14.2
Ashdown Mill: Pollution Control Record (continued)

Treatment	Overall Evaluation	Equipment
Plant Emissions:		
To water:		
BOD	16,500 lb/day (about 41 lb/ton of pulp)	
Suspended solids	N.A.	
To air:	N.A.	

Pollution Control Expenditures:
$2 million for water pollution control.

Legal Status:
(Water) Secondary treatment required by end of 1972.
(Air) In compliance.

Future Plans:
The mill will be expanded and a company representative said there will be an "addition of air pollution control equipment and aeration of lagoons." Odor control is "being studied."[1]

Profile:
"When we built the Ashdown mill," the company proudly announced in its 1969 Annual Report, "we spent $1.8 million on facilities to more than meet state requirements for water and air pollution control."
     Nekoosa failed to mention that the state has no odor standards and that the company did not bother to put in these very basic controls on its own initiative. In fact, Nekoosa did the very minimum required. The sum required to provide an odor control system is a very small part of the capital cost of a new mill, and very simple to install at original construction. The omission is evidence of disregard for the citizens who must put up with the noxious kraft mill odors.
     The mill has both primary and secondary water treatment facilities. However, the BOD load of 41 pounds per ton of pulp indicates that further secondary treatment is needed. This is being required by the state by 1972.

     CEP Estimate of Minimum Necessary Pollution Control Expenditures Not in Company Budget: $700,000 total, of which $500,000 is for an odor control system and $200,000 is for additional water treatment (aeration equipment).

Cedar Springs Mill

Location:
Cedar Springs, Georgia

Date Built:
1963

Process:
1700 TPD unbleached kraft
300 TPD NSSC

Other Production:
1700 TPD unbleached kraft linerboard and mottled white board
300 TPD corrugating medium

Total Water Use:
26 MGPD

Source:
Chattahoochee River

Discharge:
Chattahoochee River

Table 14.3
Cedar Springs Mill: Pollution Control Record

Treatment	Overall Evaluation	Equipment
Water		
Primary	✓	300-foot diameter clarifier (installed, 1965)
Secondary	✓	150-acre aerated lagoon with thirteen 50 hp aerators (installed, 1967); 25-acre stabilization basin returns effluent at the same rate as the river's flow. (NSSC liquor burned and cross-recovered in the kraft mill on the site)

Table 14.3
Cedar Springs Mill: Pollution Control Record (continued)

Treatment	Overall Evaluation	Equipment
Tertiary	X	None
Air		
Particulate (fuel)	√	Fuel: oil, coal, and bark. Mechanical fly ash collectors on coal and bark burners, 99% efficient
Particulate (production)	√	Recovery furnace with precipitator, 99% efficient; lime kilns with scrubbers, 90% efficient; demisters
Gas and odor	X	None
Plant Emissions		
To water:		
BOD	17,000 lb/day (8 lb/ton of pulp; this is the maximum specified on the state permit)	
Suspended solids	N.A.	
To air:	N.A.	

Pollution Control Expenditures:
$2.5 million for water treatment.

Legal Status:
(Water) In compliance with a permit; working with the state on further reduction in waste loads.
(Air) Not in compliance; required to submit program by 1971.

Future Plans:
Great Northern would say nothing more definite than that it intends to improve its present air and water pollution controls.

Profile:
Cedar Springs, Great Northern's only southern facility, is the newest of the

company's mills. In contrast to the two older mills which have virtually no water treatment facilities, this mill has a clarifying and aerating system that is excellent. The effluent discharged from the system into the Chatahoochee River has a BOD load of only 8 pounds per ton of pulp.

The mill's air pollution is not adequately controlled. Devices cover all stacks but mechanical collectors are not sufficient for coal burners and the lime kiln scrubber is only 90 percent efficient. Also, there is no odor control whatsoever. According to public relations director Robert Vivian, "like everyone else in the industry, we are working on the kraft odor problem, but as yet we have found no suitable solution." This answer is true as far as it goes; there is no way of totally removing the kraft odor. But many mills built before and after 1963 have found that black liquor oxidation or non-direct contact evaporation systems go a long way toward reducing it.

CEP Estimate of Minimum Necessary Pollution Control Expenditures Not in Company Budget: $500,000 total, for an odor control system.

East Millinocket Mill

Location:
East Millinocket, Maine

Date Built:
N.A.

Process:
920 TPD groundwood

Other Production:
1,080 TPD newsprint and groundwood specialty paper

Total Water Use:
20 MGPD

Source:
Penobscot River

Discharge:
Penobscot River

Table 14.4
East Millinocket Mill: Pollution Control Record

Treatment	Overall Evaluation	Equipment
Water		
Primary	X	None
Secondary	X	None
Tertiary	X	None
Other	—	Screening equipment
Air		
Particulate (fuel)	—	Fuel: hydroelectric and oil. No equipment needed
Particulate (production)	—	None needed
Gas and odor	—	None needed
Plant Emissions		
To water:		
BOD	25,000 lb/day (27 lb/ton of pulp)	
Suspended solids	N.A.	

Pollution Control Expenditures:
N.A.

Legal Status:
(Water) In compliance; no facilities required until 1976.
(Air) No regulations yet set.

Future Plans:
$3 to $4 million for a clarifier.

Profile:
Until 1969, this chemi-groundwood mill, located only 7 miles from the Millinocket mill, was responsible for 10 percent of Great Northern's pollution of

the Penobscot River. The decision was not made until the mid-1960s to change the chemi-groundwood process to a strictly mechanical one.

In chemi-groundwood pulping the chips are partially cooked to soften them before grinding. Although the liquor is reused rather than discharged, some dissolved organic materials and chemicals are washed from the pulp, creating a high BOD load. Discontinuation of the chemical stage considerably reduced these wastes in the mill's effluent. However, there is a large remaining problem of solid wastes such as fibers, dirt, and uncooked chips still being discharged. This should be largely solved by the installation of a clarifier in 1972.

CEP Estimate of Minimum Necessary Pollution Control Expenditures Not in Company Budget: None.

Millinocket Mill

Location:
Millinocket, Maine

Date Built:
1900, rebuilt 1967-1969

Process:
700 TPD groundwood
600 TPD magnesium bisulfite (company figure)

Other Production:
900 TPD newsprint, supercalendered papers, coated and uncoated papers
35 TPD wrapper

Total Water Use:
30 MGPD

Source:
Penobscot River

Discharge:
Penobscot River

Table 14.5
Millinocket Mill: Pollution Control Record

Treatment	Overall Evaluation	Equipment
Water		
Primary	X	None
Secondary	X (partial)	No equipment, but spent liquor not discharged (MgO recovery system reclaims 80-90% of the cooking chemicals and burns the waste solids)
Tertiary	X	None
Other	—	Savealls, reject refiner, fiber reclaim systems (for repulping rejected paper), paper machine close-up
Air		
Particulate (fuel)	X	Fuel: oil, bark and hydroelectric. No control equipment
Particulate (production)	✓	Recovery boiler with 3 venturi scrubbers, 99% efficient
Gas and odor	—	None needed
Plant Emissions		
To water:		
BOD	68,000 lb/day (52 lb/ton of pulp)	
Suspended solids	112,000 lb/day (86 lb/ton of pulp)	
To air:	N.A.	

Pollution Control Expenditures:
$15 million for conversion from sodium sulfite to magnesium bisulfite pulping. In addition, the mill spent $2.1 million between 1962 and 1970 for in-plant equipment such as savealls and fiber reclamation systems to reduce fiber losses.

Legal Status:
(Water) In compliance; no facilities required until 1976.
(Air) No regulations yet set.

Future Plans:
This year, preliminary to construction of primary water treatment facilities, the mill will build an interceptor sewer to collect all effluent requiring clarification, and a separate sewer system for sanitary wastes. In 1971, a $3 million clarifier and sludge handling system will be installed. These are expected to remove 95 percent of the settleable solids and reduce the BOD by 25-30 percent (approximately 10,000 pounds). The primary treatment planned for both this and the East Millinocket mill should enable Great Northern to meet the established river quality standards at best in 1973, "and certainly before the 1976 deadline."

Great Northern is also phasing out the traditional means of transporting timber to the mill (the river log drive) because the large amounts of bark which are scraped off contribute to the pollution. Trucks and railroads will be used as soon as roads are constructed.

Profile:
Until 1969, this pulp and paper plant was one of the major sources of the pollution in the Penobscot River, a vital state waterway in Maine. In that year alone, the mill discharged over 132 million pounds of BOD and 23,000 tons of suspended solids into the river (company figure). The company realized that this situation would not long be tolerated, and in the early 1960s began to investigate solutions to the enormous water pollution problem. After several years of research Great Northern arrived at the obvious conclusion that the pulping process would have to be changed to one which allowed recovery of the spent liquor. In 1967 the mill began conversion from sodium sulfite pulping to a magnesium sulfite process. A $10 million MgO recovery furnace, "the largest of its kind in the world," was constructed, which reclaims 80-90 percent of the spent chemicals for reuse and incinerates the 1.6 million pounds of solids in the waste liquor. The system also produces 300,000 pounds of steam per hour, thus reducing the mill's fuel consumption.

The Millinocket MgO recovery furnace was completed in 1969 and has already effected a major improvement in the quality of the Penobscot River water. The daily BOD discharge was reduced from 361,799 pounds to 68,000 pounds. The dissolved oxygen level of the water at the mill, which varied from 2 to 8 parts per million (ppm) in the first half of 1965, measured a steady 12 ppm in the first quarter of 1970. The revival of the river was most emphatically borne out in June, 1970, when "approximately 2 million square yards of surveyed spawning grounds re-opened for the first time in this century to Atlantic salmon inhabitants using the Penobscot River."[2] In celebration of this, Great Northern, in conjunction with the state, is building salmon ladders on its part of the river.

CEP Estimate of Minimum Necessary Pollution Control Expenditures Not in Company Budget: $1.75 million total, of which $1.5 million is for secondary water treatment and $250,000 is for improved power boiler particulate control.

Nekoosa Mill

Location:
Nekoosa, Wisconsin

Date Built:
1894-1895

Process:
340 TPD bleached kraft (company figure)

Other Production:
400 TPD watermarked and unwatered bond, safety check, lace papers and others

Total Water Use:
30 MGPD

Source:
Nepco Lake, Wisconsin River

Discharge:
Wisconsin River

Table 14.6
Nekoosa Mill: Pollution Control Record

Treatment	Overall Evaluation	Equipment
Water		
Primary	X	None
Secondary	X	None

Table 14.6
Nekoosa Mill: Pollution Control Record (continued)

Treatment	Overall Evaluation	Equipment
Tertiary	X	None
Other	—	Savealls

Air

Particulate (fuel)	X	Fuel: coal, bark. Three boilers in use equipped with dust collectors
Particulate (production)	X	Recovery furnace: precipitator with 85% actual efficiency (installed, 1968; needs to be rebuilt). No lime kiln
Gas and odor	X	No odor control system

Plant Emissions

To water:

BOD	27,420 lb/day (80 lb/ton of pulp) from pulping. 32,620 lb/day total
Suspended solids	21,540 lb/day (63 lb/ton of pulp) from pulping. 45,740 lb/day total

To air: N.A.

Pollution Control Expenditures:
$2.1 million, of which $300,000 was for the precipitator and the remainder for process equipment changes, screens, water recirculation and bleach plant modifications.

Legal Status:
(Water) State reviewing plans.
(Air) State program not yet directed to pulp mills.

Future Plans:
"To meet state regulations." As yet no specific plans for primary and secondary water treatment have been made.

Profile:
This middle-sized kraft mill is in totally appalling condition from the standpoint of pollution control. Despite a total expenditure of $2.1 million, the com-
pany has managed to provide no water pollution control, no odor and gas control, and little particulate control.

The mill's water use is 30 million gallons daily, all of which, including
bleaching chemicals, is discharged untreated into the Wisconsin River. The
BOD load for the pulping operation alone is 80 pounds per ton of pulp; this is
much higher than average BOD load of completely untreated kraft mill effluent.

Nekoosa burns coal and bark for power in boilers equipped with only the
traditional mechanical dust collectors, which provide minimal control of soot
and ash. The mill doesn't even have a lime kiln, which in most kraft mills incinerates odorous sulfurous gases and reclaims chemicals.

Are company plans afoot to right this situation? Nekoosa's Public Relations director reported in a letter of November 12, 1970, a general intent to
"meet state requirements" but said there were (after 76 years of mill operation) "no specific plans for primary and secondary water treatment." Probably
the same goes for odor and particulate controls.

CEP Estimate of Minimum Necessary Pollution Control Expenditures Not
in Company Budget: $4.8 million ($12.8 million with new furnace) total, of
which $500,000 is for primary water treatment; $2 million is for secondary
water treatment; $300,000 is for either scrubbers or precipitators for power
boilers (or conversion of fuel source to natural gas); $1 million is for lime
kiln; $250,000 is for lime kiln scrubber; $500,000 is for an odor control system; and $250,000 is for rebuilding recovery furnace precipitator. Besides
this, a new kraft recovery furnace, costing $8 million, is probably needed at
this old mill.

Port Edwards Mill

Location:
Port Edwards, Wisconsin

Date Built:
1894

Process:
230 TPD bleached magnesium sulfite base

Other Production:
360 TPD of a great variety of papers

Total Water Use:
12 MGPD

Source:
Nepco Lake

Discharge:
Wisconsin River

Table 14.7
Port Edwards Mill: Pollution Control Record

Treatment	Overall Evaluation	Equipment
Water		
Primary	X	None, but spent sulfite liquors "are pumped to seepage lagoons on an ... island." An estimated 20% of the BOD is removed from the wastes discharged to these facilities
Secondary	X	None
Tertiary	X	None
Other (optional)		Savealls
Air		
Particulate (fuel)	X	Fuel: coal, gas. Five boilers, three in regular use, of which one has a dust collector
Particulate (production)	—	N.A.
Gas and odor	—	N.A.

Plant Emissions

To water:

BOD   61,560 lb/day

Table 14.7
Port Edwards Mill: Pollution Control Record (continued)

Treatment	Overall Evaluation	Equipment
	(224 lb/ton of pulp) from pulping only*	
Suspended solids	8,200 lb/day (31 lb/ton of pulp) from pulping only*	
To air:	N.A.	

*Emission figures taken from the Report of the Upper Wisconsin River Pollution Investigation Survey, July, 1970, p. 13.

Pollution Control Expenditures:
$2.2 million for research and conversion to magnesium base; bleach plant modifications; etc.

Legal Status:
(Water) Plans for water treatment on file with the state agency since 1966, were just approved in 1970. Wisconsin River abatement orders are in preparation.
(Air) State program not yet directed to pulp mills.

Future Plans:
Sulfite liquor recover (evaporation and incineration): "target date mid-1973." Primary and secondary water treatment and increased in-plant water reuse will reduce the mill's BOD discharge by 90 percent.

Profile:
The measured BOD discharge from this mill indicates that, despite partial treatment, unrecovered magnesium pulping liquor is almost as bad for a river as dumped calcium liquor. Although recovery is always more trouble than this sort of disposal, expensive chemicals are being lost because of lack of recovery. Evidently, the company has felt unable to go ahead with installation of a recovery system (since the conversion to magnesium) without specific state requirement. Finally, only minimal efforts have been made to control ash emissions from the power boilers; one of the three regularly in use has a dust collector. The other two do not.

CEP Estimate of Minimum Necessary Pollution Control Expenditures Not in Company Budget: $3 million total, of which $300,000 is to improve power boiler particulate control; $\underline{\$200,000}$ is for a sulfur dioxide absorption system, to be installed after the recovery system is operating; $\underline{\$500,000}$ is for primary water treatment; and $\underline{\$2\ million}$ is for secondary treatment.

References for Chapter 14

1. Letter from Tad R. Meyers, Nekoosa-Edwards Paper Company, November 12, 1970.

2. Bangor Daily News, June 26, 1970.

Chapter 15

HAMMERMILL PAPER COMPANY
East Lake Road, Erie, Pennsylvania 16512

Major Products:
Pulp and paper, other paper products.

Major Consumer Brand:
HAMMERMILL BOND paper.

Financial Data ($ Million)	1969	1968	1967
Net Sales	353.3	319.0	262.8
Net Income	14.1	11.4	8.5
Capital Expenditures	28.8	17.9	15.3

Pollution Control Expenditures:
$6 million, 1950-1970 at Erie.

Annual Meeting:
The annual meeting is held during May in Erie, Pennsylvania.

Officers:
Chairman/President: John H. Devitt, Erie, Pennsylvania (Director: First
National Bank of Pennsylvania, American Sterilizer Company, General Public
Utilities Corp.)

Outside Directors:
William Beckett: President, Beckett Paper Co.
Ralph H. Demmler: Attorney, Reed, Smith, Shall Rogers and McClay
Bernard S. Kubale: Partner, Foley & Lardner
Walter A. Rentschler: President, Citizens Bank, Hamilton, Ohio
Louis H. Roddis, Jr.: President, Pennsylvania Electric Company
J. T. Thomas: President, Thilmany Pulp & Paper Company
Charles M. Williams: Professor of Banking, Harvard University Graduate
School of Business Administration

Breakdown of Total Company Production (tons):
"The great bulk of our business is in pulp and paper production along with
some manufacturing of lumber products."

Plants:
Hammermill has 12 domestic plants:

Pulp producing   4
Additional pulp and paper mills   5

Sawmill operations   2
Wood veneer   1

Timberholdings:
Hammermill has management responsibility for 312,000 acres of U.S. timberland: owning 162,000 in Pennsylvania and either owning or controlling another 150,000 in Alabama. Hammermill could supply about 50 percent of its
pulp requirements from these lands, but because of readily available wood
from industrial contractors, its timberlands now supply only about 20 percent.
The total 1969 value of Hammermill's holdings: $8,234,000.

Annual Pulp Production (estimated from the daily production figure of 1360
tons, for 355 operating days per year):
482,800 tons (1969)

Company Pollution Overview

Dr. Richard Brown, Vice President of Research for Hammermill, stated (regarding air pollution at two of the company's mills): "We are aware of the
odor and particulate problems [there], but the states have not yet adopted air
standards, so we will wait to see what we are required to do."
    Hammermill's history of dealing with pollution at its mills illustrates this
"wait and see" attitude: The company has four mills—two small pre-1910
mills in Pennsylvania (at Lockhaven and Erie) an old mill at Kaukauna, Wisconsin, and a new kraft mill at Selma, Alabama—which produce 1360 tons of
pulp per day and discharge 95 million gallons of waste water. The three old
mills have all been cited for water pollution and are required to provide adequate secondary effluent treatment: Lockhaven in 1968; Kaukauna in 1968 and
1969; and Erie, first in 1946, and, after ten extensions of time, again in 1966.
    The Erie mill was the subject of national attention in 1968 as a result of
a Wall Street Journal article describing the explosion of one of its waste liquor disposal wells, and a LIFE article photograph showing its pollution of
Lake Erie. Hammermill, after twenty years of state orders plus this publicity,
is now rebuilding (and expanding) the mill at a cost of $35 million. The pollu-
tion problems should be practically solved by the middle of 1971.
    At Lockhaven and Kaukauna, on the other hand, almost two years after
state orders were issued for secondary treatment, the company is still only
at the stage of doing "engineering studies." Treatment could have been provided in a year and a half with some company initiative, but the state deadlines don't arrive until the end of 1972, and Hammermill seems to be content
to take its time.
    In air pollution control, none of the four mills has adequate particulate
equipment, and the two kraft mills at Selma and Kaukauna, have no odor control. Hammermill has made minimal plans to correct this situation and will
do more, according to Dr. Brown, when "told what to do," despite the fact
that the technology is already available.

Particularly surprising is the fact that Hammermill's only new mill (at Selma) was constructed in 1967 without odor control and with admittedly inadequate particulate control at a time when pollution laws were being widely discussed and would obviously be put into effect in a number of years.

CEP estimates that beyond the $35 million now being spent at the Erie mill, Hammermill will have to invest at least another $6.5 million to control air and water pollution adequately at all four mills, bringing the total financing to $41.5 million.

Table 15.1
Hammermill Paper Company: Pulp Production, Water Use, and Pollution Control

Mill Location	Pulp Prod. (TPD)	Other Prod. (TPD)	Water Use (MGPD)	Water Pollution Control			Air Pollution Control	
				Pri.	Sec.	Tert.	Part.	Gas, Odor
Selma, Ala.	500		20	✓	✓	✗	✗	✗
Erie, Pa.	400	425	32	✗	✗	✗	✗	✓
Lockhaven, Pa.	80	370	15	✓	✗	✗	✗	✓
Kaukauna, Wisc.	380	500	28	✗	✗	✗	✗	✗
Total	1,360	1,295	95					

Riverdale Mill

Location:
Selma, Alabama

Date Built:
1964-1966

Process:
500 TPD bleached kraft

Total Water Use:
20 MGPD

Source:
Alabama River

Discharge:
Alabama River

Table 15.2
Riverdale Mill: Pollution Control Record

Treatment	Overall Evaluation	Equipment
Water		
Primary	✓	190-foot diameter (20 million gallon capacity) clarifier
Secondary	✓	274-acre, 60-day retention lagoon, 1 billion gallon capacity; 36-acre, 5-day retention lagoon, 100 million gallon capacity
Tertiary	X	None
Other	—	19-acre, 50 million gallon sludge storage; treated waste water is discharged to river through outfall with diffuser
Air		
Particulate (fuel)	✓	Fuel: oil, gas
Particulate (production)	X	One recovery furnace with venturi scrubber; one lime kiln with venturi scrubber
Gas and odor	X	No black liquor oxidation
Plant Emissions	N.A.	

Pollution Control Expenditures:
N.A.

Legal Status:
(Water) In compliance.
(Air) No state air standard implementation yet.

Future Plans:
None.

Profile:
The Riverdale mill was constructed three years ago. Hammermill, sticking to the "letter of the law," put into it no more control equipment than was strictly required by the state. As there were no standards for air, no odor control (black liquor oxidation) was included, and the particulate controls are admittedly inadequate.

The Company's Director of Environmental Control commented: "We know we need better particulate and odor control, and we do anticipate it will be required in the future, but right now Alabama is just adopting air standards, and we will see what we are required to put in."[1]

The community has suffered from this "watch and wait" attitude; it was no doubt eager to have Hammermill settle in Selma, bringing economic advantages via payrolls, taxes, etc., but some complaints have been made. One disgruntled instructor stationed at nearby Craig Air Force Base commented: "The smog so often prevents my student pilots from flying that it is difficult to complete the training course in the established amount of time."

The Hammermill 1969 annual report announced that the Selma mill was rated 400 TPD in 1967 but that it is now producing over 500 TPD, "exceeding its engineering specifications and our most optimistic expectations."[2] Such overproduction has been recognized as a serious cause of greatly multiplied odor and gas emissions.

CEP Estimate of Minimum Necessary Pollution Control Expenditure Not on Company Budget: $1 million total, of which $500,000 is for black liquor oxidation and $500,000 is for an electrostatic precipitator on the recovery furnace.

Erie Mill

Location:
Erie, Pennsylvania

Date Built:
1898, rebuilt in 1950s

Process:
400 TPD bleached NSSC pulp (company figure)

Other Production:
425 TPD fine writing and printing papers (company figure)

Total Water Use:
32 MGPD

Source:
Lake Erie

Discharge:
Lake Erie

Table 15.3
Erie Mill: Pollution Control Record

Treatment	Overall Evaluation	Equipment
Water		
Primary	X	Twin clarifiers only for 6 million gallons/day paper mill waste water
Secondary	X (partial)	No equipment; however, liquor is not discharged into Lake Erie, but is injected into two active deep limestone wells
Tertiary	X	Bleached plant waste discharged directly into the lake
Air		
Particulate (fuel)	X	Fuel: coal and bark. Mechanical dust collectors
Particulate (production)	—	None needed
Gas and odor	√	Sulfur dioxide wet scrubber

Plant Emissions

To water:

BOD   130,000 lb/day
                       (325 lb/ton of pulp;
                       1969 figure)

Suspended solids   81,000 lb/day
                       (202 lb/ton of pulp)

Color   2,400 PCU

To air: N.A.

Pollution Control Expenditures:
Since the early 50s, over $6 million has been spent on research into water treatment equipment including the building of several pilot plants that proved impractical and three deep liquor disposal wells (the first completed and put into operation in 1964).

Legal Status:
(Water) The State of Pennsylvania first ordered this mill to provide secondary water treatment on February 26, 1946. Since then, 10 extensions of time have been obtained by the company. On November 16, 1966, the state reordered the Erie mill to clean up its effluent and indicated that its permit to discharge inadequately treated waste would be revoked December 15, 1970. The mill, now with a new extension of time, is on a state-approved schedule for compliance by 1971.
(Air) State investigating complaints.

Future Plans:
In December, 1968, Hammermill announced a $35 million expansion and modernization program for the mill, including conversion to a new fully chemical pulping process (Neutracell II). The mill's production capacity will be increased 50 percent; pulp making costs will go down and product quality will be higher.
    The water pollution problem will be abated as waste process water will be piped to the city of Erie's municipal sewage treatment plant. Hammermill agreed June 18, 1969, to pay for part of the capital cost of plant expansion and to pay a suitable share of operating costs. The waste pulping liquor will be evaporated and burned and the deep wells abandoned. Treated bleach wastes will be discharged into Lake Erie through a new outfall pipe the company is building, under a permit from the U.S. Corps of Engineers and the Pennsylvania Sanitary Water Board.
    When the program is complete in mid 1971, Hammermill estimates a reduction of the BOD load from 130,000 pounds per day to 22,800 pounds; of settleable solids from 81,000 pounds per day to 14,000 pounds; and of color from 2400 to 400 PCU.

Profile:
In the early 1950s, this mill was converted to NSSC pulping and the use of local Pennsylvania hardwoods as raw materials. With this process, for the past 20 years the mill has contributed pollution to Lake Erie, via the daily discharge of its untreated bleach plant wastes. The mill injects its waste pulping liquors into deep limestone wells next to the lake.
    In 1968 considerable national public attention was directed to the mill's pollution. Use of deep wells to bury waste is a practice that has become of increasing concern to environmentalists, and in May, when one of Hammermill's wells sprung a leak, the Wall Street Journal reported: "The waste—a brown putrid broth left over from processing wood chips into pulp—had been

forced into the well under high pressure. When the trouble developed, the compressed liquid shot to the surface forming a 20 ft geyser that spilled into nearby Lake Erie. It took three weeks to cap the well. By that time, more than two million gallons had escaped. In six years (1962-1968) Hammermill has pumped 750 million gallons underground."[3]

In August 1968, LIFE magazine's special report, "The Blighted Great Lakes," included a two-page photograph of the mill and the adjacent part of Lake Erie where the untreated bleach plant wastes were being (and still are being) discharged. The caption read: "... looking like a giant glob of beer foam, pulp waste from the Hammermill Paper Co. stain Lake Erie's Pennsylvania shore. The white mass is penned up by a dike built of old tires and oil drums, but residue seeps through to foul open waters. Hammermill has promised action either by routing waste to an existing sewage plant or by building a new facility."[4]

Hammermill ran an ad to counter criticism of the mill called "Action Lake Erie." It asserted the bleach plant waste caused no harm to algae growth or threat to acquatic life, animal or human. Yet it gave a somewhat contradictory concession that the wastes were objectionable because of their foam, color and demand for dissolved oxygen from the lake waters.

Partially due to the adverse publicity as well as the cumulative impact of 20 years of abatement orders by the State of Pennsylvania and 10 granted extensions of time, Hammermill, in December, 1968, announced that the mill would be overhauled. At a cost of $35 million, it would be equipped with an entirely new pulping process with facilities for recovering the pulping liquor and with provision for piping the waste process waters to the Erie municipal sewage treatment plant.

Neutralized bleach plant wastes will still go directly into Lake Erie through the company's outfall.

Hammermill advertised this $35 million expenditure stressing the pollution control provisions. Although the mill conversion may have been initiated for environmental reasons, well over 90 percent of the investment is going into new, larger and more efficient production equipment. When the expansion is complete, the mill will be producing 210,000 tons a year instead of 140,000.

Although credit is due to the company for the fact that a clean-up should be occuring a year from now, it would be difficult to agree with the statement in Hammermill's 1969 annual report that "these expensive projects underway are evidence of Hammermill's long-stated resolve to meet the responsibilities of local, state and federal citizenship ... responsibilities which all of us must share in order to improve our environment."[5]

CEP Estimate of Minimum Necessary Pollution Control Expenditure Not on Company Budget: $250,000 total, for improved power boiler particulate control.

Lock Haven Mill

Location:
Lock Haven, Pennsylvania

Date Built:
1910 (mill acquired by Hammermill in 1965)

Process:
80 TPD bleached soda

Other Production:
370 fine writing and printing papers (company figure)

Total Water Use:
15 MGPD

Source:
Bald Eagle Creek

Discharge:
Bald Eagle Creek

Table 15.4
Lock Haven Mill: Pollution Control Record

Treatment	Overall Evaluation	Equipment
Water		
Primary	✓	Settling basins; 17.5 million gallon clarifier; sludge centrifuge
Secondary	X	None
Tertiary	X	None
Air		
Particulate (fuel)	X	Fuel: coal. Dust collectors
Particulate (production)	✓	Wet scrubber on recovery furnace and lime kiln
Gas and odor	—	No other equipment needed

Table 15.4
Lock Haven Mill: Pollution Control Record (continued)

Treatment	Overall Evaluation	Equipment
Plant Emissions		
To water:		
BOD	16,000 lb/day (200 lb/ton of pulp; 1970 state figure)	
Suspended solids	N.A.	
Color	462 PCU	
To air:	N.A.	

Pollution Control Expenditures:
N.A.

Legal Status:
(Water) On December 31, 1968 the State of Pennsylvania Sanitary Water Board ordered the mill to provide secondary treatment for its effluent by September, 1972, reducing BOD to 6500 pounds per day and its color to 175 PCU.

Future Plans:
As of October 1970, almost two years after the mill was cited, a company representative said that an aerated lagoon was being engineered, but that a "firm construction schedule had not been set yet" for better air control. Two new steam boilers will be installed by January, 1971, with high efficiency dust collectors.

    CEP Estimate of Minimum Necessary Pollution Control Expenditure Not on Company Budget: $2 million total, for secondary water treatment.

Thilmany Mill

Location:
Kaukauna, Wisconsin

Date Built:
1883 (acquired by Hammermill in 1969)

Process:
380 TPD kraft

Other Production:
500 TPD creped and pleated, bag, embossing, wrapping and other converting, waxing, galline and grease-proof specialty papers; this mill "makes perhaps a greater variety of paper products than any other mill in America"

Total Water Use:
28 MGPD

Source:
Fox River

Discharge:
Fox River

Table 15.5
Thilmany Mill: Pollution Control Record

Treatment	Overall Evaluation	Equipment
Water		
Primary	X	Screening; two sedimentation basins (inadequate)
Secondary	X	None
Tertiary	X	None
Air		
Particulate (fuel)	X	Fuel: oil, gas, some bark. Mechanical dust collectors
Particulate (production)	√	Recovery furnace with venturi scrubber; lime kiln has scrubber
Gas and odor	X	No black liquor oxidation

Plant Emissions

To water:

BOD   22,000 lb/day

Table 15.5
Thilmany Mill: Pollution Control Record (continued)

Treatment	Overall Evaluation	Equipment
	(59 lb/ton of pulp; 1969 company figure)	
Suspended solids	22,000 lb/day (59 lb/ton of pulp; 1969 company figure)	
To air:	N.A.	

Pollution Control Expenditures:
N.A.

Legal Status:
(Water) The Wisconsin Department of Natural Resources issued the first
water pollution abatement order to the mill May 14, 1968, with the requirement that a compliance schedule be submitted to the state by October 1, 1968.
No such schedule had been submitted as of April 1969 when Hammermill
bought the mill. An amended order was issued December 16, 1969, requiring
provision of adequate water treatment by December 31, 1972, reducing the
BOD load to 17,150 pounds per day (35 pounds per ton of pulp).
(Air) Laws not yet adopted.

Future Plans:
In October 1970, a company representative reported that "engineering studies"
were underway for secondary water treatment, but that no program had been
finalized. There are no finalized plans for better air control.

Profile:
Although Hammermill only acquired this mill in 1969 and "inherited" an
abatement order by the state, the company, in the year and a half it has
owned the mill, has made no solid plans for further water treatment. The
deadline isn't until the end of 1972, and Hammermill seems to be in no rush
to comply sooner.

No immediate attention is being given either to the inadequacies of the
mill's particulate or odor control. Dr. Richard Brown, Vice President of
Hammermill Research, noted (as he had in regard to the company's Selma,
Alabama mill): "We are aware of the pollution problems here, but the state
is just setting up standards, and we're waiting to see what will be required."
[6]

Local residents have also been very aware of the air pollution caused by this mill. One stated, "The mill casts a pall over the whole area and you are especially struck by it if you ever drive up to Milwaukee from Green Bay on Route 4."

It seems that only enthusiasm, not money, has been lacking for a clean-up at this mill, since in the 1969 Hammermill annual report it was pointed out that: "In the fourth quarter of 1969 Thilmany added a new hi-speed paper machine as the culmination of its own $35 million expansion and modernization program. The new machine expanded the mill's capacity by 25 percent (adding 25,000 tons to the mill's annual output)."

CEP Estimate of Minimum Necessary Pollution Control Expenditure Not On Company Budget: $2.75 million total, of which $2 million is for secondary water treatment; $500,000 is for black liquor oxidation; and $250,000 is for improved boiler particulate control.

References for Chapter 15

1. CEP interview, August 20, 1970.

2. Annual Report, 1969, p. 12.

3. Wall Street Journal, May 21, 1968.

4. Life magazine, August 26, 1968.

5. Annual Report, 1969, p. 15.

6. CEP interview, August 20, 1970.

Chapter 16

HOERNER WALDORF CORPORATION
2250 Wabash Avenue, St. Paul, Minnesota 55114

Major Products:
Container board, Kraft linerboard

Financial Data ($ Millions)	1969	1968	1967
Net Sales	237.3	160.4	139.9
Net Income	12.2	10.4	8.3
Capital Expenditures	14.0	17.9	10.1

Pollution Control Expenditures:
N.A.

Annual Meeting:
The annual meeting is held during February in St. Paul, Minnesota.

Officers:
Chairman: Alvin J. Huss (Director: W. T. Joyce, Tremont Lumber Co.)
President: John H. Myers (Director: Northwest Bancorporation, Northwestern Life Insurance Co.)

Outside Directors:
P. A. Schilling: Formerly Chairman of the Board, Waldorf Paper Products Co.
W. R. Driver: Partner, Brown Brothers, Harriman and Co.
Paul Christopherson: Sr. Partner, Faegre and Benson
H. L. Holtz: President, First Trust Company of St. Paul

Breakdown of Total Company Production (tons, 1969) [1]:

Pulp	60,000
Paper	160,000
Boxboard	105,000
Linerboard	495,000
Semi-chem	210,000

Plants:
Hoerner Waldorf has 44 domestic plants, all engaged in the manufacture of paper products:

Container plants

Consumer packaging	3
Flexible packaging	1
Bag plants	4
Mill operations [2]	4

Timberholdings:
Hoerner Waldorf has management responsibility for 293,000 acres of timberland. 185,000 acres are owned in North Carolina and Virginia, 72,000 acres
leased in North Carolina and 33,000 acres owned in upper Michigan. The total
value of these holdings is $23,825,767.

Annual Pulp Production (Estimated on 355 production days per year):
656,750 tons (1969)

Company Pollution Overview

Hoerner Waldorf, specialist producer of brown wrapping papers, paper bags
and containers, is the twentieth largest paper and pulp company in the United
States. The company has three pulp mills in the midwest—two middle-sized
NSSC mills (in St. Paul, Minnesota and Ontonagon, Michigan) and a larger
kraft mill in Missoula, Montana. These mills produce 1850 tons of pulp and
discharge 27.5 million gallons of waste water daily.
    The St. Paul mill, though an old facility, is quite clean, with both pri-
mary and secondary water treatment, and no air pollution problem, since
natural gas in the primary fuel.
    The Missoula kraft mill recovery operation is now being completely re-
built so that by the end of 1972 it should be one of the cleanest mills in the
country. The elaborate plans for this mill were certainly encouraged by five
years of increasingly angry complaints by the citizens of Missoula, whose
beautiful valley was covered by the heavy cloud of soot and sulfurous gases
emitted from the mill's stacks. Hoerner Waldorf had operated the mill for
thirteen years with no odor control and inadequate particulate controls. The
plans also probably were speeded up by the suit filed against the mill by the
Environmental Defense Fund in 1968, and the national publicity generated by
a two-page photograph in LIFE (February, 1969) showing yellowish brown
smoke covering the valley. Hoerner Waldorf had been claiming for years that
better pollution control was in the offing, but not until 1970 was the overhaul
program announced.
    The Ontonagon NSSC operation is now in need of attention, and is beginning
to receive it. In April, 1970, the State of Michigan requested company compliance to a timetable of plans and facilities to accomplish primary and secondary water treatment at Ontonagon by January 1, 1974. Hoerner Waldorf
says "studies are underway, and plans will be presented to the state mid-
January, 1971." The company also indicated that the air pollution problem
created by particulate and gas emissions from the mill's coal boiler stacks
was corrected in autumn 1970 when the fuel source was converted to natural

gas. Besides the $ 14 million dollars now budgeted at Missoula, CEP esti-
mates that an additional $ 3 million investment at Ontonagon (i. e., a total in-
vestment of $ 17 million) should solve all major pollution problems at the
Hoerner Waldorf pulp mills.

Table 16.1
Hoerner Waldorf Corporation: Pulp Production, Water Use, and Pollution
Control

Mill Location	Pulp Prod. (TPD)	Other Prod. (TPD)	Water Use (MGPD)	Water Pollution Control			Air Pollution Control	
				Pri.	Sec.	Tert.	Part.	Gas, Odor
Ontonagon, Mich.	200	200	8	X	X	X	X	—
St. Paul, Minn.	300	340	3.5	✓	✓	X	✓	—
Missoula, Mont.	1,350	950	16	✓	X	X	X	X
Total	1,850	1,490	27.5					

Ontonagon Mill

Location:
Ontonagon, Michigan

Date Built:
1956

Process:
200 TPD NSSC pulp

Total Water Use:
8 MGPD

Source:
Lake Superior

Discharge:
Ontonagon River

Table 16.2
Ontonagon Mill: Pollution Control Record

Treatment	Overall Evaluation	Equipment
Water		
Primary	X	None
Secondary	X (partial)	None; however, liquor is not discharged
Tertiary	X	None
Air		
Particulate (fuel)	✓	Fuel: natural gas. No equipment needed
Particulate (production)	X	One recovery furnace has cascade scrubber
Gas and odor	—	None needed
Plant Emissions	N.A.	

Pollution Control Expenditures:
N.A.

Legal Status:
(Water) This mill is not in compliance with Michigan's water quality standards
for the Ontonagon River, which specify water fitness for recreational uses
and aquatic life. The State of Michigan notified the mill April 27, 1970, that
it should provide secondary treatment to control waste solids, BOD, and phe-
nolic substances.

Future Plans:
Hoerner Waldorf has agreed to provide secondary treatment not later than
January 1, 1974, but no plans are yet finalized to reduce air pollution.

Profile:
Ontonagon is the least well equipped of the three Hoerner Waldorf mills for
pollution control. No primary or secondary water treatment is now provided,
and the present level of water pollution will evidently continue for another
four years until the state's 1974 deadline.

CEP Estimate of Minimum Necessary Pollution Control Expenditure Not on Company Budget: $3 million total, of which $1 million is for primary water treatment and $2 million is for secondary water treatment.

Saint Paul Mill

Location:
Saint Paul, Minnesota

Date Built:
1950

Process:
300 TPD unbleached NSSC

Other Production:
300 TPD board from recycled fibers
40 TPD de-inked paper stock

Total Water Use:
3.5 MGPD

Source:
Four deep wells

Discharge:
Through the metropolitan sanitary district sewage plant to the Mississippi River

Table 16.3
Saint Paul Mill: Pollution Control Record

Treatment	Overall Evaluation	Equipment
Water		
Primary	✓	Joint industrial municipal waste treatment at St. Paul plant
Secondary	✓	
Tertiary	✗	None
Air		
Particulate (fuel)	✓	Fuel: gas (oil: standby fuel for winter). Oil burner has scrubber

Table 16.3
Saint Paul Mill: Pollution Control Record (continued)

Treatment	Overall Evaluation	Equipment
Particulate (production)	—	N.A.
Gas and odor	—	N.A.
Plant Emissions	N.A.	

Pollution Control Expenditures:
$250,000 since 1967.

Legal Status:
(Water) In compliance.
(Air) In compliance.

Future Plans:
Increased water recycling ($200,000). Municipal plant may go to tertiary treatment.

Profile:
Hoerner Waldorf has two mills on the Saint Paul site, both small, old and quite clean. One mill makes all of its paper from recycled fiber: the other from a combination of 80 percent recycled fiber and 20 percent NSSC pulp. Because the amounts of pulping liquor used are therefore small and the mill has no bleaching operation, chemical water pollution problems are minimal. All effluent receives good primary and secondary treatment at the Saint Paul plant—the addition of tertiary treatment is planned; the mill's main fuel source, gas, is clean, and when standby oil is used, gas emissions are "scrubbed."

CEP Estimate of Minimum Necessary Pollution Control Expenditure Not on Company Budget: None.

Missoula Mill

Location:
Missoula, Montana

Date Built:
1957, capital cost $30 million

Process:
1,100 TPD unbleached kraft
150 TPD bleached kraft

Other Production:
950 TPD kraft linerboard

Total Water Use:
16 MGPD

Source:
Wells

Discharge:
Clark Ford River

Table 16.4
Missoula Mill: Pollution Control Record

Treatment	Overall Evaluation	Equipment
Water		
Primary	✓	Three million gallons/day clarifier, 200-foot in diameter
Secondary	X	Soil percolation (similar to spray irrigation)
Tertiary	X	None
Solid waste removal	—	Clarifier sludge used as landfill
Air		
Particulate (fuel)	X	Fuel: hog fuel, auxiliary oil. Mechanical collectors on boilers
Particulate (production)	X	Three old recovery furnaces with inadequate precipitators
Gas and odor	X	None
Plant Emissions	N.A.	

Pollution Control Expenditures:
$1.5 million for primary clarifier.

Legal Status:
(Water) In compliance.
(Air) Not in compliance, but on schedule.

Future Plans:
Three old recovery furnaces to be replaced with two large Babcock and Wilcox
low odor units, precipitators (99.5 percent), and two wet gas scrubbers for
hog fuel power boilers. Estimated cost $14 million.

Profile:
A kraft mill was built in Missoula, Montana in 1957, at a capital cost of over
$7 million, for production of 250 tons per day of kraft pulp. In the 13 years
since the mill began operations, it has expanded to a pulp capacity of 1350
tons per day, presenting a capital investment of over $30 million. The mill
sits in the Missoula Valley (pop. 15,000). It draws 16 million gallons of water
a day from wells and discharges its effluent into the Clark Ford River.

As the mill grew and expanded over the years, its air and water pollution
problems grew also. The mill's stacks poured out thousands of pounds of
black soot and ash per day into the air (11,000 pounds per day from its boiler
stacks alone) and sulfurous gases from uncontrolled chemical recovery operation stacks. The pall of smoke and stench sat for miles around the mill, because the Missoula Valley location is subjected to frequent temperature inversions. The company maintained that its fumes were only causing an odor,
not a health, problem. However, Dr. Lambert, a Missoula physician, said
that hospital admissions in Missoula for respiratory ailments go up dramati-
cally during the worst inversion periods.

By the mid 1960s Missoulians began to complain about the pollution from
the mill. A group of local housewives banded together, called themselves
GASP (gals against smoke and pollution), picketed the mill, and worked to get
a pollution control bill through the Montana State legislature (Montana's Clean
Air Act was finally passed in 1967).

In 1968 additional pressure was brought to bear against the mill by the
Environmental Defense Fund, a national organization working in the anti-
pollution field. EDF filed a suit to stop the mill from producing "its emission
of noxious sulfur compounds that were degrading the balance of life in the
Missoula area."

Hoerner Waldorf's first response was a reminder to the town of the economic good the mill was doing for the area: providing jobs for 438 people (a
total payroll of $4 million) and making annual purchases of over $20 million
worth of goods, services, and raw materials, and supporting local businesses.
Its assertion that it was working on better pollution control for the mill did
not satisfy the public.

In February 1969, one additional spur was added to the publicity about
the mill. LIFE published a two-page color photograph of the mill and the
great clouds of yellowish smoke it spread across the valley sky and some of
the text read: "Some Missoula residents call the mill 'Little Hiroshima.' It
is the biggest single contributor to their polluted air."

Hoerner Waldorf officials mentioned that if too much pressure was brought to bear for pollution control all at once, it might not be economically feasible and the mill might even have to close or move. However, when the blows were dealt, Hoerner Waldorf came up with a plan to completely overhaul the mill.

By the end of 1972, it should have state-of-the-art air control and practically that for water. Changes include: (1) replacement of the three existing recovery furnaces with two large new Babcock and Wilcox furnaces; (2) installation of a new indirect contact evaporator called a concentrator (capital cost, $322,000), with which the hot gases used for heating are piped through the liquor instead of floating directly over it, greatly reducing the chances of gas escaping up the stacks; (3) a new electrostatic precipitator rated 99.5 percent efficient (cost, $443,000); (4) installation of two new wet gas scrubbers (cost, $48,600) to supplement the two mechanical collectors already on the hog fuel boiler stacks.

The total $14 million cost of this conversion will be financed via federally tax exempt bonds. The county of Missoula has received permission to float these bonds under the States Industrial Development Act of 1965. This enables the company to borrow the money at as much as 2 percent below standard borrowing rates.

The Hoerner Waldorf resident manager at Missoula said, "We are very pleased. We were committed to go ahead without it. This ruling will provide a new source for pollution control."

CEP Estimate of Minimum Necessary Pollution Control Expenditure Not on Company Budget: None.

References for Chapter 16

1. The breakdown of total company production is taken from Chem 26, June, 1970, p. 20.

2. One of these locations is the Roanoke Rapids plant, which is a subsidiary of Albermarle Paper Company. The 900 TPD kraft mill at this plant is not included in the present study.

Chapter 17

INTERNATIONAL PAPER
220 East 42nd Street, New York, New York 10017

Major Products:
Paper, lumber and hospital supplies.

Major Consumer Brands:
CONFIL fabric, ICE-PAK containers, FACELLE tissue products,
FLUSHABYES disposable diapers.

Financial Data ($ Millions)	1969	1968	1967
Net Sales	1,777.3	1,574.3	1,421.4
Net Income	115.6	99.9	89.9
Capital Expenditures	197.8	145.1	238.5

Pollution Control Expenditures:
$23 million from 1965 to 1968; $7 million in 1969; estimated $101 million
from 1970 to 1974 ($45 million for water; $56 million for air).

Annual Meeting:
The annual meeting takes place during May at the New York office.

Officers:
Chairman: Frederick R. Kappell, Bronxville, New York (Director: Standard
Oil, New Jersey, Whirlpool, Chase Manhattan Bank, American Telephone
and Telegraph)
President: Edward B. Hinman, New York, New York (Director: Arizona
Chemical Co., Toronto-Dominion Bank)

Outside Directors:
William S. Brewster: Chairman of the Board and Chief Executive Officer,
United Shoe Machinery Company
Malcolm G. Chace, Jr.: Chairman, Berkshire-Hathaway, Inc.
George Champion: Former Chairman, Chase Manhattan Bank NA
John M. Kingsley: President, Bessemer Securities Corp.
Donald Lourie: Chairman, Quaker Oats Co.
W. B. Murphy: President, Campbell Soup Co.
Herman C. Nolen: Former Chairman, McKesson and Robbins, Inc.
Robert W. Stoddard: Chairman, Wyman-Gorden Co.

Breakdown of Total Company Production:	1969	1968
Market pulp, paper, and paperboard	6,994,504	6,698,607
Lumber board (thousands of board ft)	217,204	241,729
Plywood and veneer (thousands of sq ft)	325,476	365,175
Insulating and building board (thousands of sq ft)	156,317	137,757

Domestic Plants:
Health products	1
Flakeboard	2
Folding box & label	7
Paper milk containers	15
Bags	7
Shipping containers	27
Dissolving pulp	1
Newsprint	2
Book and bond paper	3
Groundwood specialties	2
Kraft pulp mills	11
Other	2

Long-Bell Division

Plywood	4
Sawmills	7
Other	3

Timberholdings:
International Paper has management responsibility for 23,366,000 acres in the United States and Canada. This is about the size of Ohio, and somewhat larger than Maine. In the United States, 6,881,600 acres (of which 6,519,000 are owned and 362,000 leased or controlled), and in Canada, 16,485,000 acres (of which 1,357,000 are owned and 15,128,000 under government license). The total (1969) value of the timberholdings was $153,762,000 ($141,763,000 in the United States; $11,999,000 in Canada). International Paper holds mineral rights on five million additional acres.

Total Annual Pulp Production (355 times the daily production):
5,857,500 tons (1969)

Company Pollution Overview

International Paper is the world's largest paper company and the second largest landholder in the United States — second only to the federal government.

Its 1969 output of primary grades of paper, paperboard, and market pulp—almost 7 million tons—represented 10 percent of the total North American production. The company's daily pulp production is equivalent to the combined output of Weyerhaeuser, Crown Zellerbach, and Boise Cascade Corporations. (In terms of pulp production, these latter companies rank 2nd, 3rd, and 10th, respectively, of those in the present study.)

As is so often the case with the largest company in an industry, this lumbering giant has been among the last to recognize its need for pollution control action. While smaller companies have innovated and developed pollution control equipment over the last 25 years, International Paper has retreated behind the "untouchability" of its power and size. This is demonstrated at the Jay, Maine, mill—the company owns 856,000 acres in the state. Not only has International Paper failed to develop or innovate any pollution control techniques; but Keith Fry, assistant to George Rand, Vice President of Air and Water Management, said, "activated sludge doesn't work well for paper wastes,"[1] although such secondary treatment systems have been used effectively in the industry since 1955. No significant portion of the vast resources available to this corporation has been used for pollution control.

International Paper operates 20 pulp mills at 15 locations in the United States, and a twenty-first is under construction at Texarkana, Texas. The core of company production is ten large southern kraft mills which account for 87 percent of total pulp production. (At four of the sites there are also NSSC or groundwood mills.) The company also operates at three locations in New York State, one in Maine, and one in Oregon. Daily pulp production is 16,500 tons, of which 14,170 are kraft, 1180 groundwood, 1080 NSSC, and 140 soda pulp.

Pollution control is totally inadequate at three locations (Natchez, Mississippi; Jay, Maine; and North Tonawanda, New York), where there is totally inadequate particulate control, totally inadequate control of gaseous emissions, and totally inadequate water treatment. And pollution control is generally inadequate at 9 more locations. Pollution control is adequate at three locations, one of which is the brand-new Ticonderoga, New York, mill which has replaced a nineteenth-century facility repeatedly cited in state and federal pollution abatement actions.

The company's economic power is reflected in its attitude toward public information on its activities. It was the only company in this study to deny that the public even has an interest in specifics concerning pollution control equipment; George Rand asserted that people should be satisfied with company statements that equipment is installed and doing the job.

International Paper refused to provide water use figures as well as details on capacity, efficiency, or type of water treatment at any mill. After many months, the company did agree to say whether a mill has "primary" or "secondary" treatment. However, it turned out that International Paper employs its own definition of "secondary" treatment: "anything beyond primary," which does not conform to accepted usage. The definition of secondary treatment used by every other company, every public and private organization,

and virtually all the literature in the field, is "80-90 percent BOD removal."
"Beyond primary" as used by International Paper can mean only 35 percent
BOD reduction, or just the use of outfall pipes. Without information indicating
efficiency in BOD removal, actual effluent discharge quality, capacity, and
adequacy for the location, CEP has had no alternative but to question rather
than accept the adequacy of the secondary treatment at International Paper's
mills.

The company has no water treatment at three locations: Natchez, Corinth,
and North Tonawanda; partial primary at Jay, Maine; and only primary treatment at nine others, including the two largest installations: Georgetown, South
Carolina, and Panama City, Florida.

Two mills, Mobile, Alabama, and the old Ticonderoga mill, have been
the subject of federal water pollution abatement actions. Mobile was cited in
1970 as the prime industrial polluter of Mobile Bay; and Ticonderoga was cited
in 1968 and 1970 for degrading the water of Lake Champlain. Ticonderoga was
also cited by New York State in 1966 for water pollution. The 1970 federal conference resulted in an order that Ticonderoga water pollution be abated by
July 1, 1970, and concluded that the sludge deposits already present in the
Lake would continue to constitute pollution even after the old mill's closing.
In December, 1970, the mill ceased operation. Ticonderoga was also the subject of a federal air pollution survey in 1965, which found that odorous emissions 33 miles downwind of the mill were ten times the threshold for percep-
tion.[2] All federal investigations and actions at Ticonderoga were made in
response to complaints from Vermont residents across Lake Champlain.

International Paper has also consistently refused to make the efficiency
of its air pollution control equipment public, and has limited itself to listing
pieces of equipment and characterizing them as "modern" or "old." When
pressed, Keith Fry admitted that "modern," when applied to electrostatic
precipitators, meant "rated at 99.5 percent efficiency." Yet the definition is
misleading, since industry experience shows rated efficiency may be as much
as 20 percent greater than the actual efficiency of particulate control equipment.

Seven of the company's 13 kraft mills, including the two largest at
Georgetown and Panama City, have no odor control equipment. These seven
mills account for 9075 tons, or 70 percent of the company's daily kraft pulp
production. A native of Panama City said, in despair, "The smell here is
awful—just terrible. They [International Paper] keep saying they're going to
do something about it, but nothing has been done." Of the six mills which have
odor control systems, only five have adequate ones; that at Jay, Maine, is
obviously outdated.

Six of the locations lack adequate particulate control; International Paper
claims that nine of its mills have "modern" particulate control equipment.
The six lacking it again include Georgetown and Panama City, as well as
Mobile, Natchez, North Tonawanda, and Gardiner, Oregon.

As is so often the case, the biggest company in this industry is backward
and riddled with old ideas. Its idea of responding to a problem—the pollution

which blights rivers, air, and lives — is to call in its public relations firm and have it taken care of by means of a large advertising and press release campaign. Until mid-1970, International Paper maintained a silence about pollution. Then, in a nationwide press campaign, it announced that it planned to spend $101 million in the next four years for pollution control equipment. The immense sum is equal to its 1969 profit or about half of the company's 1969 capital investment. By way of comparison, in 1969, the company's freight and delivery outlays totalled $132 million. Over the four-year period, annual expenditures will equal the company's 1969 interest payments. The year before the press release, 1969, corporate pollution control expenditures were a mere $7 million, considerably less than the $12 million the company made from dealings in foreign exchange. In relation to this, Weyerhaeuser, with pulp production 3/8 that of International Paper, has already spent $125 million on pollution control, and plans to spend an additional $12 million. Scott Paper, with 1/4 the pulp production of International Paper, plans to spend $85 million on pollution control. In its press campaign (estimated cost about $150,000) International Paper proclaimed its pollution control plans throughout the country, running a full page ad in every city and town paper where it operates a plant, in the New York Times, the Washington Post, and all major business and news magazines.

International Paper's President, Edward B. Hinman, trumpeted, "This plan places International Paper in the forefront of those taking positive, constructive measures to solve the problem of environmental quality."[3] The forefront, unfortunately, is long past. Companies in the forefront installed pollution control equipment when it became available, not when it became legally imperative or fashionable.

By the projected termination date of the program, International Paper will have mills at sixteen locations in the United States. The program represents an average investment of $6.3 million at each. However, the funds will be concentrated, not distributed evenly through the company, and some will go for modern recovery furnaces (basic production equipment) as part of mill expansions. Not until three years after the expiration date of most states' secondary treatment deadlines will International Paper have "secondary treatment" at all locations. And even then, gauging by the company's definition, this may well be inadequate. By 1975, to comply with the Oregon and Washington deadlines, the kraft mills will have odor and particulate controls. Thus, 20 years after odor control systems were first installed in the United States, International Paper will have such systems at all of its kraft mills.

The company refused to break down its projected expenditure by location or major pieces of equipment. Vice President George Rand stated that the "public does not need to know dollar amounts — they're irrelevant, as long as the job is getting done." This would, of course, be true if the public were not being asked to accept the company's irregular definitions of getting the job done. CEP asked why the advertising campaign, which had put only monetary expenditure in the headlines, had been undertaken if dollar amounts were irrelevant. There was no response: No response as to how the public can know the job is being done when International Paper defines treatment as it chooses; nor was there a response as to where the industry "leader" has been while pollution control equipment was being developed and installed by even some backward and smaller companies over the last 25 years. No response as to why the company could for example, invest $13 million in expansion of the Corinth mill, but was unable to install any but a third-best water treatment system — justifying its action on the basis of economic feasibility (see mill story).

Without the specific information which International Paper refused to make public, it is impossible to know how much of the announced investment will go for pollution control equipment (even narrowly defined), how much for production systems, and how much for production capacity expansions. Each time International Paper calls itself an industry leader we may well ask in which direction it is leading.

Is "International Paper a company fully aware and responsive to the social and environmental needs of its time?" Or is it the backward monolith its actions indicate?

Table 17.1
International Paper: Pulp Production, Water Use, and Pollution Control

Mill Location	Pulp Prod. (TPD)	Other Prod. (TPD)	Water Use* (MGPD)	Water Pollution Control			Air Pollution Control	
				Pri.	Sec.	Tert.	Part.	Gas, Odor
Mobile, Ala.	1,500	1,250	34.4	✓	✗	✗	✗	✓
Camden, Ark.	750	665		✓	?	✗	✓	✗
Pine Bluff, Ark.	1,550	1,600		✓	?	✗	✓	✗
Panama City, Fla.	2,050	700		✓	✗	✗	✗	✗
Bastrop, La.	1,700	1,530		✓	?	✗	✓	✓
Springhill, La.	1,725	1,300		✓	?	✗	✓	✗
Jay, Me.	675	525		✗	✗	✗	✓	✗
Moss Point, Miss.	700	700		✓	✗	✗	✓	✗

Table 17.1
International Paper: Pulp Production, Water Use, and Pollution Control
(continued)

Mill Location	Pulp Prod. (TPD)	Other Prod. (TPD)	Water Use* (MGPD)	Water Pollution Control			Air Pollution Control	
				Pri.	Sec.	Tert.	Part.	Gas, Odor
Natchez, Miss.	950	--		X	X	X	X	X
Vicksburg, Miss.	1,200	1,000		√	√	X	√	√
Corinth, N.Y.	255	500		X	X	X	√	—
North Tonawanda, N.Y.	140	250		X	X	X	X	?
Ticonderoga, N.Y.	550	--		√	√	X	√	√
Gardiner, Ore.	545	545		√	X	X	X	√
Georgetown, S.C.	2,230	1,660		√	X	X	X	X
Total	16,520	12,225						

*Company water use figures N.A. The Mobile mill figures come from the Mobile Bay pollution abatement conference.

Mobile Mill

Location:
Mobile, Alabama

Date Built:
Pre-1929

Process:
1,200 TPD kraft, of which 420 is bleached
300 TPD groundwood

Other Production:
1,250 TPD polyethylene-coated papers, wrapping and convertible papers, business papers, newsprint

Total Water Use:
34.4 MGPD (federal abatement conference data)

Source:
Mobile River, reservoir

Discharge:
Chickasaw Creek (runs into Mobile River and thence into Mobile Bay)

Table 17.2
Mobile Mill: Pollution Control Record

Treatment	Overall Evaluation	Equipment
Water		
Primary	✓	N.A.
Secondary	X	None
Tertiary	X	None
Solid waste removal	—	Incineration
Air		
Particulate (fuel)	X	Fuel: coal, bark. Fly ash collectors (inadequate)
Particulate (production)	X	Recovery furnaces: two with "modern electrostatic precipitators; two with old precipitators (inadequate)
Gas and odor	✓	Black liquor oxidation

Plant Emissions

To water:

BOD	Approx. 330,000 lb/day	

Table 17.2
Mobile Mill: Pollution Control Record (continued)

Treatment	Overall Evaluation	Equipment
	(220 lb/ton of pulp)	
Suspended solids	N.A.	
To air:	N.A.	

Pollution Control Expenditures:
N.A.

Legal Status:
(Water) The mill was cited in January 1970, at a federal abatement conference
as the largest polluter of Mobile Bay; the City of Mobile was second. The data
in the study, "Pollution Affecting Mobile Bay, Alabama," indicated the mill
contributed about 47 percent of the total coliform discharged in one Mobile
metropolitan area,[4] and 94 percent of the nonmunicipal loading.

Because of a "grandfather clause" in the state law the secondary treatment deadline for this mill is 1974, rather than the state general pollution
abatement deadline of 1972.
(Air) State standards not yet implemented.

Future Plans:
In 1970 the company began construction of secondary waste water treatment,
and a new recovery furnace with a 99.5 percent efficient electrostatic precipitator to replace the two old precipitators. The expected completion date is
1973. $6 million will be spent on modernization and a new bleach plant.

Profile:
Besides the federal abatement action, there is other evidence of water pollution from this mill. An unnamed paper mill has been implicated in three
fishkills in Chickasaw Creek: on May 14, 1967, 1039 fish were killed along
four miles of river over two days; on September 25, 1968, 5013 fish were
killed along two miles; on June 25, 1969, 4249 fish were killed. These incidents are listed in the annual fish kill report of the Department of the Interior
as caused by paper or pulp mill discharges. International Paper disclaims
responsibility. There is also a Scott Paper Company mill in Chickasaw Creek;
however, this Scott mill has a very large activated sludge secondary water
treatment system which operates during the summer months and reduces the
mill BOD to 740 pounds per day.

Finally, under a state requirement, the company is beginning construction of secondary treatment.

More efforts have been taken for air pollution control, but the equipment is not adequate. The power plant should have scrubbers, as mechanical dust collectors are not adequate for coal boilers. Particulate control should be adequate when the two old recovery furnaces are replaced with a larger new one with a "modern" electrostatic precipitator. No efficiency data was available for the odor control system.

CEP Estimate of Minimum Necessary Pollution Control Expenditures Not in Company Budget: $11.7 million total, of which $3 million is for secondary water treatment (activated sludge as at the nearby Scott mill); $200,000 is for a power boiler scrubber; $8 million is for a new recovery furnace (planned but no cost given); and $500,000 is for a precipitator for the new recovery furnace (planned but no cost given).

Camden Mill

Location:
Camden, Arkansas

Date Built:
Company information not released

Process:
750 TPD unbleached kraft

Other Production:
665 TPD unbleached and semibleached kraft wrapping and convertible paper

Total Water Use:
N.A.

Source:
Ouachita River and wells

Discharge:
Ouachita River

Table 17.3
Camden Mill: Pollution Control Record

Treatment	Overall Evaluation	Equipment
Water		
Primary	✓	N.A.
Secondary	?	Company statement that something beyond primary treatment is provided
Tertiary	X	None
Solid waste removal		N.A.
Air		
Particulate (fuel)	X	Fuel: bark, oil, gas. Fly ash collectors (inadequate)
Particulate (production)	✓	Three recovery furnaces: "modern" electrostatic precipitators. Lime kiln, "various other vents": scrubbers
Gas and odor	X	None
Plant Emissions	N.A.	

Pollution Control Expenditures:
N.A.

Legal Status:
(Water) In compliance with secondary treatment requirement.
(Air) In compliance.

Future Plans:
No new equipment was under construction in 1970.

Profile:
There is no odor control equipment at this mill, although there may be adequate control of other pollution sources. The company has no immediate plans to alleviate the odor.

CEP Estimate of Minimum Necessary Pollution Control Expenditures Not in Company Budget: $500,000 total, for an odor control system.

Pine Bluff Mill

Location:
Pine Bluff, Arkansas

Date Built:
N.A.

Process:
1,150 TPD bleached kraft
400 TPD groundwood

Other Production:
1,600 TPD bleached kraft board, newsprint, and directory and catalogue papers

Total Water Use:
N.A.

Source:
Mainly underground wells

Discharge:
Arkansas River

Table 17.4
Pine Bluff Mill: Pollution Control Record

Treatment	Overall Evaluation	Equipment
Water		
Primary	✓	N.A.
Secondary	?	Company indicated some facilities were completed in 1970 but would give no efficiency or capacity figures
Tertiary	X	None
Solid waste removal	—	Sludge from primary treatment used as landfill

Table 17.4
Pine Bluff Mill: Pollution Control Record (continued)

Treatment	Overall Evaluation	Equipment
Air		
Particulate (fuel)	X	Fuel: oil, gas, bark. Fly ash collectors (inadequate)
Particulate (production)	✓	One recovery furnace with precipitator, two with scrubbers. Lime kiln, "various other vents": scrubbers
Gas and odor	X	None
Plant Emissions	N.A.	

Pollution Control Expenditures:
$ 2 million for water treatment.

Legal Status:
(Water) In compliance with secondary treatment requirement.
(Air) In compliance.

Future Plans:
No equipment was under construction in 1970.

Profile:
As at Camden, the company has taken steps to control most pollution sources except odor. Odor control is an urgent need at such a large kraft mill. Unlike most of the mills, noncondensible gases are not incinerated here; thus, the odor problem is exacerbated.

CEP Estimate of Minimum Necessary Pollution Control Expenditures Not in Company Budget: $700,000 total of which $200,000 is for incineration of noncondensible gases and $500,000 is for an odor control system.

Panama City Mill

Location:
Panama City, Florida

Date Built:
1931

Process:
2,050 TPD kraft, of which 650 TPD is bleached

Other Production:
700 TPD containerboard

Total Water Use:
N.A.

Source:
"Mostly wells, some surface water"

Discharge:
St. Andrews Bay (to Gulf of Mexico)

Table 17.5
Panama City Mill: Pollution Control Record

Treatment	Overall Evaluation	Equipment
Water		
Primary	√	Clarifier. Specific information N.A.
Secondary	X	None
Tertiary	X	None
Solid waste removal	—	Clarified sludge incinerated
Air		
Particulate (fuel)	X	Fuel: oil, bark. Fly ash collectors (inadequate)
Particulate (production)	X	Four old recovery furnaces: scrubbers lime kiln, "various other vents": scrubbers
Gas and odor	X	None
Plant Emissions	N.A.	

Pollution Control Expenditures:
N.A.

Legal Status:
(Water) State requires secondary treatment for 90 percent BOD removal by
the end of 1972.
(Air) The mill was cited in January 1967. At present, new air pollution regulations are coming into effect, and the old ones are not being enforced in the
meantime.

Future Plans:
Some pollution control equipment is planned as part of a $40 million program
to completely rebuild the mill. Secondary water treatment is now under construction. Also planned are a new lime kiln with scrubbers, and a new 900
TPD recovery furnace with precipitators, which will replace two old recovery
units. (The company estimates that the other old recovery furnace will be replaced in 1972 by a second new unit.) Also, an odor control system is planned.
    The company announced that the cost of the air pollution control "im-
provements" will be $11 million.

Profile:
Despite the total inadequacy of environmental protection, the Panama City
mill is loyally defended by many townspeople because it started up during the
depression and pulled the town out of the economic doldrums. A photograph
of this mill, with dark smoke belching from its four big stacks, was the cover
for International Paper's 1965 Annual Report. Those smoke and gas emissions, so proudly displayed, have not gone totally unnoticed by the public or
by the state agencies, but most Panama City residents have just learned to
live with the pollution from this mill. One noted with resignation:

"It is commonly practiced here for a person calling the local drive-in movie
to ask 'what's playing and which way is the wind blowing.' because if it's
blowing from the direction of the mill, almost no one goes to the movie—the
odor is so bad."

The houses in the mill area are run-down, shabby, and unpainted because (according to another resident) the smoke contains chemical particles that ruin
the paint.
    Although the technology has been available for almost twenty years the
mill did not install black liquor oxidation to reduce the odor. The soot is bad
too, because the mill's four recovery furnaces have long had old and inefficient scrubbers on them and the fuel stacks are covered by equally inefficient
fly ash collectors. But the town has not complained much. One local government official explained:

"What do you mean are we 'interested' in the mill? We hate the damn thing but you don't last long in a political position here by chopping at the head of the animal that feeds you—the tail usually knocks you silly."

The dinosaur metaphor is not misplaced. This is International Paper's largest kraft mill.

In the spring of 1970, some Panama City youngsters did complain, decrying the pollution in a series of letters to the town paper. An editorial, expressing the view of many of the older residents was soon published:

"The officials of mills live in their communities. They like the pollution no better than we do. Give them time. The old-timers can tell you that to them that smoke smells like bacon and eggs. They can remember when things were harder before the paper mill came here. There were no tourists then; Panama City was just a whistle stop. The paper mill put it on the map."

"Give them time"—yet for nearly 40 years no company initiative was taken, despite the availability of pollution control equipment. That none would have been taken without outside pressure was virtually admitted by Mr. Tim Caldwell, Director of Public Affairs for the mill:

"Our environmental control projects are a result of the federal legislation of 1965 and subsequent implementing legislation at the state level."

The mill even stalled on state orders. Florida cited the mill in January 1967 for its air pollution and ordered a clean-up by June 1970. As of November 1970, three years after the citation and 5 months past the deadline, not one major piece of pollution control equipment has been added, and the company has just announced that installation will begin. The chemical recovery system is being rebuilt, with new recovery furnaces and lime kiln and pollution control equipment. Secondary water treatment will also be provided but at this rate it is doubtful that most of the job will be done by the 1972 deadline.

A June 1970 press release reported that $11 million would be spent on the proposed air pollution control equipment as part of a $40 million program to rebuild the mill. No information was disclosed on the cost of planned secondary treatment facilities.

CEP Estimate of Minimum Necessary Pollution Control Expenditures Not in Company Budget: None. The planned secondary treatment will probably cost $2 - $3 million.

Bastrop Mill

Location:
Bastrop, Louisiana

Date Built:
1923, kraft mill; 1945, NSSC mill added

Process:
1,100 TPD unbleached kraft (Louisiana mill)
600 TPD NSSC (Bastrop mill)

Other Production:
970 TPD kraft paper and board, machine coated board
560 TPD semichemical corrugating medium

Total Water Use:
35 MGPD approximately—not verified by company

Source:
Well water

Discharge:
Stockinghead Creek

Table 17.6
Bastrop Mill: Pollution Control Record

Treatment	Overall Evaluation	Equipment
Water		
Primary	✓	Series of lagoons. Specific details N.A.
Secondary	?	
Tertiary	✗	None
Solid waste removal		N.A.
Air		
Particulate (fuel)	✗	Fuel: oil, gas, bark. Fly ash collectors (inadequate)
Particulate (production)	✓	Two kraft recovery furnaces: "modern electrostatic precipitators." Lime kiln, "various other vents": scrubbers. Three NSSC recovery furnaces: one "modern electrostatic precipitator," two scrubbers

Table 17.6
Bastrop Mill: Pollution Control Record (continued)

Treatment	Overall Evaluation	Equipment
Gas and odor	✓	Black liquor oxidation. Incineration of noncondensible gases
Plant Emissions	N.A.	

Pollution Control Expenditures:
N.A.

Legal Status:
(Water) In compliance, has permit.
(Air) In compliance; state has not yet taken emissions inventory of pulp mills.

Future Plans:
At the NSSC mill, the company is currently constructing "additions to water treatment systems." At the kraft mill, "improvements to water treatment system and various scrubbers" are under way.

Profile:
Air pollution control except for the power plant is adequate here. This very large installation has an effluent treatment system which may be judged by the fact that the company is planning improvements to it.

    CEP Estimate of Minimum Necessary Pollution Control Expenditures Not in Company Budget: $1.2 million total, of which $1 million is for current improvements to waste water treatment and $200,000 is for a scrubber for the power boiler.

Springhill Mill

Location:
Springhill, Louisiana

Date Built:
1937

Process:
1,725 TPD kraft, of which 1,000 TPD is bleached

Other Production:
1,300 TPD kraft containerboard, bleached kraft paper and board

Total Water Use:
40 MGPD

Source:
Company-constructed 7,000-acre fresh water lake

Discharge:
Bayou Bodcaw

Table 17.7
Springhill Mill: Pollution Control Record

Treatment	Overall Evaluation	Equipment
Water		
Primary	✓	Lagoon system. Specific information N.A.
Secondary	?	
Tertiary	X	None, but pilot-plant-scale "massive lime" technique color-removal project underway for bleach plant effluent. Federal grant of $565,000; IP's share, $280,000
Particulate (fuel)	X	Fuel: oil, gas, bark. Fly ash collectors (inadequate)
Particulate (production)	✓	Three recovery furnaces: two "modern electrostatic precipitators," one scrubber. Lime kiln, "various other vents:" scrubbers
Gas and odor	X	None
Plant Emissions	N.A.	

Pollution Control Expenditures:
Company information N.A. beyond $280,000 contribution to pilot project.

Legal Status:
(Water) In compliance, has permit.
(Air) In compliance, state has not yet taken emissions inventory of pulp mills.

Future Plans:
"No construction for new equipment will begin in 1970."

Profile:
At the effluent discharge point, Bayou Bodcaw is about four feet wide. High-efficiency waste treatment facilities are thus essential to preserve any life that might be in the water.

International Paper is reported to be planning a new recovery unit for 1971 installation at which time it will decide whether or not to install the odor controls it has operated without for 33 years. If International Paper is somewhat lacking in enthusiasm and concern for adding odor control, the State of Louisiana is doing nothing to press it into action. The March 1970 state air pollution progress report commented:

"Paper mill emissions are not considered to have a high priority for action, principally because technology in odor control of this processing has not been proved practicable, and because residents of town or cities near paper mills usually work for the mill or serve mill employees, and they accept the odor as a part of their occupation."[5]

Odor control technolgy has for years been proved practicable, although not 100 percent effective. Such a weak and incorrect state assessment only frees the company to pollute the air and leaves the 7,000 citizens of Springfield (of whom 2,100 work for International Paper) with no recourse but in fact to "accept the odor as part of their occupation."

CEP Estimate of Minimum Necessary Pollution Control Expenditures Not in Company Budget: $2.25 million total, of which $1 million is for an odor control system (cost is high because mill is large); $250,000 is for a power boiler scrubber; and $1 million is for a precipitator for the third recovery furnace.

Jay Mill

Location:
Jay, Maine

Date Built:
N.A.

Process:
500 TPD kraft, of which 400 bleached
175 TPD groundwood

Other Production:
525 TPD lightweight bond machine coated book paper, carbonized paper, etc.

Total Water Use:
25.8 MGPD (unconfirmed)

Source:
Androscoggin River

Discharge:
Androscoggin River

Table 17.8
Jay Mill: Pollution Control Record

Treatment	Overall Evaluation	Equipment
Water		
Primary	X	Clarifier (mill has two sewers, one goes to clarifier, one to river)*
Secondary	X	None
Tertiary	X	None
Solid waste removal	—	Clarifier sludge to landfill
Air		
Particulate (fuel)	X	Fuel: oil, bark. Fly ash collector (inadequate)
Particulate (production)	√	Recovery furnace: "modern electrostatic precipitator." Lime kiln, "various other vents": scrubbers
Gas and odor	X	Incineration of noncondensible gases; weak black liquor oxidation (inadequate)

Table 17.8
Jay Mill: Pollution Control Record (continued)

Treatment	Overall Evaluation	Equipment
Plant Emissions		
To water:		
BOD	70,600 lb/day (105 lb/ton of pulp)	
Suspended solids	39,600 lb/day (59 lb/ton of pulp)	
To air:	N.A.	

*Data from Maine Environmental Commission, 1970.

Pollution Control Expenditures:
N.A.

Legal Status:
(Water) In compliance, state requires secondary treatment in 1976.
(Air) State standards not yet established.

Future Plans:
The company began construction in 1970 of secondary waste water treatment; a switch to heavy black liquor oxidation; and improved sludge handling and incineration. The company indicated no completion dates.

Profile:
International Paper owns 856,000 acres in the State of Maine.

An article in the Christian Science Monitor referred to the Jay Mill as one of the six so-called "untouchables" in the Maine paper industry. Professor Donaldson Koons, Chairman of the Water and Air Environmental Improvement Commission and a biologist at Colby College, told Monitor reporter Richard C. Halverson that he could sometimes smell the plant from his home 30 miles away.[6]

According to a state report, the mill was involved in a week-long fish kill in June, 1967, caused by "masses of organic matter rising from the bottom" of the river with a foul odor. On July 22, 1968, there was another fish kill near Lewiston, Maine, killing over 5,000 fish along four miles of river. The cause of this kill was a "chemical discharge accident" according to Marshall F. Burk, Executive Secretary, Natural Resource Council of Maine.

He commented that the dissolved oxygen content of the river was less than one part per million, and the temperature was above 80 degrees (F). However, International Paper denied its mill was responsible for this accident.

CEP Estimate of Minimum Necessary Pollution Control Expenditures Not in Company Budget: $2.95 million total, of which $500,000 is for primary water treatment; $2 million is for secondary water treatment; $250,000 is for a power boiler scrubber; and $200,000 is for heavy black liquor oxidation.

Moss Point Mill

Location:
Moss Point, Mississippi

Date Built:
1913, rebuilt 1969

Process:
700 TPD bleached kraft

Other Production:
700 TPD kraft wrapping and convertible paper; paper, paperboard, and machine coated bleached kraft board

Total Water Use:
50 MGPD (not confirmed by company)

Source:
Escatawpa River

Discharge:
Escatawpa River

Table 17.9
Moss Point Mill: Pollution Control Record

Treatment	Overall Evaluation	Equipment
Water		
Primary	✓	N.A.
Secondary	X	None
Tertiary	X	None

Table 17.9
Moss Point Mill: Pollution Control Record (continued)

Treatment	Overall Evaluation	Equipment
Solid waste removal	—	Sludge burned (this was the first mill to burn primary sludge)
Air		
Particulate (fuel)	X	Fuel: oil, bark. Fly ash collectors (inadequate)
Particulate (production)	✓	Three recovery furnaces: "modern electrostatic precipitators". Lime kiln, "other vents": scrubbers
Gas and odor	X	None
Plant Emissions	N.A.	

Pollution Control Expenditures:
N.A.

Legal Status:
(Water) Mill permit expired August, 1970; secondary treatment with controlled release is required by state by August, 1972.
(Air) State standards not yet developed.

Future Plans:
Company says construction is beginning in 1970 on a secondary waste water treatment system, with expected completion by the 1972 deadline.

Profile:
It is interesting that the recent rebuilding project at this large mill failed to include an odor control system. This was a serious omission. The company failed to indicate what controls, if any, are on the sludge incineration system.

CEP Estimate of Minimum Necessary Pollution Control Expenditures Not in Company Budget: $2.7 million total, of which $500,000 is for an odor control system; $2 million is for a secondary water treatment system; and $200,000 is for a power boiler scrubber.

Natchez Mill

Location:
Natchez, Mississippi

Date Built:
1950

Process:
950 TPD kraft

Other Production:
None

Total Water Use:
60 MGPD (not confirmed by company)

Source:
Groundwater wells

Discharge:
Mississippi River

Table 17.10
Natchez Mill: Pollution Control Record

Treatment	Overall Evaluation	Equipment
Water		
Primary	X	None
Secondary	X	None
Tertiary	X	None
Air		
Particulate (fuel)	X	Fuel: mainly oil, bark. Fly ash collector (inadequate)
Particulate	X	Recovery furnaces: Four have "modern electrostatic precipitators", two have scrubbers (inadequate). Lime kiln: scrubber

Table 17.10
Natchez Mill: Pollution Control Record (continued)

Treatment	Overall Evaluation	Equipment
Gas and odor	X	None
Plant Emissions	N.A.	

Pollution Control Expenditures:
N.A.

Legal Status:
(Water) Permit expired September 1970; state requires secondary treatment at this mill by December 1, 1972.
(Air) State standards not yet adopted.

Future Plans:
Construction is beginning on a new electrostatic precipitator to replace the two recovery furnace scrubbers, and on improved scrubbers.

Profile:
This mill has no water treatment and no odor control. Neither is planned as of 1970, although the ongoing improvements to the particulate control system indicate that there is no question of closing down the mill. International Paper obviously imagines that the Mississippi River and the noses of local residents have infinite waste assimilation capacities.

CEP Estimate of Minimum Necessary Pollution Control Expenditures Not in Company Budget: $5.5 million total, of which $500,000 is for an odor control system; $3 million is for primary and secondary water treatment; power boiler scrubber; and $1 million is for a precipitator to replace the scrubber.

Vicksburg Mill

Location:
Vicksburg, Mississippi

Date Built:
1967

Process:
1,200 TPD bleached and unbleached kraft

Other Production:
1,000 TPD linerboard and container board

Total Water Use:
45-50 MGPD (not confirmed by company)

Source:
Yazoo River

Discharge:
Yazoo River

Table 17.11
Vicksburg Mill: Pollution Control Record

Treatment	Overall Evaluation	Equipment
Water		
Primary	✓	Clarifier. Specific information N.A.
Secondary	✓	Lagoon. Specific information N.A.
Tertiary	✗	None
Solid waste removal	—	Clarifier sludge incinerated in new sludge plant
Air		
Particulate (fuel)	✗	Fuel: gas, oil, bark. Fly ash collectors (inadequate)
Particulate (production)	✓	Two recovery furnaces: "modern electrostatic precipitators"; lime kiln, "various other vents": scrubbers
Gas and odor	✓	Incineration of noncondensible gases. Black liquor oxidation
Plant Emissions	N.A.	

Pollution Control Expenditures:
N.A.

Legal Status:
(Water) In compliance with secondary treatment required; has nonexpiring
state discharge permit.
(Air) State standards not yet adopted.

Future Plans:
In 1970, the company began construction of additional scrubbers.

Profile:
This is quite a new mill, built only three years ago. A company representative said that it is the company's "showcase" in the South, with the "best odor
control of any International Paper mill... It also meets the latest water quality objectives."

   CEP Estimate of Minimum Necessary Pollution Control Expenditures Not
in Company Budget: $200,000 total, for a power boiler scrubber.

Corinth Mill

Location:
Corinth, New York

Date Built:
Company information not released

Process:
255 TPD groundwood (sulfite mill on site is idle)

Other Production:
500 TPD coated printing paper, label paper, mill wrapper and core paper

Total Water Use:
N.A.

Source:
Hudson River

Discharge:
Hudson River

Table 17.12
Corinth Mill: Pollution Control Record

Treatment	Overall Evaluation	Equipment
Water		
Primary	X	None
Secondary	X	None
Tertiary	X	None
Air		
Particulate (fuel)	✓	Fuel: oil. Soot collector
Particulate (production)	—	None needed
Gas and odor	—	None needed
Plant Emissions		
To water:		
BOD	N.A.	
Suspended solids	N.A.	

Pollution Control Expenditures:
N.A.

Legal Status:
(Water) Not in compliance, but plans for treatment were approved by the
state in January, 1969.
(Air) In compliance.

Future Plans:
The company says construction of primary and secondary water treatment
was begun in 1969. The plans include primary tanks, and a trickling filter
with bio-oxidating media for BOD removal.

Profile:
In 1969, International Paper completed a $13 million modernization program
at Corinth, which did not include construction of any water treatment facility.

Although the company would not release the mill's construction date, it probably has a long history of discharging untreated wastes into the Hudson. No doubt the pollution situation is better than it was when the sulfite mill on the site was operating, but it is still not good.

It was learned from an interview with Mr. Keith Fry that after state criticism of mill effluent, the company discussed treatment alternatives with the state agency. International Paper refuses to install activated sludge secondary treatment systems on principle, and there was not enough room on the site for a different, equally efficient system. Piping the effluent up the hill behind the mill to a level area where extensive lagooning could be constructed was suggested, but the company rejected that solution as being "too expensive." The third possibility is providing a smaller, less efficient treatment system located on the limited space next to the mill. This will provide consistent but less adequate BOD removal than would the two other alternatives. It is the solution International Paper chose, and the state unfortunately accepted.

CEP Estimate of Minimum Necessary Pollution Control Expenditures Not in Company Budget: $2.8 million total, of which $800,000 is for primary water treatment and $2 million is for secondary treatment.

North Tonawanda Mill

Location:
North Tonawanda, New York

Date Built:
N.A.

Process:
140 TPD bleached soda

Other Production:
255 TPD book, offset, label, tablet, and special converting paper

Total Water Use:
12.5 MGPD (approximate, not confirmed by company)

Source:
Niagara River (near end of Erie Canal)

Discharge:
Niagara River

Table 17.13
North Tonawanda Mill: Pollution Control Record

Treatment	Overall Evaluation	Equipment
Water		
Primary	X	None
Secondary	X	None
Tertiary	X	None
Air		
Particulate (fuel)	X	Fuel: coal, oil, bark. Fly ash collectors (inadequate)
Particulate (production)	X	Recovery furnace: "old electrostatic precipitator" (inadequate). Lime kiln: scrubber
Gas and odor	?	Fume collection and treatment system. Specific information N.A.

Plant Emissions

To water:

BOD	21,680 lb/day (155 lb/ton of pulp)
Suspended solids	30,820 lb/day (220 lb/ton of pulp)
To air:	N.A.

Pollution Control Expenditures:
N.A.

Legal Status:
(Water) Mill charged by the New York State Department of Environmental
Quality and International Joint Commission as major polluter of Niagara
River.
(Air) On abatement schedule for odor and particulate controls.

Future Plans:
Under construction: "corrections to sewer system; rebuilding of the precipitator; and additional scrubbers and improved fly ash collection."

Profile:
As implied by the state governmental criticism of this mill, untreated pulping wastes have always been discharged into the Niagara River. The mill does have required chemical recovery system of any soda mill, but has no waste treatment of its effluent. No water treatment seems to be planned.

Air pollution control measures are finally being taken, with a rebuilding program for the precipitator, perhaps in response to a long history of complaints by passersby on the "scenic route" to Niagara Falls. It is reported that strong mercaptan odors are caused four times every eight hours when the digesters are blown.

CEP Estimate of Minimum Necessary Pollution Control Expenditures Not in Company Budget: $3.2 million total, of which $200,000 is for scrubbers for power boilers (planned but no cost given); $500,000 is for an odor control system; and $2.5 million is for primary and secondary water treatment.

Ticonderoga Mill

Location:
Ticonderoga, New York

Date Built:
1971 (under construction)

Process:
550 TPD kraft

Other Production:
N.A.

Total Water Use:
21 MGPD (not confirmed by company)

Source:
N.A.

Discharge:
Lake Champlain

Table 17.14
Ticonderoga Mill: Pollution Control Record

Treatment	Overall Evaluation	Equipment
Water		
Primary	✓	Clarifier (capacity N.A.)
Secondary	✓	Aerated lagoon, 21 million gallon capacity (17-acre area). Sedimentation tanks (area of whole secondary system, 200 acres)
Tertiary	✗	None
Other	—	Outfall diffusers
Air		
Particulate (fuel)	?	Fuel: N.A.
Particulate (production)	✓	Recovery furnace: "modern precipitator;" lime kiln: scrubber
Gas and odor	✓	Incineration of noncondensible gases, indirect contact evaporation system as at American Can at Halsey, Ore.
Plant Emissions	N.A.	

Pollution Control Expenditures:
Of the total mill capital cost of $80 million, $6.4 million is attributed to pollution control.

Legal Status:
(Water) Expected to be in compliance.
(Air) Expected to be in compliance.

Future Plans:
None; mill is brand-new.

Profile:
This mill will replace an 88-year old kraft mill in the town of Ticonderoga.

The old mill was the subject of considerable state, federal, and citizen criticism for both air and water pollution.

An article in the New York Times reported "Visitors are surprised... that such a smell could come from a village in such an idyllic spot at the foot of the Adirondacks."[7] Water pollution problems were similar:

"Water at one edge of Ticonderoga rushes crystal clear from Lake George, tumbling down the creek into the paper mill. From there it emerges black and murky and heavy with suspended residue."[8]

Complaints by citizens across on the Vermont side of Lake Champlain in the path of the prevailing wind from the mill, led to a federal air pollution investigation, published in November, 1965.[9] The report said the pulp mill was the only area source of the distinctive "rotten egg" and "rotten cabbage" odors characteristic of kraft pulp mills, and that significant concentrations of the odorous emissions occurred at relatively great distances from the source.

Meanwhile, there was also criticism of water pollution due to the mill. In 1966, the State of New York objected to mill discharges because of their high content of settleable and floating solids, the sludge deposits they caused, the low dissolved oxygen conditions they created; and because emissions of "chlorine compounds were sufficient to prevent fish life."

In November and December of 1968, a Federal Water Pollution Abatement Conference took place at the request of the Vermont Department of Water Resources; in 1970, the conference was reconvened. There are massive sludge deposits in the lake, which the governments have been trying to get International Paper to remove. The company disclaims responsibility, citing as the source the many sawmills which long ago also operated on the shores of Lake Champlain. The 1970 session also expressed concern about possible mercury deposits in the sludge, and ordered a study on the problem.

In the midst of this enforcement activity, in December of 1967, International Paper announced that it would build a new mill a few miles from Ticonderoga and close the old mill. Enchanted with this prospect, the town sold the land which had been planned for an airport to the company for $40,000. This will be the site of the new mill.

In December 1970, the old mill was closed for good. International Paper has announced that all the modern pollution control equipment will be used at the new mill, including water treatment facilities capable of treating far more effluent than the mill will originally produce—an unprecedented step for the company.

CEP Estimate of Minimum Necessary Pollution Control Expenditures Not in Company Budget: None; mill is being built with needed equipment.

Gardiner Mill

Location:
Gardiner, Oregon

Date Built:
1964

Process:
545 TPD unbleached kraft

Other Production:
545 TPD container board

Total Water Use:
N.A.

Source:
Surface water from small tributaries of Umpqua River

Discharge:
Pacific Ocean

Table 17.15
Gardiner Mill: Pollution Control Record

Treatment	Overall Evaluation	Equipment
Water		
Primary	✓	Specific information N.A.
Secondary	X	None
Tertiary	X	None
Other	—	Ocean outfall
Air		
Particulate (fuel)	X	Fuel: oil (mainly), bark. Fly ash collector (inadequate)
Particulate (production)	X	Recovery furnaces: one with "modern precipitator," one with scrubber (inadequate).

Table 17.15
Gardiner Mill: Pollution Control Record (continued)

Treatment	Overall Evaluation	Equipment
		Lime kiln, "various other vents": scrubbers (inadequate)
Gas and odor	√	Heavy black liquor oxidation

Plant Emissions

To water: N.A.

To air:

TRS	3,000 lb/day (6 lb/ton of pulp)
Particulates	10,760 lb/day (20 lb/ton of pulp)

Pollution Control Expenditures:
N.A.

Legal Status:
(Water) Mill is operating with a state discharge permit; but, according to state officials, water treatment "needs some improvement."
(Air) Mill is on compliance schedule for kraft mills; does not now meet future TRS or particulate emissions limits.

Future Plans:
In a letter to CEP, the company said it is beginning construction in 1970 of turpentine recovery systems (to keep it out of the effluent). Incineration of noncondensible gases is also planned.

Profile:
This is another mill built relatively recently with inadequate effluent treatment — consisting of solids control and ocean dumping — sanctioned by the state.

Significant expenditures will probably be required in order to comply with the state's 1975 particulate and TRS emissions limits. The present particulate emissions are quite high, and the TRS emissions are three times the ultimate standard.

CEP Estimate of Minimum Necessary Pollution Control Expenditures Not in Company Budget: $2.7 million total, of which $2 million is for secondary water treatment; $200,000 is for a power boiler scrubber; and $500,000 is for an electrostatic precipitator for recovery furnace lacking one.

Georgetown Mill

Location:
Georgetown, South Carolina

Date Built:
N.A.

Process:
1,750 TPD kraft, of which 400 TPD is bleached

Other Production:
480 TPD NSSC
1,660 TPD bleached kraft board, unbleached kraft container board, semi-chemical corrugated medium

Total Water Use:
N.A.

Source:
Pee Dee River

Discharge:
Sampit River (into Winyar Bay Estuary)

Table 17.16
Georgetown Mill: Pollution Control Record

Treatment	Overall Evaluation	Equipment
Water		
Primary	√	N.A.
Secondary	X	None
Tertiary	X	None
Solid waste removal	—	Primary sludge burned

Table 17.16
Georgetown Mill: Pollution Control Record (continued)

Treatment	Overall Evaluation	Equipment
Air		
Particulate (fuel)	X	Fuel: N.A. Fly ash collectors (inadequate)
Particulate (production)	X	Recovery furnaces: one with "modern precipitator," two with scrubbers (inadequate). Lime kiln, "various other vents": scrubbers.
Gas and odor	X	None
Plant Emissions	N.A.	

Pollution Control Expenditures:
N.A.

Legal Status:
(Water) Not in compliance. State cited mill for causing deterioration of stream quality.
(Air) State standards not implemented yet.

Future Plans:
International Paper plans to begin construction in 1970 of secondary water treatment and to add heavy black liquor oxidation and more scrubbers for "various vents." No completion date was given.

Profile:
This very large mill is the only major industry on the small, narrow, Sampit River, and thus may be considered the major pollution source. State measurements taken below the mill show very low summer dissolved oxygen contents. Over a period of time, 11 out of 17 measurements showed a zero oxygen content, five varied between zero and 2.4 parts per million and the highest was 2.4 parts per million. Dissolved oxygen readings above the mill generally showed a higher oxygen content. There is also no odor control and inadequate particulate control here.

    CEP Estimate of Minimum Necessary Pollution Control Expenditures Not in Company Budget: $3.7 million total, of which $2 million is for secondary water treatment (planned but no cost given); $500,000 is for an odor control system (planned but no cost given); $200,000 is for a power boiler scrubber; and $1 million is for two precipitators for the recovery furnaces lacking them.

References for Chapter 17

1. Keith Fry, in an interview with CEP.

2. National Emissions Standards Study (Report of the Secretary of Health, Education, and Welfare), March, 1970, pp. 65-66.

3. New York Times, May 14, 1970.

4. Pollution Affecting Shellfish Harvesting in Mobile Bay, Alabama, U.S. Department of the Interior Federal Water Pollution Control Administration, Southeast Water Laboratory, January 1970, p. 37.

5. The Louisiana Air Control Program, Louisiana Air Control Commission, Louisiana State Department of Health, Air Control Section, March 2, 1970, p. 11.

6. Christian Science Monitor, December 27, 1969.

7. New York Times, July 13, 1969.

8. Ibid.

9. Reprinted in the National Emissions Standards Study, March, 1970, pp. 65-66.

Chapter 18

KIMBERLY-CLARK CORPORATION
North Lake Street, Neenah, Wisconsin 54956

Major Products:
Industrial printing and sanitary paper products, newsprint.

Major Consumer Brands:
KLEENEX, DELSEY toilet tissue, KOTEX, FEMS sanitary napkins,
KIMWIPE, TERI-TOWEL towels, KIMLON disposable garments and sheets,
KIMBIES disposable diapers, MARVALON adhesive coverings.

Financial Data ($ Millions)	1969	1968	1967
Net Sales	834.7	720.5	733.3
Net Income	49.9	40.2	43.4
Capital Expenditures	68.0	86.7	96.4

Pollution Control Expenditures:
1970 company estimate, $5 million.

Annual Meeting:
The annual meeting is held during April in Neenah, Wisconsin.

Officers:
Chairman and Chief Executive Officer: Guy M. Minard, Neenah, Wisconsin
(Member, The Conference Board; Director of the American Paper Institute)
President: Darwin E. Smith, Neenah, Wisconsin (Director, First National
Bank, Neenah, Wisconsin; Director, Cutler-Hammer, Milwaukee, Wisconsin)

Outside Directors:
Edmund B. Fitzgerald: President, Cutler-Hammer, Inc., Milwaukee, Wis-
consin
J. George Harrar: President, The Rockefeller Foundation
William A. Mulholland, Jr.: President and Chief Executive, British New-
foundland Corporation, Ltd.
Louis Quarles: Quarles, Herriott, Clemons, Teschner & Noelke, Milwaukee,
Wisconsin
James S. Rockefeller: Retired Chairman, First National Bank, New York
William P. Schweitzer: Retired Consultant
John S. Sensenbrenner: Retired Vice President
John R. Kimberly: Director, Member of the Executive Committee and
Chairman of the Finance Committee of the Board of Directors

Breakdown of Total Company Production (Percent of Sales):

Consumer products   42
Pulp and paper   38
Foreign   20

In 1969 Kimberly-Clark produced 982,000 tons of paper and 419,000 tons of
tissue [1].

Plants:
Kimberly-Clark has 34 plants in 14 states. Twenty-four of these produce pulp,
paper, and paperboard.

Timberholdings:
Kimberly-Clark owns over 1.3 million acres of land in the United States and
leases 9 million acres in Canada. The total 1969 value of Kimberly-Clark's
holdings (at cost): $50,934,000.

Annual Pulp Production (estimated on a basis of 355 production days per year)
year):
716,745 tons

Company Pollution Overview

Kimberly-Clark is the largest U.S. manufacturer of facial tissues, yet it
operates no domestic sulfite pulp mills. According to Mr. R. M. Billings,
the company director of environment control, all the old sulfite mills were
shut down because of the "tremendous pollution problems" and the fact that
only a limited variety of woods may be pulped by the acid process. He explained that Kimberly-Clark has been able to produce all its quality papers
and tissue products using recycled paper, kraft pulp, and a small amount of
purchased sulfite pulp.
     Kimberly-Clark operates 5 pulp mills: 3 old groundwood mills at Niagara
and Kimberly, Wisconsin and Niagara Falls, New York; a large, middle-aged
kraft and groundwood mill at Coosa Pines, Alabama; and a relatively new
kraft and groundwood mill at Anderson, California. These mills produce
2019 tons of pulp and 2480 tons of paper daily using 83.4 million gallons of
water.
     Although Kimberly-Clark seems to take little initiative in controlling its
mills' pollution, it appears to be most amenable and prompt about installing
whatever the states require. In an "Earth Day" speech at Wisconsin State
University, Mr. Billings outlined pollution control plans at the company's
24 pulp, paper, and paperboard mills. He said that by the end of 1972, all
water pollution abatement facilities will be completed: 16 mills now meet or
better, state legal standards. Six are constructing facilities, and two are at
the planning stage. With regard to air pollution, he said that of the eight mills
which have problems, four are installing control equipment, two have plans,

and two are "under study." He expects that all the air pollution problems will be corrected by 1975.[2]

The general tone of this evaluation of Kimberly-Clark's position was borne out, for the most part, by CEP's study of the company's five pulp mills. Although none is presently outstanding, planned equipment will bring all but the two 80-year old groundwood mills up to a high pollution control standard. These two (Kimberly and Niagara, Wisconsin) will require additional water treatment. Coosa Pines, Niagara Falls, and Anderson are concentrating on improving their air emissions.

Obviously, Kimberly-Clark has accepted pollution control as a condition of doing business. It has planned expenditures of $11,450,000 and CEP estimates that beyond this, only $2 million need be spent for complete secondary treatment at the Niagara and Kimberly mills.

Table 18.1
Kimberly-Clark Corporation: Pulp Production, Water Use, and Pollution Control

Mill Location	Pulp Prod. (TPD)	Other Prod. (TPD)	Water Use (MGPD)	Water Pollution Control			Air Pollution Control	
				Pri.	Sec.	Tert.	Part.	Gas, Odor
Coosa Pines, Ala.	1,520	1,200	40	√	√	X	X	X
Anderson, Calif.	200	150	11	√	√	X	X	√
Niagara Falls, N.Y.	100	315	8.5	√	√	X	X	—
Kimberly, Wisc.	80	440	15	X	X	X	X	—
Niagara, Wisc.	119	375	8.9	X	X	X	X	—
Total	2,019	2,480	83.4					

Coosa Pines Mill

Location:
Coosa Pines, Alabama

Date Built:
1949 (expanded 1957, 1967)

Process:
620 TPD bleached and semibleached kraft
900 TPD groundwood (company figures)

Other Production:
1,200 TPD newsprint

Total Water Use:
40 MGPD

Source:
Coosa River

Discharge:
Coosa River

Table 18.2
Coosa Pines Mill: Pollution Control Record

Treatment	Overall Evaluation	Equipment
Water		
Primary	✓	274-foot diameter clarifier (50 MGPD capacity)
Secondary	✓	Natural stabilization lagoon on 370 acres (capacity, 1.4 billion gallons; 4 week retention)
Tertiary	X	None
Solid waste removal	—	Landfill
Air		
Particulate (fuel)	X	Fuel: bark. Cyclone fly ash systems, 96% efficient (inadequate)
Particulate (production)	X	Two recovery furnace precipitators, 80% efficient (inadequate). Two lime kilns: wet scrubbers; one is 98% efficient; one is "old and erratic"

Table 18.2
Coosa Pines Mill: Pollution Control Record (continued)

Treatment	Overall Evaluation	Equipment
Gas and odor	X	None

Plant Emissions

To water:

BOD	36,000 lb/day (24 lb/ton of pulp; company figure)
Suspended solids	900 lb/day (0.6 lb/ton of pulp; company figure)
To air:	N.A.

Pollution Control Expenditures:
$4 million including $2.5 million for water treatment and $225,000 a year for operating costs.

Legal Status:
(Water) In compliance.
(Air) State standards presently being set under a new Alabama air pollution control law.

Future Plans:
Kimberly-Clark has budgeted $4.5 million to improve air and water pollution control at this mill. Between 90 percent and 95 percent of the expenditures will be for particulate and odor control.

Profile:
Coosa Pines, Kimberly-Clark's largest mill, has good water treatment although the BOD discharge could be further reduced.

Air pollution control is not adequate and Abbott Byfield, public affairs director, stated that "we anticipate the Alabama air standards will require 98 percent or better [efficiencies] by 1976, so we will have to have some improvement."

Although the mill has planned expenditures of $4.5 million, which will certainly be sufficient to improve and augment existing controls to that level, Kimberly-Clark seems in no rush to spend the budgeted money, but will wait six more years to clean up its air, although the equipment could easily be installed in two.

CEP Estimate of Minimum Necessary Pollution Control Expenditures Not in Company Budget: None.

Anderson Mill

Location:
Anderson, California (Shasta mill)

Date Built:
1964

Process:
150 TPD bleached kraft
50 TPD groundwood

Other Production:
150 TPD coated printing papers

Total Water Use:
11 MGPD

Source:
Deep wells

Discharge:
Sacramento River

Table 18.3
Anderson Mill: Pollution Control Record

Treatment	Overall Evaluation	Equipment
Water		
Primary	√	100-foot diameter mechanical clarifier; two impoundment basins with 5 MGPD capacity
Secondary	√	Activated sludge, 50 million gallon capacity; im-

Table 18.3
Anderson Mill: Pollution Control Record (continued)

Treatment	Overall Evaluation	Equipment
		poundment basins, capacity 4 million gallons
Tertiary	X	None
Solid waste removal	—	Landfill
Other	—	Spray irrigation (pilot project)
Air		
Particulate (fuel)	✓	Fuel: gas. No equipment needed
Particulate (production)	X	Two recovery furnace scrubbers 85%-95% efficient (inadequate). Lime kiln scrubber 85%-95% efficient (inadequate)
Gas and odor	✓	Evaporator condensate treatment; black liquor oxidation

Plant Emissions	
To water:	
BOD	3,170 lb/day (15 lb/ton of pulp)
Suspended solids	1,980 lb/day (9.9 lb/ton of pulp)
To air:	N.A.

Pollution Control Expenditures:
$3 million.

Legal Status:
(Water) In compliance.
(Air) In compliance.

Future Plans:
$1 million will be spent over the next five years to improve the efficiencies of the existing equipment. $850,000 of this expenditure will be for air pollution control; the rest for water treatment.

Profile:
Anderson is Kimberly-Clark's newest pulp mill. It has excellent water treatment because, according to Abbott Byfield, "construction of this mill wouldn't have been permitted by the state if we hadn't agreed to put in the very best treatment. The BOD load has been low from the start." Even the state concurred, saying, "Kimberly-Clark has a history of being a most cooperative company as far as water pollution abatement is concerned."[3]

Beyond having excellent treatment, the Anderson mill is also doing interesting research in the applicability of using sludge, which contains traces of nitrogen, potassium, and phosphorous, as fertilizer.[4] Early results show that the crops benefit and the water retention of the soil increases. Although the pilot project has not solved all the problems involved in turning this increasingly troublesome waste material into a saleable product, environmental engineer Sandy Narum said that conceivably it could be done. "But it would mean that the market for commercial fertilizers would have to change drastically."[5]

Although the Anderson mill cleaned its water in the customary manner by installing treatment facilities, it took a sorry approach to solving the air pollution problems. When the monthly particulate fallout exceeded the 50 tons per square mile limitation set by the state, Kimberly-Clark simply bought the adjoining property which was most affected by the dustfall. This kind of "ingenuity" must surely be noted as typical of evasive attitudes which have kept industry "dirty" for so many decades.

CEP Estimate of Minimum Necessary Pollution Control Expenditures Not on Company Budget: None.

Niagara Falls Mill

Location:
Niagara Falls, New York

Date Built:
1921

Process:
100 TPD groundwood

Other Production:
195 TPD coated and uncoated book paper
120 TPD cellulose wadding and tissue

Total Water Use:
8.5 MGPD

Source:
Niagara River

Discharge:
City Sewage Plant

Table 18.4
Niagara Falls Mill: Pollution Control Record

Treatment	Overall Evaluation	Equipment
Water		
Primary	✓	Joint municipal treatment facility provides primary and secondary treatment for all effluent
Secondary	✓	
Tertiary	✗	None
Air		
Particulate (fuel)	✗	Fuel: coal. No equipment
Particulate (production)	—	None needed
Gas and odor	—	None needed
Plant Emissions	N.A.	

Pollution Control Expenditures:
N.A.

Legal Status:
(Water) In compliance.
(Air) In compliance.

Future Plans:
The mill will convert to oil for fuel in 1971 at a cost of $350,000.

Profile:
As this is a small groundwood and paper mill, pollution problems are minimal. Water treatment is adequate and by the end of 1971, air control will be
also.

CEP Estimate of Minimum Necessary Pollution Control Expenditures Not
on Company Budget: None.

Kimberly Mill

Location:
Kimberly, Wisconsin

Date Built:
1889

Process:
80 TPD groundwood (company figure)

Other Production:
440 TPD coated and uncoated book paper

Total Water Use:
15 MGPD

Source:
Fox River

Discharge:
Fox River

Table 18.5
Kimberly Mill: Pollution Control Record

Treatment	Overall Evaluation	Equipment
Water		
Primary	X	None
Secondary	X	None
Tertiary	X	None

Table 18.5
Kimberly Mill: Pollution Control Record (continued)

Treatment	Overall Evaluation	Equipment
Air		
Particulate (fuel)	X	Fuel: coal, bark, gas. Coal boilers have dust collectors, 93% efficient (inadequate); bark emissions cleaned in gas boiler
Particulate (production)	—	None needed
Gas and odor	—	None needed
Plant Emissions		
To water:		
BOD	19,400 lb/day (242 lb/ton of pulp)	
Suspended solids	33,200 lb/day (415 lb/ton of pulp)	
To air:	N.A.	

Pollution Control Expenditures:
$535,400 since 1967.

Legal Status:
(Water) An abatement order was issued in December 1969 requiring either on-site or joint municipal-industrial treatment to reduce the suspended solids load to less than 20 pounds per ton of paper produced by January 1973.
(Air) New state standards define compliance deadline of 1973.

Future Plans:
$4 million will be spent to provide primary and partial secondary waste water treatment by the end of 1972. The mill also plans to convert all power boilers to gas fuel in order to meet the new state air standards.

Profile:
In the spring of 1968, Kimberly-Clark shut down a 100 TPD calcium sulfite

mill on this site because, according to the company, "it was obsolete and un-
competitive in light of new pollution control standards." Although this action
considerably reduced Kimberly-Clark's water pollution, in February, 1970,
the Wisconsin Department of Natural Resources ordered the remaining groundwood pulp and paper mill to abate its pollution by January, 1973.

As a consequence, primary and partial secondary treatment facilities are
now being planned which are expected to reduce the daily BOD discharge to
12,000 pounds, or 150 pounds per ton of pulp, and the suspended solids load
to 2,400 pounds. This would be adequate primary treatment, but is only a 40
percent BOD reduction. Full secondary treatment would result in an 80 percent to 90 percent reduction in BOD.

CEP Estimate of Minimum Necessary Pollution Control Expenditures Not
on Company Budget: $1 million total, for additional secondary treatment.

Niagara Mill

Location:
Niagara, Wisconsin

Date Built:
1889

Process:
119 TPD groundwood (company figure)

Other Production:
375 TPD coated and uncoated book paper

Total Water Use:
8.9 MGPD

Source:
Menominee River

Discharge:
Menominee River

Table 18.6
Niagara Mill: Pollution Control Record

Treatment	Overall Evaluation	Equipment
Water		
Primary	X	None
Secondary	X	None
Tertiary	X	None
Solid waste removal	—	Landfill of 900 tons per year of fibers and chips
Other	—	"White water settlers," "groundwood rejects system," "reclaimed water system"
Air		
Particulate (fuel)	X	Fuel: coal. No equipment
Particulate (production)	—	None needed
Gas and odor	—	None needed
Plant Emissions		
To water:		
BOD	17,540 lb/day (148 lb/ton of pulp)*	
Suspended solids	N.A.	
To air:	N.A.	

*State Department of Natural Resources data for 1969.

Pollution Control Expenditures:
$190,000 for in-plant equipment to reduce fiber and chip loss.

Legal Status:
(Water) Operating under a state permit.
(Air) New state standards define deadline for compliance as 1973.

Future Plans:
Kimberly-Clark is planning a $1.3 million water treatment facility which will include a 150-foot diameter reactor clarifier and lagoons for dewatering the sludge. The first stage (a sewer system to collect and centralize all mill effluent) was completed in 1969. The company will also spend $300,000 to convert three of the four coal boilers to natural gas by 1972.

Profile:
This 80-year-old groundwood and paper mill was cited at the 1963 Interstate Michigan-Wisconsin Water Pollution Conference as a principal source of organic wastes and solids in the Menominee River, and was required to stop discharging wood chips and fibers. Partial control was achieved by installation of in-plant equipment; and a clarifier is now under construction.

The company estimates that primary treatment will reduce the mill's BOD discharge to 11,000 pounds per day or 92 pounds per ton of pulp. This is definitely not adequate and secondary treatment will most probably be required in the future.

CEP Estimate of Minimum Necessary Pollution Control Expenditures Not on Company Budget: $1 million total, for secondary water treatment.

References for Chapter 18

1. Chem. 26, June, 1970, p. 32.

2. Speech at Wisconsin State University, Oshkosh, Wisconsin, "Earth Day," April 22, 1970.

3. Questionnaire from Shasta County, California.

4. Pulp and Paper, April, 1970, pp. 117-118.

5. Ibid.

Chapter 19

MARCOR, INCORPORATED (Container Corporation of America)
One First National Plaza, Chicago, Illinois 60670

Major Products:
Retailing, paperboard and paper bags, folding cartons, finished packaging products.

Major Consumer Brands:
MONTGOMERY WARD products.

Financial Data ($ Millions)	1970	1969	1968
Net Sales	2715.2	2500.7	2352.3
Net Income	66.9	53.8	37.4
Capital Expenditures	136.3	99.7	96.0

Pollution Control Expenditures:
$ 19,000,000 from 1960-1970

Annual Meeting:
The annual meeting is held during May in Chicago.

Officers:
Chairman: R. E. Brooker, Winnetka, Illinois
President: Leo H. Schoenhofen, Barrington, Illinois (Director: Banker's
Life Co., Honeywell, Inc., Northern Trust Co., U.S. Gypsum Co.)

Breakdown of Total Company Sales (Percent, 1970):

Shipping containers	46
Cartons, cans, bags and plastics	37
Paper, pulp, waste products and other	17

(Container Corporation accounted for 18.8 percent of corporate sales.)

Plants:
Container Corporation of America, one component of Marcor, has 82 domestic plants. (Montgomery Ward, the other component of Marcor, operates nearly 500 stores around the nation.)

Timberholdings:
Container Corporation owns, leases, or has cutting rights to 779,000 acres of timberland of which 74,000 acres were bought or leased in 1969. In addition, the company has equity of 49 percent in the T. R. Miller Co., Inc.,

which owns 194,000 acres. The total value (1970) of Container Corporation's holdings is $51,837,000.

Annual Pulp Production:
670,950 tons (1969)

Company Pollution Overview

Of Container Corporation's 82 U.S. manufacturing plants, 14 are paperboard mills. Ten of these use only waste paper, and another—at Carthage, Indiana— is currently being converted to use only waste paper. Fresh pulping is carried on at only three locations: two large kraft mills in the South (Brewton, Alabama and Fernandina Beach, Florida), and one medium-sized NSSC mill in the Midwest (Circleville, Ohio). These three mills have a total daily pulp production of 1,890 tons and use 70 million gallons of water per day.

All three mills have been cited for water pollution by the state or federal pollution control boards: Circleville and Fernandina Beach in 1967, Brewton in 1970. None has yet installed the required level of secondary water treatment. Fernandina Beach will not even have primary treatment until 1971.

Not one of the mills now has adequate air pollution control either, but no state citations have been issued, as Alabama, Ohio, and Florida have not even established air standards yet. Nonetheless, Container Corporation has begun extensive overhaul of both kraft mill recovery operations; plans include new furnaces, new precipitators, and indirect contact "low odor" evaporators. The company has budgeted $24 million for the new systems, which should be completed by 1972 and should provide excellent air pollution control.

The Circleville mill is fueled by high sulfur coal and oil which result in heavy smoke and sulfur dioxide emissions. The company doesn't plan to attack this problem for another two years; the laws will not be in force until then. A company representative added that no source of low-sulfur fuel is presently available at a feasible cost.

It is somewhat difficult to understand how Container Corporation has been able to embark on a $70 million expansion of the Fernandina Beach mill, which is, according to the 1969 annual report, the largest single capital expenditure in the company's history, yet be willing to leave an air pollution problem such as the one at Circleville hanging for another two years.

The company has budgeted $24 million for pollution controls at its three pulp mills. The Council estimates that an additional $4.25 million will be required, bringing the total to $28.25 million.

Table 19.1
Marcor, Incorporated (Container Corporation of America): Pulp Production,
Water Use, and Pollution Control

Mill Location	Pulp Prod. (TPD)	Other Prod. (TPD)	Water Use (MGPD)	Water Pollution Control			Air Pollution Control	
				Pri.	Sec.	Tert.	Part.	Gas, Odor
Brewton, Ala.	800	800	38	✓	✓	X	X	X
Fernandina Beach, Fla.	850	830	30	X	X	X	X	✓
Circleville, O.	240	240	2	✓	X	X	X	X
Total	1,890	1,870	70					

Brewton Mill

Location:
Brewton, Alabama

Date Built:
1957

Process:
800 TPD bleached kraft

Other Production:
800 TPD kraft board, bleached food board, and unbleached linerboard

Total Water Use:
38 MGPD

Source:
Conecuh River

Discharge:
Conecuh River (which becomes the Escambia River in Florida and flows into
Escambia Bay off the Gulf of Mexico)

Table 19.2
Brewton Mill: Pollution Control Record

Treatment	Overall Evaluation	Equipment
Water		
Primary	✓	Clarifier and settling lagoon
Secondary	✓	Series of natural oxidation lagoons where free oxygen is added to the waste water. System added in 1970 achieves 80% removal of organic wastes (state findings)
Tertiary	X	None
Air		
Particulate (fuel)	✓	Fuel: gas. No equipment needed
Particulate (production)	X	Recovery furnaces have old and inefficient electrostatic precipitators
Gas and odor	X	None
Plant Emissions	N.A.	

Pollution Control Expenditures:
$50,000 for the free oxygen water treatment system.

Legal Status:
(Water) A January, 1970, federal abatement conference requested by the State of Florida cited the mill as one of three major contributors of carbonaceous water pollution to interstate waters; and required 90 percent removal of its carbonaceous (organic) waste and improvement of effluent color by December 31, 1972.
(Air) No state regulations implemented yet.

Future Plans:
A $10 million expenditure is planned which will include installation of a new recovery furnace with a "recently designed dual chamber (dry bottom construction) electrostatic precipitator that will... effectively remove 99.5 percent of the particulate matter in the atmospheric discharge..." and a Swenson indirect contact evaporator. A company representative said that this equipment will begin operation in mid-1972.

Profile:

The summary of the first session of the conference on Pollution of the Waters of the Escambia River Basin (Alabama-Florida) January 20-21, 1970 states:

"The entire upper section of Escambia Bay north of the Louisville and Nashville Railroad Company bridge is in a state of accelerated eutrophication as shown by unstable dissolved oxygen variations resulting from algal activity; high carbon, nitrogen, and phosphorous concentrations; and oxygen demanding sludge deposits."

Container Corporation of America is cited as one of the three major dischargers of carbonaceous (organic) wastes to the river and is required, under Provision 10 of the Federal Water Pollution Control Act, to reduce these wastes by 90 percent and to provide complete removal of settleable solids by no later than December 31, 1972.

Container Corporation is also cited for discoloring the Escambia River with its discharge of treated wastes, and is instructed that the "color in the Escambia River ... shall be reduced to levels meeting Alabama, Florida, and Federal Standards."

CEP Estimate of Minimum Necessary Pollution Control Expenditure Not on Company Budget: $500,000 total, for improved effluent treatment.

Fernandina Beach Mill

Location:
Fernandina Beach, Florida

Date Built:
1937

Process:
850 TPD unbleached kraft

Other Production:
830 TPD unbleached kraft liner and container board

Total Water Use:
30 MGPD

Source:
Wells

Discharge:
Amelia River

Table 19.3
Fernandina Beach Mill: Pollution Control Record

Treatment	Overall Evaluation	Equipment
Water		
Primary	X	None
Secondary	X	None
Tertiary	X	None
Air		
Particulate (fuel)	X	Fuel: bark and oil. Cyclone collectors
Particulate (production)	X	New recovery furnace with electrostatic precipitator installed 1970, 99.5% efficient; one old furnace with an old precipitator
Gas and odor	✓	"Uni-Tech" indirect contact evaporators
Plant Emissions		
To water:		
BOD	20,000 lb/day (24 lb/ton of pulp; state data)	
Suspended solids	N.A.	
To air:	N.A.	

Pollution Control Expenditures:
N.A.

Legal Status:
(Water) This mill, as well as a nearby one owned by ITT Rayonier, was cited by the Florida State Department of Air and Water Pollution Control in November 1967 for water pollution. It was required to provide 90 percent BOD removal by March 1971. The mill has submitted plans, now approved by the state.

(Air) New state regulations under development.

Future Plans:
A 250-foot diameter clarifier with aerators is now being built. This and in-plant changes to increase the amount of water reused should improve the quality of the effluent.

The mill is currently undergoing a major expansion that mill manager Franklin Jones defined as the largest which Container Corporation has ever embarked upon. It will involve an expenditure of over $70 million, of which, according to the 1969 report, $14 million (20 percent) will be invested in equipment for air and water pollution control. The equipment includes: a new recovery furnace and precipitators, a new lime kiln, a water recycling system, the 250-foot aerated clarifier, fiber and chemical recovery systems, and a water cooling tower. Of this $14 million, perhaps as much as $10 million will be spent on the new recovery furnace and lime kiln, which are basic production equipment.

Profile:
In a WJXT News Special Report aired on April 2, 1970, in Jacksonville, Franklin Jones outlined the future actions of his mill. "We plan to install certain in-plant treatments of [the] water to remove suspended material. In that plant we expect to take care of certain spills that occur periodically and reclaim them." In response to a question about the condition of the river, he said:

"The condition of the river could stand improvement ... I suspect that on some occasions there have been some effects on marine life along the river by people who live along the river. I don't know to what degree that this has been done, because we have really no yardsticks to compare with." [1]

It is interesting to compare this statement of the water quality in the Amelia River with that written by William Curtis Lovelace in a letter to the U.S. Corps of Engineers.

"This river has suffered several major fish kills in recent months and marine biologists have stated it is useless to take oxygen samples from the water." [2]

In another WJXT broadcast, aired April 3, 1970, shrimp fisherman Linton Cook described the condition of the marine life in the river. "I've caught fish and you couldn't eat them for the turpentine in them." Another shrimper, Tony Tringali, explained the conflict of interests in the town, and in fact, in the nation.

"Well naturally the people that's working at the mills will say that it's not hurting the shrimp or the fish, but they're working there so they got to protect their job. You know. But as far as the shrimp fishermen, we know that the pollution has hurt the fish and the shrimp here." [3]

CEP Estimate of Minimum Necessary Pollution Control Expenditure Not on Company Budget: $500,000 total, for improved power boiler particulate control.

Circleville Mill

Location:
Circleville, Ohio

Date Built:
1882

Process:
240 TPD NSSC

Other Production:
240 TPD corrugating medium

Total Water Use:
2 MGPD

Source:
Wells

Discharge:
Scioto River

Table 19.4
Circleville Mill: Pollution Control Record

Treatment	Overall Evaluation	Equipment
Water		
Primary	✓	2 million gallon capacity clarifier
Secondary	X (partial)	No equipment, but spent liquor is not discharged (in-plant recovery of liquor and fibers). Liquor is treated, using the organic wastes as fuel and isolating the inorganic contents for reuse by Container Corporation kraft mills and local glass manufacturers. (This system was developed independently by Container Corporation)

Table 19.4
Circleville Mill: Pollution Control Record (continued)

Treatment	Overall Evaluation	Equipment
Tertiary	X	None
Solid waste removal	—	Sludge used as landfill
Air		
Particulate (fuel)	X	Fuel: high sulfur coal and oil. Scrubber (inadequate)
Particulate (production)	√	Recovery furnace, two-stage scrubbing.
Gas and odor	—	None needed
Plant Emissions	N.A.	

Pollution Control Expenditures:
$2 million (for liquor disposal system).

Legal Status:
(Water) This mill was cited by the Ohio Pollution Control Board in 1967 for high BOD level, high settleable solids level, and objectionable color. Its current permit requires improved water treatment.
(Air) No state standards yet. However, the State Control Board reported receiving complaints from local residents and health department for smoke and sulfur dioxide emissions.

Future Plans:
Container Corporation is working with the city of Circleville to plan a joint water treatment facility which will provide secondary water treatment. An activated sludge system for the mill is also being investigated. There are no definite plans for air pollution control despite the problems from high sulfur emissions because, "the state hasn't got air control laws yet," so, "we won't have to come up against that one until 1972." A company representative added, "low-sulfur fuels are not presently available to us for use at Circleville at costs that would permit continued operation of the mill. When they do become available it is our stated position to implement appropriate action to alleviate the problem."

    CEP Estimate of Minimum Necessary Pollution Control Expenditures Not on Company Budget: $3.5 million total, of which $3 million is for secondary water treatment (activated sludge or joint plant) and $500,000 for improved power boiler particulate control.

References for Chapter 19

1. WJXT News Special Report, Jacksonville, Florida, aired April 2, 1970, p. 2 of script. (Script sent to CEP by the Southeastern Environmental Council, P.O. Box 31278, Yukon Branch, Jacksonville, Florida 32230.)

2. Letter sent to Colonel A. S. Fullerton, Engineer, U.S. Corps of Engineers, from Wm. Curtis Lovelace, member of the Environmental Advisory Committee, U.S. House of Representatives, June 24, 1970.

3. WJXT News Special Report, p. 4.

Chapter 20

THE MEAD CORPORATION
118 West First Street, Dayton Ohio 45402

Major Products:
White paper, paperboard, educational products, pipes and fittings.

Major Consumer Brands:
WESTAB (manufactured by a subsidiary) notebooks, book covers; MONTAG stationary.

Financial Data ($ Millions):	1969	1968	1967
Net Sales	1,031.7	952.7	633.1
Net Income	36.0	34.0	20.4
Capital Expenditures	61.3	62.6	29.3

Pollution Control Expenditures:
Over $ 31 million, 1960-1970, plus about $5 million per year in operating costs.

Annual Meeting:
The annual meeting is held during March in Dayton, Ohio.

Officers:
Chairman/President: James W. McSwiney, Dayton, Ohio (Director: Winters National Bank and Trust, Gem City Savings Association, British Columbia Forest Products, Ltd.; Director and Senior Vice President, Northwood Pulp Ltd.; President and Director, Brunswick Pulp and Paper)

Outside Directors:
Vernon R. Alden, President, Ohio University
Newton H. DeBardeleben, President, First National Bank of Birmingham
Arthur L. Harris, President, Scripto, Inc.
Alfred W. Jones, Chairman of the Board, Sea Island, Inc.
R. Stanley Laing, President, National Cash Register
John J. McDonough, Former Chairman of the Board, Georgia Power Co.
Paul F. Miller, Jr., President, Drexel, Harriman, Ripley, Inc.
George H. Pringle, Retired President, The Mead Corp.
William M. Spencer III, President, Owens Richards Co.
C. William Verity, Jr., President, Armco Steel Corp.
John M. Walker, Managing Partner, G. H. Walker and Co.
H. E. Whitaker, Retired Chairman of the Board, The Mead Corp.

Breakdown of Total Company Production (Percent of Sales, 1969):

Paper	20
Paperboard and converting	22
Wholesale paper and paper products	23
Educational products	11
Metals, construction materials and precision castings	20
Interior furnishings	4

Plants:
Mead has 117 domestic production facilities:

Paper	13
Board and packaging and container	43
Educational products and school supplies	21
Metals, construction materials and precision casting	30
Interior furnishings	10

Timberholdings:
Mead and its affiliates (Brunswick Pulp and Paper Co., Northwood Pulp Ltd., and British Columbia Forest Products, Ltd.) own and lease more than 2.25 million acres of timberland as well as holding cutting rights to more than 7 million acres of Canadian Crown lands. The total 1969 value of Mead's holdings was $38,556,851 (including land and mineral reserves).

Annual Pulp Production (estimated on a basis of 355 operating days per year): 594,625 tons (1969)

Company Pollution Overview

Mead Corporation has six pulp mills: four medium-sized, middle-aged NSSC and sulfite mills in the Southeast (Tennessee, Virginia and North Carolina), and a small old groundwood mill and new, larger kraft mills in the Midwest (Escanaba, Michigan and Chillicothe, Ohio). In addition it is joint owner with Scott, of a 1,100 ton per day kraft mill in Brunswick, Georgia (see Scott profile). These mills produce 2225 tons of pulp and discharge 130.9 million gallons of effluent daily. Georgia-Kraft Corporation, Mead's board-making subsidiary, operates three kraft mills in the South which have not been included in this study.

From the fall of 1959 through the spring of 1970, Mead Corporation battled the citizens of Escanaba, Michigan for authorization to build a $65 million kraft mill in the town. The citizens determinedly kept the permit from going through until every possible legal requirement for optimal pollution control was agreed upon in writing. Mead now proudly claims that the new mill will be one of the cleanest, if not THE cleanest, in the United States. When the company decided that growth and profit necessitated the new Escanaba facility, management found the millions of dollars to finance it. When the company found the construction obstructed, the technical director focused his attention on pollution control. The results will provide a rather strong contrast to the conditions of the company's other mills where no such energy has been channelled into cleaning up the pollution.

At its existing mills, the company's greatest effort seems to have gone into publicizing the difficulty and expense of its few accomplishments. At the March 1970 Annual Meeting, Mead President James McSwiney dwelt on his company's "long-standing identification with public concern" (about pollution control), and the $31 million spent during the past decade. He also explained, "It is essential to recognize that however much we might like to attack all aspects of pollution at once, technological knowledge and economic consideration will force a timetable and an ordering of pollution abatement goals." Technology clearly is not the problem, and Mr. McSwiney might well reassess the impressiveness of an annual pollution control expenditure of $3.1 million in comparison with the $65 million summoned up for the new kraft mill.

Mead's old mills are obviously not top priority. Only one of these — the small groundwood mill in Michigan — has no air pollution problem. The other five all emit excessive gas and particulate matter from coal and oil boilers which are controlled only by mechanical dust collectors. Chillicothe does have black liquor oxidation to reduce odors. However, the kraft mill Mead owns jointly with Scott at Brunswick, Georgia, does not. (See the Scott company analysis for more details on this mill.)

As for water treatment, the little groundwood mill in Escanaba, Michigan is the only one with both primary and secondary treatment. However, when these facilities broke down last year, the mill continued to produce. As a result, parts of the Escanaba River now have no life in them at all. Two of the other mills have primary treatment — Chillicothe, Ohio, and Kingsport, Tennessee — and the Chillicothe mill has constructed a second aerated lagoon to provide adequate secondary treatment. The facility, required by the state in 1967, will be financed with tax-free state government revenue bonds via "a new concept in pollution financing which Ohio is pioneering." The state will assume ownership of the property during the construction period so that the company payments may be tax exempt.

Three other mills — Sylva, North Carolina, Lynchburg, Virginia, and Harriman, Tennessee — have virtually no water treatment, although pollution at Sylva and Lynchburg has been reduced by installation of spent liquor recovery systems. After 1967 state recommendations for improve treatment at all three mills, and follow-up implementation hearings, Mead reports only that studies are underway, but no plans have been finalized. A relevant comparison can be drawn between the three years which have elapsed in these cases and the brief two year period which Mead has alloted for complete construction of its new mill.

Company Public Relations Director Richard Lowe wrote CEP "I want you to know that all our mills are either meeting pollution control standards or

are on schedule with regulatory approved programs to attain the standards set
in the very near future." This statement, in light of the "mill facts," is not
reassuring, but rather, indicates that the states, with their obliging issuance
of permits and flexible deadlines, are following the company's lead: two steps
forward and one step back.

Mead has budgeted about $14 million for pollution control; $12 million at
Escanaba and $2 million at Chillicothe. CEP estimates that an added $17 million will be required to clean up all six mills (including half of the $10.25 mil-
lion needed at Brunswick, Georgia) for a total of $31 million.

Table 20.1
The Mead Corporation: Pulp Production, Water Use, and Pollution Control

Mill Location	Pulp Prod. (TPD)	Other Prod. (TPD)	Water Use (MGPD)	Water Pollution Control			Air Pollution Control	
				Pri.	Sec.	Tert.	Part.	Gas, Odor
Brunswick,* Ga.	550	125	37.5	✓	X	X	?	X
Escanaba, Mich.	100	280	10	✓	✓	X	✓	—
Sylva, N.C.	300	300	5.5	X	X	X	X	—
Chillicothe, O.	600	600	38	✓	✓	X	X	✓
Harriman, Tenn.	185	190	7.9	X	X	X	X	—
Kingsport, Tenn.	300	550	20	✓	X	X	X	—
Lynchburg, Va.	190	375	12	X	X	X	X	—
Total	2,225	2,420	130.9					

*Half owned by Scott, half owned by Mead; production divided weekly between
Scott and Mead (the water use and production figures for this plant are divided
between the two companies). For mill profile, see Scott (Chapter 25).

Escanaba Mill

Location:
Escanaba, Michigan

Date Built:
1912

Process:
100 TPD groundwood

Other Production:
280 TPD coated converting and printing paper

Total Water Use:
10 MGPD

Source:
Escanaba River

Discharge:
Escanaba River

Table 20.2
Escanaba Mill: Pollution Control Record

Treatment	Overall Evaluation	Equipment
Water		
Primary	✓	180-foot diameter clarifier
Secondary	✓	30-acre aerated lagoon
Tertiary	X	None
Solid waste removal		Clarifier sludge used for landfill
Air		
Particulate (fuel)	✓	Fuel: gas. No equipment needed
Particulate (production)	—	No chemicals used; none needed

Table 20.2
Escanaba Mill: Pollution Control Record (continued)

Treatment	Overall Evaluation	Equipment
Gas and odor	—	None needed
Plant Emissions	N.A.	

Pollution Control Expenditures:
Total of $2 million, 90 percent spent in the last three years.

Legal Status:
(Water) In compliance.
(Air) In compliance.

Future Plans:
In mid-1969 Mead announced plans to build a kraft mill at Escanaba. The citizens of the town rose to arms. A group calling itself the Delta County Citizen's Committee to Save our Air sent representatives to the meetings of the State Pollution Control Board and the company. Mead considered the addition of the new mill vital, as it would supply all the pulp needed for Escanaba's paper production, which heretofore had to be largely purchased from Canadian mills.

Mead promised to install the most effective air control equipment available, but the citizen's group was not satisfied. At one of the hearings, the group showed samples of the Escanaba River to justify its suspicions concerning Mead's good intentions. Group leader John Walbridge, a local businessman, explained: "Mead touted a new water pollution control system installed last fall, but it broke down when the river froze, and still the mill goes on operating without it. Not even a sludge worm could live in that river now." He questioned what would happen to the air next if Mead were allowed to build the kraft mill.

Mead guaranteed that its new mill would have as good air control as the new American Can Company mill in Halsey, Oregon, considered one of the best in the United States. Mead, in fact, claimed that the Halsey mill was virtually odor free. Still, when the group sent a representative to inspect the mill, he returned saying it was good but certainly not "odor free." The odor was proving especially annoying to residents living downwind from the mill.

Mead then said that it would "do better." Delta County citizens, still skeptical of promises, before agreeing to the construction, pushed through a special local Air Pollution Ordinance, creating a nine-member control commission to review all construction plans for a new proposed plant in the county and to enforce, without any allowable industrial variances, air standards higher than those of the state. At this mill, an "alert" limit was set at three parts per million of hydrogen sulfide in the recovery furnace stack gases. If the gas emissions ever reach 10 ppm and cannot be reduced within six hours, the furnace responsible would be shut down until the company is able to operate it within regulations again. Continuous monitoring of particulate and gas emissions is also required.

On April 7, 1970, when Mead agreed to sign a pledge to comply with the new local standards, the company finally received permission to build. $12 million of the total $65 million capital cost for the mill will go into pollution control equipment (half for air, half for water). The kraft recovery operation will include a Babcock and Wilcox furnace which assures hydrogen sulfide emissions under 1 ppm, with an indirect contact evaporator, a dual chamber dry bottom electrostatic precipitator (if one chamber malfunctions, the other will provide a temporary replacement), and high efficiency scrubbers. The effluent will be screened, and then go through a clarifier and a ten-day retention aerated lagoon. It will then be measured before discharge into the Escanaba River.

Mead Vice President said the mill will be a "model, a showplace," which indeed the citizens have insured it must be if it is to be allowed to operate.

Sylva Mill

Location:
Sylva, North Carolina

Date Built:
1928

Process:
300 TPD unbleached ammonia based sulfite

Other Production:
300 TPD paperboard for corrugating

Total Water Use:
5.5 MGPD

Source:
Scotts Creek

Discharge:
Tuckaseegee River

Table 20.3
Sylva Mill: Pollution Control Record

Treatment	Overall Evaluation	Equipment
Water		
Primary	X	None
Secondary	X (partial)	None. However, spent pulping liquor is burned and not discharged into river
Tertiary	X	None
Air		
Particulate (fuel)	X	Fuel: coal, bark, some oil. Two boilers, both have mechanical dust collectors; liquor burned in these boilers
Particulate (production)	—	None needed
Gas and odor	—	None needed
Plant Emissions	N.A.	

Pollution Control Expenditures:
In 1968 the mill converted from sodium to ammonia base sulfite liquor, and
$3 million was spent for liquor burning equipment, to avoid discharging waste
liquor into the Tuckaseegee River.

Legal Status:
(Water) The February 9, 1967 hearings [1] of the North Carolina State Stream
Sanitation Committee cited the mill for "still discharging untreated or sub-
stantially untreated paper mill waste," and indicated that the mill had been
given a schedule for developing primary and secondary treatment.
(Air) Law brand-new, not implemented yet.

Future Plans:
The mill is reducing water use to 3 million gallons per day by in-plant changes.
Not treatment plans are finalized.
    As at Mead's Lynchburg mill, the water pollution problem at Sylva has
been substantially reduced by changing pulping liquor to ammonia which can
be disposed of by burning. However, the mill, despite a 1967 state order,
has no external water treatment or plans to provide it, nor does it have adequate soot and gas controls on its fuel boilers.

CEP Estimate of Minimum Necessary Pollution Control Expenditures Not on Company Budget: $2.75 million total, of which $250,000 is for improved power boiler particulate controls; $1 million is for primary water treatment; and $1.5 million is for secondary treatment.

Chillicothe Mill

Location:
Chillicothe, Ohio

Date Built:
1963

Process:
600 TPD bleached kraft

Other Production:
600 TPD book (coated and machine coated), duplicator, ledger, mimeograph, sales book, bond, writing papers

Total Water Use:
38 MGPD

Source:
Underground wells

Discharge:
Paint Creek (flowing two miles into Scioto River)

Table 20.4
Chillicothe Mill: Pollution Control Record

Treatment	Overall Evaluation	Equipment
Water		
Primary	✓	210- by 80-foot clarifier, 80% solids removal
Secondary	✓	Aerated lagoon
Tertiary	X	None
Solid waste removal		Clarifier sludge used for landfill

Table 20.4
Chillicothe Mill: Pollution Control Record (continued)

Treatment	Overall Evaluation	Equipment
Air		
Particulate (fuel)	X	Fuel: coal. Three boilers with mechanical collectors
Particulate (production)	X	Two recovery furnaces with electrostatic precipitators (one over 90% efficient, one under 90%)
Gas and odor	✓	Black liquor oxidation; incineration of digester gases
Plant Emissions	N.A.	

Pollution Control Expenditures:
Mead indicates a total of $4.5 million spent on controls, including $1 million for a clarifier and $1 million for black liquor oxidation.

Legal Status:
(Water) In 1967 the state required improved secondary water treatment. Mill now in compliance.
(Air) No relevant standards implemented yet.

Future Plans:
A second ten-acre aerated lagoon is to be built by 1974 at an estimated cost of $2 million.

Profile:
To "lighten the company's financial burden" of adding the required water treatment at this mill, Mead arranged to transfer ownership of the proposed 10-acre lagoon to the Ohio Water Development Authority. Its construction may then be financed via government revenue bonds, saving Mead the high interest rate the company would have to pay. Once the lagoon is built, ownership will be transferred back to Mead.
Mead's 1969 annual report stated: "The new water facility will be the first to take advantage of a new concept in pollution financing which Ohio is pioneering." Even with the new water facilities, Chillicothe will not be perfect. The mill has a very tall stack for dispersion of gaseous emission, but particulate controls—both on the fuel boilers and recovery furnaces—still need considerable improvement.

CEP Estimate of Minimum Necessary Pollution Control Expenditures Not on Company Budget: $1 million total, of which $500,000 is for improved power boiler particulate control and $500,000 is for an electrostatic precipitator.

Harriman Mill

Location:
Harriman, Tennessee

Date Built:
1929

Process:
185 TPD NSSC pulp

Other Production:
190 TPD paperboard for corrugating

Total Water Use:
1.9 MGPD

Source:
Emory River

Discharge:
Tennessee River

Table 20.5
Harriman Mill: Pollution Control Record

Treatment	Overall Evaluation	Equipment
Water		
Primary	X	Joint partial primary treatment plant with the city of Harriman; waste liquor discharged in effluent to plant
Secondary	X	None. But pilot project for joint secondary treatment underway with FWQA grant
Tertiary	X	None
Solid waste removal	—	Screening of effluent before it leaves mill

Table 20.5
Harriman Mill: Pollution Control Record (continued)

Treatment	Overall Evaluation	Equipment
Other	—	Pipe to Tennessee River
Air		
Particulate (fuel)	✕	Fuel: oil. No equipment
Particulate (production)	—	No liquor burning, therefore no air pollution problems here
Gas and odor	—	None needed
Plant Emissions	N.A.	

Pollution Control Expenditures:
N.A.

Legal Status:
(Water) At the February 9, 1967 hearings, [2] the Tennessee Stream Pollution Control Board noted that the mill's waste, consisting of spent liquor and amounts of bark and wood fiber, was only being screened at the mill, and then given partial primary treatment at the city sewage plant. There were also complaints resulting from "wastes spilling from the effluent line due to leaks." The board recommended that city and mill officials begin planning for secondary treatment.
(Air) Location not yet given area air classification. So far, state only has Ringleman standard.

Future Plans:
Water treatment study underway.

Profile:
The Harriman mill is the main industry in this community. The town and the mill have had problems in transporting waste water from the Harriman sewage plant to the Tennessee River because the hydrogen sulfide in the pulp mill effluent has damaged and corroded the pipes. No speedy action has been taken since the 1967 hearings. Three years later Mead reported only that an FWQA pilot project was underway to determine design criteria for building a full-scale joint secondary treatment plant.

CEP Estimate of Minimum Necessary Pollution Control Expenditure Not on Company Budget: $2.25 million total, of which $250,000 is for improved power boiler particulate control and $2 million is for improved primary and secondary treatment.

Kingsport Mill

Location:
Kingsport, Tennessee

Date Built:
1917

Process:
300 TPD bleached soda pulp

Other Production:
550 TPD fine white printing paper, machine finish book, tablet, envelope, bond, lithograph. "Second largest papermaking operation in the company"

Total Water Use:
20 MGPD

Source:
Holston River

Discharge:
Holston River — South Fork (interstate stream)

Table 20.6
Kingsport Mill: Pollution Control Record

Treatment	Overall Evaluation	Equipment
Water		
Primary	√	Clarifier installed in 1965, achieves 96% solids removal and 30% BOD reduction
Secondary	X	None
Tertiary	X	None

Table 20.6
Kingsport Mill: Pollution Control Record (continued)

Treatment	Overall Evaluation	Equipment
Air		
Particulate (fuel)	X	Fuel: coal. Power boilers (built in 1948) equipped with dust collectors
Particulate (production)	X	Recovery furnace has old (1948) precipitator and scrubber (over 90% efficient)
Gas and odor	—	—
Plant Emissions	N.A.	

Pollution Control Expenditures:
$150,000 spent for electrostatic precipitator (1948) and $500,000 for clarifier (1965).

Legal Status:
At the February 1967 public hearings,[3] the Tennessee State Stream Pollution Control Board named the mill as one of the major users of Holston River water and stated that the mill was "still discharging considerable amounts of waste water characterized by high organic strength and high levels of suspended and dissolved solids, in spite of 1961 state recommendations." The board indicated more water treatment was needed by the mill.

At the November 1969 implementation and enforcement meetings, the board labelled Mead's water treatment "unsatisfactory," and required a plan for secondary water treatment by April 1970, beginning of construction October 1970, and operation by April 1972, to achieve a BOD limit of 6,000 pounds (20 pounds per ton of pulp).
(Air) N.A.

Future Plans:
None finalized (as of October 1970).

Profile:
At the 1967 hearings a Mead representative said, "I want to assure you that we believe in stream improvement and mean to continue as we have in the past to exert every effort we can within the limits set for us by the economics of our industry and our company to help accomplish it." Nevertheless, after the 1967 recommendation, and a 1969 state implementation order, the mill

has made no plans for the required treatment and is now months behind the state deadlines.

The old 1948 particulate controls on the coal boilers and recovery furnaces are inadequate.

CEP Estimate of Minimum Necessary Pollution Control Expenditures Not on Company Budget: $2.75 million total, of which $250,000 is for improved power boiler particulate control; $500,000 is for an electrostatic precipitator for a recovery furnace; and $2 million is for secondary water treatment.

Lynchburg Mill

Location:
Lynchburg, Virginia

Date Built:
1929

Process:
190 TPD unbleached ammonia-based sulfite

Other Production:
375 TPD paperboard for corrugating, laminated board, polyethylene coated board

Total Water Use:
12 MGPD

Source:
James River

Discharge:
James River

Table 20.7
Lynchburg Mill: Pollution Control Record

Treatment	Overall Evaluation	Equipment
Water		
Primary	X	None
Secondary	X (partial)	None; however, all waste liquor is burned and not discharged into river

Table 20.7
Lynchburg Mill: Pollution Control Record (continued)

Treatment	Overall Evaluation	Equipment
Tertiary	X	None
Air		
Particulate (fuel)	X	Fuel: coal. Boiler also used to burn waste liquor; equipped with mechanical dust collector
Particulate (production)	—	None needed, since liquor is burned in fuel boiler
Gas and odor	—	
Plant Emissions	N.A.	

Pollution Control Expenditures:
Twelve years ago, the mill installed a chemical recovery system (pulp washer, evaporator and furnace) for $3 million. The mill was pulping with sodium liquor, and during burning, the chemicals converted to corrosive sulfuric acid, which gradually ate the furnace away. In 1966 the mill converted to use of ammonia-based liquor and invested another $1 million for a new furnace. Liquor is now burned, not discharged.

Legal Status:
(Water) The mill has been ordered by the state to have both primary and secondary water treatment by July, 1972. Currently the city and the mill are working on a joint treatment plan. A company representative said, "We are behind on our schedule, but we are committed to meeting the '72 deadline."
(Air) State agency policy is to provide no information to the public.

Future Plans:
The company gave no details on the water treatment facilities planned, but did say it would meet state deadlines.

Profile:
Although the Mead Corporation has invested $4 million in spent liquor recovery at this mill, still no effluent treatment is provided, and particulate control on the fuel boiler is inadequate.

CEP Estimate of Minimum Necessary Pollution Control Expenditures Not on Company Budget: $3.25 million total, of which $250,000 is for improved power boiler particulate control; $1 million is for primary water treatment; and $2 million is for secondary treatment.

References for Chapter 20

1. Public Hearings on Water Uses, Stream Standards, and Implementation Plans for Interstate Streams for Tennessee, Vol. IV, Hearing No. 5, Knoxville, Tennessee, p. 45.

2. Ibid., pp. 5, 7.

3. Ibid., Vol. II, pp. 10, 18.

Chapter 21

OWENS-ILLINOIS
405 Madison Avenue, Toledo, Ohio 43604

Major Products:
Glass containers, paper, paper dinnerware.

Major Consumer Brands:
LIBBEY glasses and tableware, LILY paper cups, plates and bowls.

Financial Data ($ Millions):	1969	1968	1967
Net Sales	1,294.4	1,152.2	1,085.2
Net Income	69.7	55.1	52.8
Capital Expenditures	83.9	99.5	145.8

Pollution Control Expenditures:
$15 million for four container board mills.

Annual Meeting:
The annual meeting is held during April in Toledo, Ohio.

Officers:
Chairman: R. H. Mulford
President: Edwin D. Dodd

Outside Directors:
E. C. Arbuckle: Chairman, Wells Fargo Bank
Walter J. Bergman
Arthur S. Bowes
Harry E. Collin: Associate, Clark, Dodge and Co.
G. A. Costanzo: Executive Vice President, First National City Bank, New
York City
Max M. Fisher: Chairman, Fisher-New Center Co. and Safran Printing Co.
A. J. Hettinger, Jr.: Partner, Lazard, Frères and Co.
John A. Hill: Management Consultant, Hartford, Connecticut
C. E. Janssen: Vice Chairman, Societe Generale de Banque S.A.
Robert H. Levis II: Chairman, First National Bank and Trust Co.
Harris McIntosh
Kenneth S. Pitzer: President, Stanford University
R. L. Snideman

Breakdown of Total Company Production (Dollar Value and Percent of Sales, 1969):

Glass Container	$ 589 million	46%
Forest Products	241	19
International Division	173	13
Consumer and Tech. Products Division	147	11
Lily-Tulip Division	138	11
Plastics Products	60	5
Other	27	2
	$1,294	100%*

*Figure arrived at after subtracting $81 million and 7% due to interdivisional sales eliminated in consolidation.

Plants:
Owens-Illinois has 102 domestic facilities.

Timberholdings:
Owens-Illinois owns, leases or has cutting rights to 1.4 million acres in Florida, Georgia, Louisiana, Michigan, Texas, Virginia and Wisconsin (1969 Annual Report).

Annual Pulp Production:
1,004,650 tons (1969)

Company Pollution Overview

Owens-Illinois operates four pulp mills—three in the South and one in Wisconsin. It is the only company in this study to have adequate pollution control equipment at every mill. In addition, interest in pollution control at Owens-Illinois dates back at least 20 years. Much of the equipment was developed internally, and most was installed prior to state requirements. As mills were built or rebuilt, Owens-Illinois installed state-of-the-art pollution control equipment. As the state of the art has changed, the company has duly upgraded this equipment.

Daily pulp production is about 2,830 tons, of which 1,770 is unbleached kraft pulp (produced at Valdosta, Georgia and Orange, Texas) and 1,060 is NSSC pulp (Big Island, Virginia and Tomahawk, Wisconsin). Daily water use is much less than would be expected—totalling 41 million gallons—because Owens-Illinois reuses a far higher percentage of water than most pulp and paper manufacturers.

The NSSC mills had the industry's first NSSC liquor recovery systems in the mid 1950s. Everyone else said it couldn't be done. The Orange, Texas kraft mill has a new odor control technique, which should be the most effective ever. Water pollution control systems at the mills consist of large, extended lagooning systems, a technique which Owens also pioneered, at the Valdosta mill in 1954. In 1954, paper and pulp mills did not generally have even primary treatment; but from inception, this system has provided very good secondary treatment also.

Although the company operates only four mills, none is small, none is brand-new, and none lacks adequate pollution control. Credit for this unusual situation is given to corporate policy and to William Webster, a former engineer and then vice-president with Owens-Illinois. He not only designed the water treatment system at Valdosta, but was also one of the first (in the early 1940s) to work on odor control systems.

Owens-Illinois does not attempt to hide its light under a barrel. In 1969, the company entered its pulp mills in a competition for the best company-wide pollution control program, and, deservedly, won the annual Gold Medal of the Sports Foundation, Inc., for water pollution control.

Over time, the company has not only developed pollution control equipment, but also a pollution measurement system, and operates a touring laboratory to sample and calculate air and water pollution loads at each of its four mills. "... The pollution load unit expresses in a single factor the total measurement of a plant's gaseous and liquid discharges. When this measurement is applied to a pollution load chart, management can quickly see a profile of the plant's performance over the years, in relation to production, pollution abatement controls and regulatory requirements."[1]

These facts give validity to company statements such as the following "Policy on Pollution," issued by (then President, now Chairman) R. H. Mulford in August of 1966:

"... We attach the same importance to air and water pollution abatement that we attach to quality, safety, fire prevention, and operating efficiencies. In constructing new facilities, we will, within the bounds of technical and economic feasibility, provide the best pollution control equipment available to meet or exceed community criteria."

A recent issue of Owens-Illinois Outlook comments: "As a manufacturer, we 'borrow' billions of gallons of water and enormous quantities of air during a year's time. They have been borrowed before — countless times — and not always by responsible users."[2] It is a pleasure to define Owens-Illinois as a responsible user.

Although pollution control is now adequate at all mills, improvements in particulate and water controls are planned for mid-1971. These additions should cost the company no more than $3.9 million.

Over time, taking the initiative seems to have been profitable. Owens estimates pollution control expenditures of $15 million for the four mills. That sum could, today, easily be the cost of upgrading one mill, had it been left until today.

Table 21.1
Owens-Illinois: Pulp Production, Water Use and Pollution Control

Mill Location	Pulp Prod. (TPD)	Other Prod. (TPD)	Water Use (MGPD)	Water Pollution Control			Air Pollution Control	
				Pri.	Sec.	Tert.	Part.	Gas, Odor
Valdosta, Ga.	870	825	12	✓	✓	✗	✓	✓
Orange, Tex.	900	900	10	✓	✓	✗	✓	✓
Big Island, Va.	510	510	12	✓	✓	✗	✓	—
Tomahawk, Wisc.	550	620	7	✓	✓	✗	✗	✓
Total	2,830	2,855	41					

Valdosta Mill

Location:
Valdosta, Georgia

Date Built:
1953

Process:
870 TPD unbleached kraft pulp

Other Production:
825 TPD linerboard

Total Water Use:
12 MGPD

Source:
Wells

Discharge:
Withlacoochee River (Florida)

Table 21.2
Valdosta Mill: Pollution Control Record

Treatment	Overall Evaluation	Equipment
Water		
Primary	✓	Seven holding ponds, total capacity, 1 billion gallons
Secondary	✓	60-day retention (1954), 90%-95% BOD reduction
Tertiary	✗	None
Solid waste removal	—	Landfill
Air		
Particulate (fuel)	✓	Fuel: gas, bark. Mechanical collectors, 93%-94% efficient
Particulate (production)	✓	Recovery furnace: precipitator, 99.5% efficient. Lime kiln: venturi scrubber, 95%-97% efficient
Gas and odor	✓	Strong black liquor oxidation
Plant Emissions		
To water:		
BOD	N.A.	
Suspended solids	N.A.	
To air:		
TRS	2 ppm (company data)	
Particulates	N.A.	

Pollution Control Expenditures:
$4.2 million was spent between 1967 and 1970 for additional ponding, improved aeration, scrubbers on two lime kilns, and the rebuilding of one recovery furnace and precipitator.

Legal Status:
(Water) Pond system discharges in Florida; in compliance with a Florida permit.
(Air) Shortest compliance schedule of any mill in the state under recently established regulations.

Future Plans:
Collectors for bark burners; recovery furnace scrubber; 175-foot diameter clarifier (for 1971); company designed color removal system.

Profile:
As is evident from the low odorous gas emissions (TRS), the odor control system at this mill is excellent. The company's calculated pollution load chart shows that the pollution load would have quadrupled since 1967 as production rose from 500 to 900 tons a day without abatement measures. Instead, the pollution load units at the mill are half of the 1967 level.

In other words, Owens-Illinois planned the capacity increase and a pollution control increase simultaneously; this is what all companies should do. The mill was built with then state-of-the-art pollution control equipment. It has been updated as attainable equipment efficiency has increased. Owens did not install a liquid oxygen odor control system here as it did at Orange, Texas, because there was no local supply of molecular oxygen, but instead supplemented the original weak black liquor oxidation with strong black liquor oxidation.

The water treatment system, installed when the mill was built, was the first massive lagooning setup in the industry, and has consistently provided a BOD reduction of more than 90 percent over the years. (Such a system is only feasible for certain locations. In order to set it up, Owens-Illinois purchased the part of the stream — Jumping Gully Creek — which is in Georgia.) The mill site, in fact, was selected by company engineer William Webster in order to accommodate the ponding system he had designed. Webster then met with the Georgia and Florida health departments (there were no pollution control boards) and the system was jointly approved. At the time, there were no clarifiers at mills, so ponds were built to deal with removal of both solids and BOD. The company comments that everyone made fun of the "nature ponds" then, but they don't any longer. Here, as elsewhere, Owens-Illinois uses relatively little water.

CEP Estimate of Minimum Necessary Pollution Control Expenditure Not on Company Budget: $3 million total, of which $2.5 million is for water treatment and $500,000 is for improved power boiler particulate control.

Orange Mill

Location:
Orange, Texas

Date Built:
1967

Process:
900 TPD unbleached kraft pulp

Other Production:
900 TPD kraft linerboard

Total Water Use:
10 MGPD

Source:
Sabine River

Discharge:
Sabine River

Table 21.3
Orange Mill: Pollution Control Record

Treatment	Overall Evaluation	Equipment
Water		
Primary	✓	Clarifier, 15 million gallon capacity
Secondary	✓	1.8 billion gallon, 37-acre aeration basin; 12 rain guns reduce foam problem and lower carbon dioxide. Six-month holding pond
Tertiary	X	None
Air		
Particulate (fuel)	✓	Fuel: gas. None needed
Particulate (production)	✓	Two recovery furnaces, both with 99.5% efficient precipitators. Demisters on lime kiln

Table 21.3
Orange Mill: Pollution Control Record (continued)

Treatment	Overall Evaluation	Equipment
Gas and odor	✓	Molecular-oxygen liquor oxidation system, over 95% efficient

Plant Emissions

To water:

BOD	25 ppm (about 2 lb/ton of pulp; company figure)
Suspended solids	N.A.
COD	200 ppm (company figure)

To air:

TRS	2-5 ppm (company figure)

Pollution Control Expenditures:
N.A.

Legal Status:
(Water) In compliance.
(Air) In compliance.

Future Plans:
None.

Profile:
This mill is very close to state-of-the-art pollution control. "We developed a new technology for odor control at this mill using molecular oxygen instead of air, to oxidize black liquor. It's unique—no one else we know of is using molecular oxygen—the MOBLO system," said a company spokesman.

Note that even at this level of control there will be a detectable odor, since mercaptans and hydrogen sulfide can be smelled in a concentration of a few parts per billion parts of air.

Water pollution control technology at this mill was also worked on by the company. A common and very difficult problem in the use of the aeration basin at southern mills—foam which built up, dried, and then was blown about by any wind—was solved by spraying the aerators frequently with water from rain guns. This action killed most of the foam and kept the rest wet. Without the rain gun, a blanket of foam built up a foot or two thick and floated in the wind as far as 5-10 miles from the mill. An intensive company study in 1969 showed that the 12 rain guns alone could do a better job of raising the oxygen content in the basin than 10 aerators. The aerators have been taken out. The BOD discharge is very low.

Texas is one of few states to have a standard of COD (chemical oxygen demand) on industrial effluent. The mill reports COD emissions of 200 parts per million. This is down from 300 ppm before the rain guns and from the average mill's 400-800 ppm level.

CEP asked a member of the State Water Quality Control Department about the mill (September 21, 1970), who said: "I do know that it is not on my list of things to be concerned about this week—nor on the list of things to be concerned about this month." That is a nice testimonial from a very strict state agency.

CEP Estimate of Minimum Necessary Pollution Control Expenditure Not on Company Budget: None.

Big Island Mill

Location:
Big Island, Virginia

Date Built:
Pre-1930

Process:
510 TPD NSSC pulp

Other Production:
510 TPD corrugated board

Total Water Use:
12 MGPD

Source:
James River

Discharge:
James River

Table 21.4
Big Island Mill: Pollution Control Record

Treatment	Overall Evaluation	Equipment
Water		
Primary	✓	Airfloat clarifier (aerated)
Secondary	✓	Lagoon with 20 million capacity and 8-day retention of effluent; NSSC spent pulping liquor is recovered
Tertiary	X	None
Air		
Particulate (fuel)	X	Fuel: bark, low sulfur coal. None
Particulate (production)	✓	Own design recovery furnace, venturi scrubber
Gas and odor	—	None needed
Plant Emissions		
To water:		
BOD		8,000 lb/day (15 lb/ton of pulp; company figure). Company monitoring shows dissolved oxygen in the James at the sag point* of 3-8 ppm.
Suspended solids		N.A.
To air:		N.A.

*The biological decomposition of wastes does not take place immediately after they are discharged, but, in a flowing stream, occurs from 2 to 5 miles down-stream of the discharge. This point, where the effluent is demanding the most oxygen, is called the sag point.

Pollution Control Expenditures:
N.A.

Legal Status:
(Water) In compliance (laws new in 1970).

(Air) State agency policy is to provide no information to the public.

Future Plans:
Increased secondary treatment; 98 percent efficient dust collectors being installed.

Profile:
This mill has the second (1956) NSSC recovery system installed in the United States. The technique was developed by Owens-Illinois and provides a significant measure of pollution control. The recovery system is probably adequately serviced by the scrubber for air pollution control.

Since 1967, this mill has been seeking to build a joint-treatment plant; but found no interest in the area. Virginia adopted water pollution control regulations meeting with the approval of the Department of Interior only in 1970. The new state standards for dissolved oxygen in the James — average concentration of five parts per million, minimum concentration of four parts per million — will require increased secondary treatment for all process effluent.

The company commented that "this is a continuing program in that we anticipate more stringent regulations in future requiring complete secondary and probably tertiary treatment." In the years before 1970, the mill provided good water treatment; probably it will continue to do so. Needed particulate controls for the power boilers are being installed.

CEP Estimate of Minimum Necessary Pollution Control Expenditure Not on Company Budget: $700,000 total, of which $500,000 is for aerating the secondary treatment system and $200,000 is for improved power boiler particulate control.

Tomahawk Mill

Location:
Tomahawk, Wisconsin

Date Built:
1920s; expanded 1958

Process:
550 TPD unbleached NSSC pulp

Other Production:
620 TPD semichemical corrugated medium

Total Water Use:
7 MGPD for pulping, 5 MGPD for papermaking

Source:
Wisconsin River

Discharge:
Wisconsin River

Table 21.5
Tomahawk Mill: Pollution Control Record

Treatment	Overall Evaluation	Equipment
Water		
Primary	✓	Settling pond
Secondary	✓	Aeration pond; 7-day retention lagoon with 34 million gallon capacity; NSSC spent pulping liquor is recovered
Tertiary	X	None
Solid waste removal		Landfill
Other		Water run over dam after treatment for additional aeration and some power for mill operations
Air		
Particulate (fuel)	X	Fuel: 33% of power is purchased, 10%-15% is hydroelectric; 40% is low-sulfur coal, rest is bark and gas
Particulate (production)	X	Kraft-type recovery furnace with precipitator, 92% actual efficiency
Gas and odor	✓	Non-direct contact evaporation system

Plant Emissions

To water:

BOD   10,000 lb/day (state figure); 11,600 lb/day
                       (21 lb/ton of pulp; company figure). 3-8 ppm
                       dissolved oxygen at sag point.

Table 21.5
Tomahawk Mill: Pollution Control Record (continued)

Treatment	Overall Evaluation	Equipment
To air:		
TRS	Hydrogen sulfide, 2 ppm or less; sulfur dioxide also very low (company data)	
Particulates	N.A.	

Pollution Control Expenditures:
N.A.

Legal Status:
(Water) In compliance.
(Air) Applicable regulation not yet implemented.

Future Plans:
In process of rebuilding precipitator for over 99 percent efficiency.

Profile:
This was the first American NSSC mill (in 1953) to have a spent cooking liquor
recovery system. Needless to say, this was before any state pressure for
water pollution control was applied. Owens-Illinois spent a considerable sum
developing the system, and initiated a study with the Institute of Paper Chemistry in Appleton, Wisconsin in 1952. Both this system and the one at the Big
Island, Virginia, mill have been working ever since they were put in.

Despite liquor recovery and equivalent primary and secondary water
treatment, the 1970 Wisconsin River Report found "heavy slimes were in
evidence on the rocks and debris ... the majority of organisms were tolerant
or very tolerant. The unbalanced condition observed at this location [below
the Wisconsin Dam at Tomahawk ... below Owens-Illinois, Inc.] indicates
that critical dissolved oxygen levels do occur periodically."

In order to correct the depletion of dissolved oxygen presently caused by
mill effluent, Owens will try to completely recycle water here. This would
be one of the first such systems in the industry. Presently, Green Bay Packaging Corp., under an FWQA grant, is setting up the first such system.

CEP Estimate of Minimum Necessary Pollution Control Expenditure Not
on Company Budget: $200,000 total, for improvement of the precipitator.
Costs of the closed-loop water system under investigation have not been established.

References for Chapter 21

1. Design News, June 8, 1970.

2. Owens-Illinois Outlook, July, 1970, p. 1.

Chapter 22

POTLATCH FORESTS, INC.
One Maritime Plaza, Golden Gateway Center, San Francisco, California 94119

Major Products:
Lumber, plywood, lumber products, pulp, paper, paperboard, relocatable buildings.

Consumer Brands:
See Appendix 4.

Financial Data ($ Millions):	1969	1968	1967
Net Sales	337.1	318.1	274.3
Net Income	14.9	16.4	11.8
Capital Expenditures	27.3	27.3	18.8

Pollution Control Expenditures:
$ 12 million from 1950-1970
$ 21 million (estimated) from 1970-1975

Annual Meeting:
The annual meeting is held during May in San Francisco.

Officers:
President: Benton R. Cancell, San Francisco, California (Director: First Security Corp., Bank of California)

Outside Officers:
Samuel F. Bowlby: Consultant, Bank of California
Frederick W. Davis: Investment Manager
George F. Jewett, Jr.
Harry T. Kendall, Jr.
John J. Pascoe: Chairman, J. W. Clement Co.
Edward L. Palmer: Executive Vice President, First National City Bank
Willis S. Hutchinson: Economist and Market Management Researcher
Hans Schneider: President, Reemehrs Schneider Co., Inc.
Langdon S. Simons, Jr.: Chairman Financial Committee, Laird Norton Co.
Frederick T. Weyerhaeuser, Manufacturing Executive

Breakdown of 1969 Company Forest Products Production:
Lumber	545,460,000 board feet
Plywood	416,169,000 square feet
Paperboard and tissue	283,434 tons
Printing and business papers	256,341 tons

Plants:
Potlatch has 34 domestic operations:
Wood products   15
Paperboard and packaging   15
Printing and business papers   2
Manufactured buildings   2

Timberholdings:
Potlatch owns 1,366,000 acres of timberland in the United States and has cutting rights to "many hundreds of thousands of acres" overseas in Colombia
and an exclusive right to cut and manufacture timber on both government and
forest lands in Western Samoa. Domestic holdings consist of 240,000 acres in
Minnesota, 595,000 acres in Idaho and Washington, and 595,000 acres in Arkansas. The total 1969 value of Potlatch's holdings, including timber, was
$40,262,109.

Annual Pulp Production:
445,525 tons (1969)

Company Pollution Overview

Potlatch Forests, Inc. was the fourth-largest lumber producer in the United
States in 1968, and in 1969 supplied 3.5 percent of the total plywood market
in the country and 6.2 percent of the high grade bleached paperboard market.
A subsidiary, Northwest Paper, operates two pulp mills; one at Lewiston,
Idaho on the Washington-Idaho border, and one at Cloquet, Minnesota. A
second subsidiary, Swanee Paper Corp. operates a paper mill. The two pulp
mills produce 1,270 tons of pulp using 52 million gallons of water daily.
        In a speech before an audience of securities analysts, Potlatch President
Benton Cancell stated: "Being endowed with a wealth of natural resources,
located close to manufacturing plants, our job is to fully utilize such resources." The company's pollution control record indicates that "fully utilize"
is closer to "fully use up" where air and water are concerned. The Cloquet
mill has only a five million gallon capacity clarifier to handle 22 million gallons of bleached calcium sulfite and unbleached kraft effluent per day and its
air pollution controls are similarly inadequate. Lewiston has no secondary
treatment and its air pollution control equipment is so overloaded as to be almost useless. The mill emits over 6,100 pounds of particulate matter and
between 7,000 and 9,000 pounds of sulfurous gases daily into the atmosphere.
        Needless to say, neither mill is in compliance with state water or air
pollution regulations and Lewiston is the only industrial facility in the nation
to have been the subject of two separate federal pollution abatement conferences; water pollution in 1964, and air pollution in 1967. In fact, the enforcement division of the Department of the Interior Water Quality Administration
is reconvening the abatement conference because there has not been satisfactory improvement in the water quality of the Snake River.

Despite the fact that there is no mention of pollution, pollution control, environment, ecology, etc., in Potlatch's annual report, the company is aware of the problem. To illustrate its concern, Potlatch ran a nationwide advertisement showing a picture of the sparkling clean Clearwater River. The text read, "It cost us a bundle, but the Clearwater still runs clear." The message was clear, but the medium was a bit muddy. The photograph was taken many miles upstream of the Lewiston mill, which doesn't even discharge into the Clearwater, but pipes its effluent to the Snake River. If the Clearwater is clean, it didn't cost Potlatch "a bundle," it cost the "price" of the Snake River and the public must pay the bill.

Not surprisingly, Potlatch ranks high in lack of cooperation with CEP. It seems to have been waiting for CEP to go away just as it has waited for its pollution to go away. One of the company's responses to a question about air pollution equipment at Cloquet reflects its attitude: "You know, you don't just pick up this stuff off the shelf in a grocery store." From the condition of its pulp mills, it is clear that Potlatch hasn't even window-shopped for pollution control equipment. Apparently production equipment is more easily obtainable, however, because the company was able to spend $27 million in 1969 on a new off-machine coater at Cloquet and a new continuous digester and bleach plant at Lewiston.

Some may recall that the word "potlatch" comes from an American Indian festival during which an annual bonfire was fueled with all of one's possessions to demonstrate both wealth and the confidence of replenishing it within the next year. Potlatch Forests seems to feel this way about our air and water resources.

CEP estimates that, in addition to the $21.6 million which Potlatch has reportedly budgeted for pollution control at its two pulp mills, between $6.4 and $15.4 million more will be required to thoroughly clean both air and water, making a total expenditure of between $28 and $37 million.

Table 22.1
Potlatch Forests, Inc.: Pulp Production, Water Use, and Pollution Control

Mill Location	Pulp Prod. (TPD)	Other Prod. (TPD	Water Use (MGPD)	Water Pollution Control			Air Pollution Control	
				Pri.	Sec.	Tert.	Part.	Gas, Odor
Lewiston, Id.	830	685	32	√	X	X	X	X
Cloquet, Minn.	440	330	20	X	X	X	X	X
Total	1,270	1,015	52					

Lewiston Mill

Location:
Lewiston, Idaho

Date Built:
1950

Process:
830 TPD bleached kraft

Other Production:
640 TPD bleached board, manila paper, bleached speciality grade papers,
and polyethylene-coated food board
45 TPD tissue and towelling

Total Water Use:
32 MGPD

Source:
Clearwater River

Discharge:
Snake River

Table 22.2
Lewiston Mill: Pollution Control Record

Treatment	Overall Evaluation	Equipment
Water		
Primary	✓	35 million gallon Eimco clarifier installed in 1968 (removes 20 TPD of settleable solids)
Secondary	X	None
Tertiary	X	None
Solid waste removal	—	Sludge to landfill
Other	—	Outfall pipe to Snake River

Table 22.2
Lewiston Mill: Pollution Control Record (continued)

Treatment	Overall Evaluation	Equipment
Air		
Particulate (fuel)	X	Fuel: gas, bark. Mechanical dust collectors (inadequate)
Particulate (production)	X	Three recovery furnace precipitators 95% efficient (overloaded). Smelt dissolver scrubbers. Lime kiln scrubbers.
Gas and odor	X	Weak black liquor oxidation (overloaded); digester gases burned in lime kiln

Plant Emissions

To water:

BOD	90,000 lb/day (108 lb/ton of pulp)
Suspended solids	70,000 lb/day (84 lb/ton of pulp)

To air:

TRS	7,000-9,000 lb/day (8-10 lb/ton of pulp; 1969)
Particulates	6,136 lb/day (over 7 lb/ton of pulp; 1969)

Pollution Control Expenditures:
$ 3 million.

Legal Status:
(Water) Not in compliance with state regulations; secondary treatment is required. A federal water pollution abatement conference in 1964 recommended that primary treatment be installed by 1968. At present, the FWQA is planning reconvention of the conference.
(Air) The State Air Pollution Control Commission said: "Our state agency has had very little to do with Potlatch Forests, Inc.'s mill in Lewiston since the Lewiston-Clarkston region has been under a federal abatement program. Potlatch is making engineering studies in preparation for several major projects in connection with air and water pollution abatement. These projects will be designed to exceed current regulations and meet those that will be

proposed in the near future."[1] U.S. Public Health Service air pollution
survey took place in the winter of 1961-1962, and a PHS abatement conference, initiated December, 1965, convened in February, 1967.

Future Plans:
On September 22, 1970, Standard and Poor's reported that $9.6 million "will
be spent for the first stage of an air and water quality improvement program"
at Lewiston. This will include secondary "and possibly tertiary water effluent
treatment... Further recovery boiler modifications to reduce sulfurous gas
emissions, sulfur dioxide emissions, hydrogen sulfide emissions as well as
mercaptan emissions. Further particulate emission reductions are being engineered." Potlatch reported that pilot plant studies are underway and secondary treatment will be completed in 1974.

Profile:
This 20-year-old mill has received more federal abatement attention than any
industrial facility in the United States and is the only pulp mill to have been
the subject of both air pollution and water pollution abatement conferences.
Federal agencies first focused attention on Lewiston in 1961; nine years later,
the first stage of an air and water quality "improvement program" at the mill
was announced. The recorded history of Lewiston's water pollution begins in
1964 when a federal abatement conference traced 94 percent of the waste in
the area to the mill. It was found that "a rotten cabbage odor prevails in the
Snake River below Potlatch Forests, Inc.'s waste discharge point"[2] and
that "the waste discharged from the pulp and paper mill is vividly evident as
a deep, brown, odorous, and turbid section in the Snake River."[3] The pipe
which diverts effluent into the Snake instead of the Clearwater is also a continual source of trouble, because, according to a spokesman for the Idaho
pollution control agency, it was not laid on grade and "accumulates slime and
ruptures" about once a year. He added that "people get excited" when the pipe
overflows and the effluent gushes into the Clearwater.
    The mill was also reported to be adding 320 tons of solids daily to the
Snake River, although there was some question about this figure, because it
was the result of only one sample, taken in the middle of the river at the end
of the outfall pipe. At the conference, the data was presented along with the
explanation that sampling was not permitted on mill property and that mid-
river was too perilous to allow another attempt to sample water quality
there.[4] The result of the conference was a recommendation that Potlatch
install primary treatment by 1968. This was done within the generous four
year timetable. At present, Idaho reports that the problems of odor and discoloration in the river have "not been solved, but have diminished" and it is
expected that the odor will disappear when the mill provides secondary water
treatment.
    A Corps of Engineers dam project has often been cited as a reason for
Potlatch's failure to construct water pollution control facilities. A dam would
affect the flow and possibly the location of the river. However, Potlatch could

probably provide the water treatment now needed and be better prepared to meet future requirements.

The effects of mill air pollution are equally far reaching. Prevailing winds carry the emission plume across the Idaho border to Clarkston, Washington, but are not strong enough to carry it out of the valley. The geographical situation and frequent inversions compound the problem. The severe odor problems in Clarkston were first documented in 1962 by a HEW study which found measured concentrations of hydrogen sulfide and mercaptans ranged from nine to 44 parts per billion. (Nine parts per billion is twice the annoyance level for these compounds.)

Between 1962 and the February 1967 air pollution abatement conference, the mill's only odor control, weak black liquor oxidation, was discontinued because of operation problems, further increasing emissions of odorous gases from the recovery furnace and evaporation system. A 1967 report stated, "... total hydrogen sulfide emissions may reach a rate of 5,400 pounds per day from all three recovery furnaces. This rate of ... emissions from recovery furnaces of a mill this size having no facility for controlling (them) is not extraordinary."[5] Further, these levels do not compare favorably with the limits set by Washington and Oregon of two pounds a day per ton of pulp, or a 1660-pound total for a mill this size.

At the abatement conference, Potlatch reported that it had finally planned improvements to control air pollution at the mill. Between February 1967 and July 1968, it installed a precipitator on the number three recovery furnace, a collection system for incineration of evaporator off-gas, and reinstalled the black liquor oxidation system. In November, 1968—almost two years after the conference and seven years after the beginning of the initial investigation—personnel from the National Air Pollution Control Administration toured the area and found that there had been some improvement which they expected to continue "... as the company applies newer techniques to contain the effluent; provided there is no overall increase in production."[6] However, since then, production has increased and, with it, so have odorous emissions.

The Lewiston mill has been the subject of federal interest for all these years because geography, meteorology, and lack of control measures have spread its air and water pollution through Lewiston, Idaho, and into Clarkston, Washington. It is this interstate feature which has enabled the State of Idaho to minimize its involvement by enlisting federal help in dealing with the area's largest employer. The documentation of pollution from this mill has been voluminous. The concentration of odorous gases in the two towns has actually increased in the decade; and the company has shown an apparent disregard for environmental considerations in any of its production activities. This is despite the fact that the relative newness of the mill should simplify design and installation of pollution control equipment.

CEP Estimate of Minimum Necessary Pollution Control Expenditures Not in Company Budget: $3.7-$12.7 million total, of which $1.5-$10 million is for an odor control system (the higher figure includes a new recovery furnace); $200,000 for improved power boiler particulate control; and $2 million for a tertiary (color removal) system, recommended by the State of Idaho.

Cloquet Mill

Location:
Cloquet, Minnesota

Date Built:
Sulfite, 1915; kraft, 1924

Process:
110 TPD bleached calcium sulfite
330 TPD bleached kraft

Other Production:
330 TPD of a wide variety of coated and uncoated papers

Total Water Use:
20 MGPD (4 MGPD for paper mill)

Source:
Lake Superior and St. Louis River

Discharge:
St. Louis River

Table 22.3
Cloquet Mill: Pollution Control Record

Treatment	Overall Evaluation	Equipment
Water		
Primary	X	80-foot diameter, 5 MGPD clarifier (for paper mill effluent only)
Secondary	X	None
Tertiary	X	None
Solid waste removal	—	Landfill on own property
Other	—	Screening, savealls; lagooning of lime mud

Table 22.3
Cloquet Mill: Pollution Control Record (continued)

Treatment	Overall Evaluation	Equipment
Air		
Particulate (fuel)	X	Fuel: coal, gas, oil. Mechanical dust collectors rated 90% efficient (inadequate)
Particulate (production)	X	Recovery furnace scrubber, 90% efficient (inadequate)
Gas and odor	X	None
Plant Emissions		
To water:		
BOD	77,000-118,000 lb/day (175-268 lb/ton of pulp)	
Suspended solids	25,000-32,000 lb/day (57-72 lb/ton of pulp)	
To air:		
TRS	2,200 lb/day (5 lb/ton of pulp)	
Particulates	10,000 lb/day (22.7 lb/ton of pulp)	

Pollution Control Expenditures:
$4.1 million from 1950 to 1969 of which $500,000 was spent since 1967.

Legal Status:
(Water) Not in compliance; company has signed stipulation to have secondary treatment (80 percent BOD removal) by November, 1973. Until then, the mill is operating under a permit.
(Air) Not in compliance: company has filed program with the state.

Future Plans:
$12 million will be spent between 1971 and 1975 as part of a $45.4 million investment "to increase the capacity of the kraft mill and to bring both current and additional kraft pulping operations up to required air and water quality standards."[7] The company indicated that it is presently operating a pilot secondary treatment plant to evaluate the best method of treatment with the work to be completed in September. Engineering and planning are underway for construction of secondary treatment facilities, revision of mill

effluent sewers, spill catch basins, odor-free recovery boiler, vaporsphere and other related equipment."[8]

Profile:
This is quite an old facility at which 425 tons of bleached calcium sulfite and kraft pulp are produced without adequate air or water pollution controls. There is strong evidence of the damage caused by the daily discharge of highly oxygen-demanding spent sulfite liquor and bleaching chemicals. In 1969, two fish kills in the St. Louis River were attributed to mill effluent: the first, on July 25, was "moderate"; it affected one mile of the river for one day. The second, August 21, was considered to be "heavy"; the water in 15 miles of river killed aquatic life for a duration of two weeks. This water pollution situation is rather bad, yet it will continue unabated for three more years until the planned water treatment goes into operation and the sulfite mill is closed down. This is scheduled just in time to meet the 1973 Minnesota deadline for a BOD reduction to 59 pounds per ton.

The air pollution situation is simularly bad. The control equipment that exists is old and inadequate. There is no odor control and the lack of a lime kiln indicates that those gases which are incinerated in most kraft mills are simply being released into the atmosphere. Northwest Paper has been effusive in announcing the planned pollution control expenditures at Cloquet, yet it is just another company being forced by the state to take some action after decades of irresponsibility, and is doing so as part of a major expansion of production facilities.

CEP Estimate of Minimum Necessary Pollution Control Expenditures Not in Company Budget: $2.7 million total, of which $500,000 is for complete primary water treatment (at present the mill has only partial treatment facilities); $200,000 is to improve the coal boiler particulate control; and $2 million is for secondary treatment.

References for Chapter 22

1. Letter to CEP, July 28, 1970.

2. Summary of the Conference in the Matter of Pollution of the Interstate Waters of the Snake River and Its Tributaries, January 15, 1964, Lewiston, Idaho, U.S. Department of Health, Education, and Welfare, p. 27.

3. Ibid, p. 28.

4. Ibid. Material abstracted from pp. 50-54.

5. Lewiston, Idaho/Clarkston, Washington Air Pollution Abatement Activity, U.S. Department of Health, Education, and Welfare, Public Health Service, National Center for Air Pollution Control, February, 1967, p. 24.

6. National Emissions Standards Study (Report of the Secretary of Health, Education, and Welfare), April 27, 1970, pp. 17-19.

Potlatch's (Northwest Paper Company) Cloquet mill is the worst environmental offender of the four pulp mills in Minnesota. The noxious fumes usually drift over the city in the early evening, around supper-time. Photo by Mike Zerby, *Minneapolis Tribune;* © The Minneapolis Star and Tribune Company.

7. Standard and Poor's Daily News, September 22, 1970, p. 6678.

8. Letter to CEP, September 18, 1970.

Chapter 23

RIEGEL PAPER CORPORATION
260 Madison Avenue, New York, New York 10016

Major Products:
Paper, paperboard packaging products.

Financial Data ($ Millions):	1969	1968	1967
Net Sales	184.0	165.2	148.1
Net Income	7.8	5.0	6.6
Capital Expenditures	14.7	7.2	11.7

Pollution Control Expenditures:
Approximately $5 million from 1960 to 1970; $13.5 million (estimated) from 1970 to 1975.

Annual Meeting:
The annual meeting is held on the final Tuesday in April.

Officers:
President: William J. Scharffenberger

Outside Directors:
J. S. Armstrong: Executive Vice President, U.S. Trust Co. of New York
German H. H. Emory: Chairman, Riegel Textile Corp.
Charles E. Hartford: Retired Senior Vice President, Riegel Paper Corp.
E. V. Hogan: President, Richmond County Bank
F. S. Leinbach: Retired President, Riegel Paper Corp.
A. D. Lewis: Sr. Partner, F. S. Smithers and Co.
W. M. McFreely: Retired Vice President, Riegel Textile Corp.
Theodore Riegel: Retired Vice President, Riegel Textile Corp.
William E. Reig: President, Riegel Textile Corp.
J. A. Stratka: Retired Chairman of the Board, Cheseborough-Pond, Inc.

Breakdown of Company Production (Dollar Value and Percent of Sales):

	$ Millions	Percentages
Paper Division (Manufacture of pulp and paper for printing, packaging and technical uses)	112.5	61
Packaging Division (manufacture of folding cartons and packaging)	63.5	34

	$ Millions	Percentages
Industrial Products Division	7.5	5
(manufacture of industrial films, metallic		
yarns and security items—credit and identi-		
cation cards)		

Plants:
Riegel has plants at 22 domestic locations:

Packaging	10
Paper Division	7
Forest Products and Real Estate Division	1
Industrial Products Division (one belongs to an affiliated	4
company, Britains-Riegel, Ltd.)	

Timberholdings:
Riegel owns or controls over 345,000 acres of timberlands in North and South Carolina. During 1969, the company cleared and planted over 11,000 acres in these two states. The total 1970 value of Riegel's holdings: $7,624,000.

Annual Pulp Production (Estimated on a basis of 355 production days per year):
390,500 tons (1969)

Company Pollution Overview

Riegel Paper Company operates only one pulp mill—a large 1,100 TPD kraft mill in Riegelwood, North Carolina, which has good water treatment, inadequate particulate control, and no odor control at all.

In the 1969 Annual Report, Riegel's President stated the company's pollution control position: "... management is fully aware of all pollution problems associated with its operations and ... is committed to the development of programs which will minimize air and water pollution to the maximum extent technically feasible in the shortest possible time."[1]

The air pollution problems at the Riegelwood mill and the company's average annual pollution control expenditure of only $500,000 over the past decade possibly throw some doubt on the truth of this statement. However, it is made more creditable by the fact that Riegel—the 24th largest paper company [2]—developed the now widely used surface aeration system for secondary water treatment.

Perhaps, when North Carolina's air standards for pulp mill emissions are formulated, Riegel will show similar inventiveness in solving its air pollution problems.

Riegel has budgeted $13.5 million for pollution control at all its facilities. The $11.5 million which has been allocated for improvements at Riegelwood should be sufficient to clean it up.

Table 23.1
Riegel Paper Corporation: Pulp Production, Water Use, and Pollution Control

Mill Location	Pulp Prod. (TPD)	Other Prod. (TPD)	Water Use (MGPD)	Water Pollution Control			Air Pollution Control	
				Pri.	Sec.	Tert.	Part.	Gas, Odor
Riegelwood, N.C.	1,100	725	40	✓	✓	✗	✗	✗

Riegelwood Mill

Location:
Riegelwood, North Carolina

Date Built:
1951; expanded 1956

Process:
1,100 TPD bleached and semibleached kraft (company figure)

Other Production:
725 TPD coated and uncoated bleached kraft board for folding, food, ice cream, and other cartons

Total Water Use:
40 MGPD

Source:
Cape Fear River

Discharge:
Cape Fear River

Table 23.2
Riegelwood Mill: Pollution Control Record

Treatment	Overall Evaluation	Equipment
Water		
Primary	✓	60 million gallon clarifier

Table 23.2
Riegelwood Mill: Pollution Control Record (continued)

Treatment	Overall Evaluation	Equipment
Secondary	✓	Aerated stabilization lagoon with 29 high-efficiency surface aerators (230-acre, 550 million gallon capacity, built in 1964)
Tertiary	X	None
Solid waste removal	—	Landfill on company property
Air		
Particulate (fuel)	X	Fuel: coal, oil, gas; 80% efficient precipitator (inadequate)
Particulate (production)	X	Three recovery furnaces: precipitators rated 90% efficient, actually 80%. Two lime kilns: scrubbers, 90% efficient (system overloaded)
Gas and odor	X	None
Plant Emissions	N.A.	

Pollution Control Expenditures:
$4.5 million, primarily for the aerated lagoon.

Legal Status:
(Water) The mill has a permit, which expires June 1, 1971, requiring more aeration facilities and studies to determine whether additional aeration will be sufficient.
(Air) The state is planning to establish specific emission standards for pulp mills, but has not done so yet.

Future Plans:
By the end of 1970, Riegel will finalize plans for expansion of this mill. The company's environmental control director indicated that the three recovery furnaces may be replaced by two new large ones, and a new evaporator may be included. Riegel is now considering odor control alternatives (black liquor oxidation or a nondirect contact evaporator). Of the company-wide $13 million pollution control investment projected over the next five years, $11.5 million will be spent at this pulp mill.

Profile:

This large kraft mill is Riegel's only pulping facility. It produces 1,100 tons
of bleached and semibleached pulp and 725 tons of paper daily. The mill has
good effluent treatment. Its surface-aerated stabilization system, installed in
1964, was the first surface aeration system in the country and Riegel received
an award from the North Carolina Wildlife Federation for developing it. It has
served as a model for more than 30 similar systems installed throughout the
country. Although the system is good, the state evidently thinks it should be
better and has required Riegel to provide more aeration by July 1971.

Air pollution control at Riegelwood is totally inadequate. The mill has
operated for 17 years with no odor control at all, and the environmental director conceded that it has been producing beyond the 1,000 tons per day rated
equipment capacity. This overproduction has caused particulate control equipment efficiency to fall from 90 percent to 80 percent.

Now that Riegel is planning a major mill expansion, initiative will probably be taken to install adequate particulate and odor control. This is particularly likely since the state is now formulating air emissions standards
for pulp mills.

CEP Estimate of Minimum Necessary Pollution Control Expenditures Not
in Company Budget: None.

References for Chapter 23

1. Annual Report, 1969, p. 3

2. Chem. 26, June, 1970, p. 35.

Chapter 24

ST. REGIS PAPER COMPANY
150 East 42nd Street, New York City, New York 10017

Major Products:
Paper, lumber and plywood products, packaging machinery, food processing machinery.

Major Consumer Brands:
WHIRLPOOL Trash Masher bags, NIFTY school supplies, SUPERIOR paper plates & cups, MURRAY bags.

Financial Data ($ Millions):	1969	1968	1967
Net Sales	867.8	786.2	721.8
Net Income	41.2	35.3	31.7
Capital Expenditures	35.6	61.8	99.8

Pollution Control Expenditures:
$9.94 million, 1967-1969
$65-$70 million (estimated), 1970-1975.

Annual Meeting:
The annual Meeting is held during April in New York.

Officers:
Chairman: R. K. Ferguson, Oyster Bay, New York (Chairman: Norwood and St. Lawrence Railroad)
President: W. R. Adams, New Canaan, Connecticut (Director: Greyhound Corp., Norwood and St. Lawrence Railroad Co., University of Maine Pulp and Paper Foundation; President: North Western Pulp and Power Ltd., Trustee: Union College)

Outside Directors:
A. D. Pace
Lawrence S. Pollock: President, Pollock Paper Co.
Curtis E. Neldner: Partner, White, Weld, Inc.
J. C. Pace: Former Vice President, St. Regis Paper Co.
J. P. Lewis: Affiliated with J. P. Lewis Corp.-Latex Fiber Ind.
J. H. Laeri: Executive Vice President, First National City Bank of N.Y.
T. M. McClellan: President, Nifty Manufacturing Co.
P. Neils: Former President, J. Neils Lumber Co.
George P. Jenkins: Financial Vice President, Metropolitan Life Insur. Co.
A. C. Leiby: Partner, LeBoeuf, Lamb, Leiby and MacRae
Baldwin Maull: President, Marine Midland Corp.

W. I. Osborne, Jr.: President, Pullman, Inc.
Henry D. Schmidt: Former President, Schmidt and Ault Paper

Breakdown of Total Company Production (1969):

Kraft Production and specialty papers	1,767,000 tons
Printing and fine paper	492,000 tons
Pulp	2,082,000 tons
Paper and paperboard converted	990,000 tons
Lumber	377,329 thousands of board ft
Plywood	118,044 thousands of square ft (3/4-inch basis)

Plants:
St. Regis has 117 domestic plants:

Pulp and paperboard	6
Boxboard using secondary fiber	4
Corrogated box	23
Other paper products	62
Lumber and wood	8
Metal culvert	6
Food processing systems	4
Chemical	1
Prestressed concrete	3

Timberholdings:
St. Regis has management responsibility for 8,234,000 acres of timberland in
North America, of which 2,700,000 acres (primarily in the United States) are
owned outright or on long-term lease. The total 1969 value of St. Regis'
holdings: $70,670,000.

Annual Pulp Production:
2,082,000 tons (1969)

Company Pollution Overview

St. Regis, although a leader in paper making technology, has totally lacked
initiative in pollution control. The company's new mill, in Monticello, Mississippi, considered the most advanced paper making facility in the country,
is the only one of nine mill locations with excellent air or water control. The
other eight are all painfully far behind; only two have any control equipment
for pulping effluent at all.
    St. Regis' problems in making pollution control decisions are multiplied

and complicated by the fact that five of its mills are very small and old, built before 1930 — marginal operations at best. Two of these are in the East, at Bucksport, Maine and Deferiet, New York; three are in the Midwest at Sartell, Minnesota and Cornell and Rhinelander, Wisconsin. The company also has four large kraft mills, three in the South (Pensacola and Jacksonville, Florida, and Monticello, Mississippi) and one in Tacoma, Washington.

These nine mills produce a total of 5,655 tons of pulp per day, using and discharging 222.3 million gallons of water in the process.

Eight of the St. Regis mills have been approached by state authorities to clean up their effluent; three were cited at government abatement conferences — Bucksport in 1967, Pensacola and Rhinelander in 1970. Despite these citations, and numerous public complaints, as of October, 1970, only Monticello and Pensacola had any water treatment.

The air pollution control situation is not much better. Four of the nine mills need improved particulate control and three of the four kraft mills still spew out completely uncontrolled odorous sulfur gases. Full plans for solving these problems have been set only at the Tacoma mill.

This is a dismal pollution control record, but St. Regis is finally taking some positive steps, encouraged by pressure from the government and the public. One focus is the large kraft mill at Tacoma, Washington. By the end of 1971, after an $18.5 million expenditure for a new recovery system and primary water treatment, the mill should be in fairly good shape. Another $5.8 million has been budgeted for installation of primary treatment at the Bucksport and Sartell mills, and $11.5 million for secondary treatment at Pensacola, Rhinelander, and Jacksonville.

In no case is the company rushing to finish the job before state deadlines. For example, primary treatment at Bucksport, Maine, which should take no more than a year to plan and build, will not even begin to be constructed until 1974 because the state water pollution standards don't apply until 1976. (An aside — 52 percent of the land in Maine is owned by the paper and pulp industry.)

In other cases the company is stalling because the costs of cleaning up may be more than the value of the mills. No plans exist for two of the oldest, smallest mills — Deferiet, New York and Cornell, Wisconsin — which have been pouring untreated groundwood and bleaching wastes into the Chippewa and Black Rivers for 70 years. When the state deadline arrives, St. Regis will decide whether to invest or close down, a decision which already has been made at Rhinelander, Wisconsin. The company recently announced that in light of the new Wisconsin water standards, the chemical pulping operations there would be shut down, probably at the end of 1972.

Though St. Regis is trying to catch up in pollution control and was extremely honest in discussing its pollution problems, its lax history is making the job much more difficult. The company anticipates that within two years, five of the nine locations will have secondary treatment and all will have primary, but the planned total pollution control expenditure of $36,050,000 will still only half clean up all the mills. CEP estimates that an additional

$23-$43 million will have to be invested before the job is done, bringing the total to between $59 and $79 million.

Table 24.1
St. Regis Paper Company: Pulp Production, Water Use, and Pollution Control

Mill Location	Pulp Prod. (TPD)	Other Prod. (TPD)	Water Use (MGPD)	Water Pollution Control			Air Pollution Control	
				Pri.	Sec.	Tert.	Part.	Gas, Odor
Jacksonville, Fla.	1,370	1,300	78	X	X	X	✓	X
Pensacola, Fla.	900	910	28	✓	X	X	X	X
Bucksport, Me.	270	600	12	X	X	X	X	—
Sartell, Minn.	125	200	5.8	X	X	X	✓	—
Monticello, Miss.	1,620	1,500	20	✓	✓	X	✓	✓
Deferiet, N.Y.	300	375	18	X	X	X	✓	X
Tacoma, Wash.	900	610	30	X	X	X	✓	X
Cornell, Wisc.	50	200	5.5	X	X	X	X	—
Rhinelander, Wisc.	120	240	25	X	X	X	X	—
Total	5,655	5,945	222.3					

Jacksonville Mill

Location:
Jacksonville, Florida

Date Built:
1955

Process:
1,370 TPD kraft

Other Production:
1,300 TPD wrapping paper, shipping sacks, linerboard, gumming paper

Total Water Use:
78 MGPD (60 MGPD for cooling)

Source:
Wells; Broward River (cooling water)

Discharge:
St. John's River

Table 24.2
Jacksonville Mill: Pollution Control Record

Treatment	Overall Evaluation	Equipment
Water		
Primary	X	None
Secondary	X	None
Tertiary	X	None
Air		
Particulate (fuel)	√	Fuel: bark, 1.5% sulfur oil. 95% efficient multiple cyclone collectors on boilers; one installed in 1967, one in 1969
Particulate (production)	√	Recovery furnace has precipitator; two of three lime kilns have scrubbers
Gas and odor	X	None
Plant Emissions		
To water:		
BOD	71,000 lb/day (50 lb/ton of pulp; company estimate)	
Suspended solids	N.A.	
To air:	N.A.	

Pollution Control Expenditures:
$2.3 million, including the cost of the two lime kiln scrubbers ($250,000); the
multiple cyclones ($250,000); the conversion of some power boilers from bark
to low sulfur oil ($100,000); and the renovation of the bark boilers ($350,000).
Also included are the costs of monitoring equipment, rebuilding and overhaul-
ing the recovery furnaces and precipitators, and $10,000 for a three-year
pilot program to develop an adequate scrubber for bark boilers.

Legal Status:
(Water) Not in compliance; will be in 1972. State suit pending for damages
from an oil spill.
(Air) Not in compliance with old regulations; new ones just developed.

Future Plans:
This mill is beginning a $6 million water quality program which is expected
to remove 90 percent of the oxygen-demanding pollutants in the mill process
water effluent. In 1971 a $3 million, 300-foot diameter clarifier with a capac-
ity of 20 million gallons will be installed to provide primary treatment. A
$1.4 million aerated lagoon system with a capacity of 200 million gallons will
go into operation by July 1, 1972.
    The mill is also planning to reduce its particulate emissions with a third
multiple cyclone dust collector on the power boilers and a scrubber on the
lime kiln. An odor control system is "under study."

Profile:
The St. Regis mill in Jacksonville is one of the largest industrial plants in
Florida and has long been a major concern of conservation-minded citizens,
particularly the Southeastern Environmental Council, Inc. Sulfur and carbon
emissions have significantly contributed to the city's air pollution problems
and the mill was described by newsman Charles Thompson as "Duval County's
largest water polluter," discharging "90 million gallons of mill wastes into
the St. John's River each day."[1]
    This effluent, combined with the 15 million gallons of raw sewage con-
tributed by the city, has rendered the river so dirty that the shellfish are in-
edible.
    Mill manager Denholm Smith, who is also a member of Duval County's
Water Quality Control Board, expressed a less than enthusiastic philosophy
about a thorough "clean up." On a WJXT News Special Report Smith said: "I
think it highly unlikely that St. Regis or any company would do this [under-
take a six million dollar investment in pollution control] on a voluntary basis
[without state enforcement] unless it could be assured that other companies
in the same business were going to do so—otherwise it would put us at a competitive disadvantage."[2] Thus admitting that state laws were the determining
factor in inspiring any pollution control program, he complained in a newspaper interview: "Standards for both air and water quality appear to be changing almost daily—and always higher—making our job of correcting our pollu-

tion problems doubly difficult. It has been our recent experience that no sooner than we spend a large amount of investors' money for pollution control equipment the rules change and we are in violation of new standards."[3]

By "large amounts of investors' money," Mr. Smith was evidently referring to the $2.3 million spent in the last few years for the mill's first major pieces of air control equipment—lime kiln scrubbers, a recovery furnace precipitator, and conversion of some power boilers from bark to low sulfur oil as fuel. He also may have had in mind the $6 million planned expenditure for a clarifier and aerated lagoon system being built to satisfy a new state requirement of 90 percent BOD reduction by January 1973.

Even when all the planned equipment has been completed, however, St. Regis will have some very basic unsolved pollution problems which the state is bound to notice. For example, the mill will still have no odor control; 60 million gallons of cooling water will still be returned to the St. Johns River 8-9 degrees warmer than when it left; and 18 million gallons of fresh water will still be withdrawn daily from what scientists fear is a dwindling water table.[4]

CEP Estimate of Minimum Necessary Pollution Control Expenditure Not on Company Budget: $800,000 total, of which $500,000 is for an odor control system and $300,000 is for a cooling tower.

Pensacola Mill

Location:
Pensacola, Florida

Date Built:
1941, 1955

Process:
900 TPD kraft from two mills (300 TPD bleached)

Other Production:
580 TPD natural kraft shipping sack and liner board
250 TPD bleached board specialties, drinking cups, food wrap and bag papers
80 TPD boards for food containers, tag boards

Total Water Use:
28 MGPD

Source:
Deep wells

Discharge:
Eleven Mile Creek (flows into Perdido Bay)

Table 24.3
Pensacola Mill: Pollution Control Record

Treatment	Overall Evaluation	Equipment
Water		
Primary	✓	Settling pond
Secondary	X	Natural lagoon for 3.5-day retention (inadequate)
Tertiary	X	None
Solid waste removal	—	Landfill on the mill site
Other	—	Pilot program for waste water treatment with activated carbon (aimed at reuse of water). Partially financed by FWQA grant of $878,472
Air		
Particulate (fuel)	✓	Fuel: gas and bark. Precipitator on boiler
Particulate (production)	X	Three of six recovery furnaces have 90% efficient precipitators
Gas and odor	X	"Cooking liquor controls"
Plant Emissions		
To water:		
BOD		55,000 lb/day (61 lb/ton of pulp; FWQA estimate)
Suspended solids		30,000 lb/day (33 lb/ton of pulp; company estimate)
To air:		N.A.

Pollution Control Expenditures:
$1.2 million plus the company share of the pilot activated carbon project which was $583,090 direct and $2 million indirect expenditure.

Legal Status:
(Water) State abatement action on January 1, 1967, required 90 percent removal of BOD. The mill will be in compliance when the lagoon expansion is complete at the end of 1971.

(Air) Not in compliance. Regulations in the process of revision and the old ones are not being enforced.

Future Plans:
By the end of 1971, the secondary treatment lagoon will be expanded to retain effluent for 4.5 days. The company is also studying the possibility of consolidating the two mills on the site and then installing an odor control system at some time in the future. (At present, the only budgeted figure is $1.5 million for the lagoon expansion.)

Profile:
In January 1970, a federal water pollution abatement conference on Perdido Bay concluded that the inadequately treated effluent from the St. Regis mill is the major cause of the low dissolved oxygen level, unsightly foam, excessive sludge deposits, and increased lignin in Perdido Bay and River, as well as the degraded water quality in Eleven Mile Creek. The mill discharges 98 percent of the total BOD in Perdido Bay, the equivalent of the wastes of 330,000 people.

The conference recommended that St. Regis reduce the amounts of carbonaceous materials discharged by 90 percent so that the BOD of the effluent does not exceed 8,800 pounds per day; remove foam-causing materials; and improve the color of the effluent so that it meets the water quality standards of Florida, Alabama, and the federal government. In addition, the company should make a feasibility study of an essentially closed system involving recirculation, treatment, and reuse of its process water by January 1, 1971.

The conference said that abatement facilities to satisfy these recommendations should be in operation by December 31, 1972 and the company has planned its $1.5 million lagoon expansion for compliance.

CEP Estimate of Minimum Necessary Pollution Control Expenditure Not on Company Budget: $3.5 million total, of which $3 million is for new precipitators on recovery furnaces and $500,000 is for black liquor oxidation. Should St. Regis plan to rebuild the mill's kraft recovery operation with a large new furnace, a "low odor" indirect contact evaporator, a new electrostatic precipitator and accessory equipment (as at the company's Tacoma mill), an investment of $10 to $20 million might be called for.

Bucksport Mill

Location:
Bucksport, Maine

Date Built:
1930

Process:
270 TPD groundwood (75 TPD bleached)

Other Production:
600 TPD coated publication grade paper

Total Water Use:
12 MGPD

Source:
Lakes

Discharge:
Penobscot River (estuary)

Table 24.4
Bucksport Mill: Pollution Control Record

Treatment	Overall Evaluation	Equipment
Water		
Primary	X	None
Secondary	X	None
Tertiary	X	None
Air		
Particulate (fuel)	X	Fuel: 2.25% sulfur oil. No equipment
Particulate (production)	—	None needed
Gas and odor	—	None needed
Plant Emissions		
To water:		
BOD	16,000 lb/day (59 lb/ton of pulp)	
Suspended solids	N.A.	
To air:	N.A.	

Pollution Control Expenditures:
$500,000 for in-plant collection and separation of sanitary wastes.

Legal Status:
(Water) On compliance schedule, but state has not yet approved plans.
(Air) No state standards implemented yet.

Future Plans:
A $3 million primary treatment clarifier with a capacity of 12 million gallons
is now being planned and will be in operation in 1975. By 1972 sanitary wastes
will be piped from the mill to a municipal treatment plant now under construction in the town. $1 million will be spent on piping and process rearrange-
ment for this.

Profile:
Until 1968, the Bucksport site also had a sulfite mill. This was closed for
"economic and pollution abatement reasons" after an April 1967 federal water
pollution abatement conference on the quality of the Penobscot River. This
conference recommended an 85 percent reduction in waste sulfite liquor effluent and treatment for reducing BOD and suspended solids. Closing down
the sulfite mill satisfied these requirements and alleviated the pollution of the
river considerably.

    However, the groundwood mill and paper plant which remain in operation
at Bucksport have no water treatment facilities and the quality of the effluent
is poor. The BOD load from groundwood pulp ordinarily averages 30-60 pounds
per ton; at this mill it is 59 pounds per ton. The figure is so high, presumably, because of the bleaching and paper mill wastes in the effluent. St. Regis
is not rushing to clean up this mill. Maine laws conveniently require almost
nothing of industry until 1976, so it seems that it will be four years before
construction of any water control equipment will begin.

    CEP Estimate of Minimum Necessary Pollution Control Expenditure Not
on Company Budget: $2.25 million total, of which $2 million is for secondary
water treatment and $250,000 is to improve power boiler particulate control.

Sartell Mill

Location:
Sartell, Minnesota

Date Built:
1906

Process:
125 TPD bleached groundwood

Other Production:
200 TPD catalogue and magazine papers

Total Water Use:
5.8 MGPD

Source:
Mississippi (headwaters)

Discharge:
Mississippi (headwaters)

Table 24.5
Sartell Mill: Pollution Control Record

Treatment	Overall Evaluation	Equipment
Water		
Primary	X	None
Secondary	X	None
Tertiary	X	None
Air		
Particulate (fuel)	✓	Fuel: gas, bark, 2.2% sulfur coal. 85% efficient multiple-cyclone scrubbers on boilers
Particulate (production)	—	None needed
Gas and odor	—	None needed
Plant Emissions	N.A.	

Pollution Control Expenditures:
$130,000 (for particulate control equipment); $1.8 million is being spent on clarifier now under construction.

Legal Status:
(Water) Not in compliance, but the company has agreed to construct adequate sewage facilities. The mill has until May 31, 1973, to install and have in operation secondary water treatment facilities.
(Air) In compliance. State permit system does not begin until 1971.

Future Plans:
Secondary treatment facilities by May 31, 1973.

Profile:
This is another very old mill, which, as it produces groundwood, does not present very serious or hard-to-handle pollution problems — bits of wood, fibers, and bleaching chemicals. Adequate control is being provided on the slowest possible timetable — perhaps because St. Regis has so many larger mills that lack adequate controls and take priority.

CEP Estimate of Minimum Necessary Pollution Control Expenditure Not on Company Budget: $1 million total, for secondary water treatment.

Ferguson Mill

Location:
Monticello, Mississippi

Date Built:
1969

Process:
1620 TPD unbleached kraft

Other Production:
1,500 TPD natural kraft speciality papers and linerboard

Total Water Use:
20 MGPD
16 MGPD discharged

Source:
Pearl River

Discharge:
Pearl River

Table 24.6
Ferguson Mill: Pollution Control Record

Treatment	Overall Evaluation	Equipment
Water		
Primary	√	12.5 million gallon clarifier
Secondary	√	850 million gallon aerated lagoon; 20-day retention pond

Table 24.6
Ferguson Mill: Pollution Control Record (continued)

Treatment	Overall Evaluation	Equipment
Tertiary	X	None
Air		
Particulate (fuel)	✓	Fuel: gas and bark. No equipment needed
Particulate (production)	✓	Recovery furnace: precipitator, 99% efficient. Lime kiln: scrubber, 99% efficient
Gas and odor	✓	Heavy black liquor oxidation

Plant Emissions

To water:

BOD   5,000 lb/day* (3.3 lb/ton)

Suspended solids   N.A.

To air: N.A.

*This represents a 90% reduction of the 50,000 pound BOD load that enters the clarifier.

Pollution Control Expenditures:
$4.84 million of the $100 million total capital cost of the mill. $3 million for water; $1.84 million for air.

Legal Status:
(Water) In compliance with a nonexpiring discharge permit.
(Air) No relevant state standards yet.

Future Plans:
None needed

Profile:
This brand new mill is the star of St. Regis and shows that the company is capable of providing excellent pollution control equipment. It is in compliance with state water standards and has even prepared for the time when Mississippi has relevant air quality standards.

Deferiet Mill

Location:
Deferiet, New York

Date Built:
1900

Process:
200 TPD bleached and unbleached groundwood
100 TPD magnesium bisulfite

Other Production:
375 TPD coated and uncoated book paper; catalogue, directory, offset and
printing, and opaque papers

Total Water Use:
18 MGPD

Source:
Black River

Discharge:
Black River

Table 24.7
Deferiet Mill: Pollution Control Record

Treatment	Overall Evaluation	Equipment
Water		
Primary	X	None
Secondary	X	None. All waste pulp and bleaching liquor discharged directly into Black River
Tertiary	X	None
Air		
Particulate (fuel)	✓	Fuel: 1.8% sulfur coal. 94% efficient multiple cyclone collectors on boiler

Table 24.7
Deferiet Mill: Pollution Control Record (continued)

Treatment	Overall Evaluation	Equipment
Particulate (production)	X	None
Gas and odor	X	None

Plant Emissions	
To water:	
BOD	84,000 lb/day (280 lb/ton of pulp)
Suspended solids	N.A.
To air:	N.A.

Pollution Control Expenditures:
$80,000 for the multiple cyclone dust collectors.

Legal Status:
(Water) State abatement order in February, 1966, cited "excessive floating and settleable solids, too little dissolved oxygen, and sludge deposits." The combination of no water clarifying facilities and direct discharge of waste liquor accounts for the inordinately high BOD load. The state deadline for adequate water treatment is December, 1972.
(Air) Power boiler complies with state regulations. The state is reviewing the situation of other emissions.

Future Plans:
St. Regis is only planning primary treatment here to reduce the BOD to the still very high 24,000 pounds per day. No decision has been made on what equipment will be installed or when.

Profile:
St. Regis converted to magnesium pulping liquor in 1966. Although chemical recovery is feasible with this liquor base, the mill has continued to discharge all its liquor into the Black River for the past four years. It is clear that no chemical recovery system has been built because, for this very small old mill, the cost would be prohibitively high. The mill will probably continue to operate as is, stalling until it is face to face with the December, 1972, New York State deadline—when it will be forced to make a decision to invest in the necessary pollution control equipment or close down its chemical pulping operations.

CEP Estimate of Minimum Necessary Pollution Control Expenditure Not
On Company Budget: $12 million total, of which $10 million is for a chemical
recovery system and $2 million is for primary and secondary water treat-
ment.

Tacoma Mill

Location:
Tacoma, Washington

Date Built:
1928, 1960

Process:
900 TPD bleached, semi-bleached, and unbleached kraft (company figure)

Other Production:
610 TPD of a wide variety of specialty papers, including sack and wrapping
papers, tag and cup stock, and linerboards

Total Water Use:
30 MGPD

Source:
Puyallup River

Discharge:
Puyallup River

Table 24.8
Tacoma Mill: Pollution Control Record

Treatment	Overall Evaluation	Equipment
Water		
Primary	X	Clarifier treating only 3 million gallons of barking waste water
Secondary	X	None
Tertiary	X	None
Other	—	Outfall pipe

Table 24.8
Tacoma Mill: Pollution Control Record (continued)

Treatment	Overall Evaluation	Equipment
Air		
Particulate (fuel)	X	Fuel: 1.25% sulfur oil, hog fuel. No equipment
Particulate (production)	✓	Three recovery furnaces each with 99% efficient precipitator
Gas and odor	X	None
Plant Emissions		
To water:		
BOD	44,000 lb/day (49 lb/ton of pulp)	
Suspended solids	N.A.	
To air:	N.A.	

Pollution Control Expenditures:
$1.9 million; $1.75 million is now being spent on a clarifier to reduce the
BOD to 40,000 pounds per day and remove 90 percent of suspended solids.

Legal Status:
(Water) Not in compliance now. The present permit requires installation of —
or provision for — secondary treatment facilities, but the state is not pressing
for this.
(Air) Not in compliance now. Company aims to meet part of the odor deadline
in 1971 and the standards for noncondensible gases and recovery boiler emissions by 1974.

Future Plans:
St. Regis is now making extensive expenditures for air pollution control in
order to comply with the Washington State laws. A $17 million system which
includes nondirect contact evaporation to reduce odor, and equipment for the
collection and burning of noncondensible gases will be completed in 1971. The
mill will then be in very good shape in terms of particulate and gas emissions.

Profile:
This mill has made improvements in air pollution control; however, the liquid

effluent will still be very dirty, even after completion of the new 1.7 million gallon clarifier. When mill manager Robert F. Lynch announced the planned clarifier in February, 1970, he stressed that his would "put the finishing touches on a water control program which has maintained high standards of quality through the years."[5] It would be hard to draw that conclusion for a mill that even now will only be providing adequate primary treatment.

CEP Estimate of Minimum Necessary Pollution Control Expenditure Not on Company Budget: $2.25 million total, of which $2 million is for secondary water treatment and $250,000 is to improve power boiler particulate control.

Cornell Mill

Location:
Cornell, Wisconsin

Date Built:
1900

Process:
50 TPD groundwood (and some recycled waste paper)

Other Production:
200 TPD of a wide variety of boards, including manila folding box, book match, and milk bottle cap boards

Total Water Use:
5.5 MGPD

Source:
Chippewa River

Discharge:
Chippewa River

Table 24.9
Cornell Mill: Pollution Control Record

Treatment	Overall Evaluation	Equipment
Water		
Primary	X	None
Secondary	X	None

Table 24.9
Cornell Mill: Pollution Control Record (continued)

Treatment	Overall Evaluation	Equipment
Tertiary	X	None
Air		
Particulate (fuel)	X	Fuel: coal, gas, hydroelectric. No equipment
Particulate (production)	—	None needed
Gas and odor	—	None needed
Plant Emissions		
To water:		
BOD	4,000 lb/day (80 lb/ton of pulp)	
Suspended solids	N.A.	
To air:	N.A.	

Pollution Control Expenditures:
None seem to have been made.

Legal Status:
(Water) Plan for treatment facilities pending at the state pollution control authority.
(Air) State program not yet directed to pulp mills.

Future Plans:
N.A.

Profile:
Another on the St. Regis roster of old uncontrolled mills. There is no bleaching here, so pollution problems are minimal.

CEP Estimate of Minimum Necessary Pollution Control Expenditure Not on Company Budget: $600,000 total, of which $100,000 is for a clarifier or settling pond; $500,000 is to improve power boiler particulate control.

Rhinelander Mill

Location:
Rhinelander, Wisconsin

Date Built:
1913

Process:
120 TPD bleached and unbleached calcium sulfite

Other Production:
240 TPD of a wide variety of treated papers including wax paper, glassine, greaseproof, transparent, tracing papers

Total Water Use:
25 MGPD; 18 MGPD discharged

Source:
Upper Wisconsin River

Discharge:
Upper Wisconsin River

Table 24.10
Rhinelander Mill: Pollution Control Record

Treatment	Overall Evaluation	Equipment
Water		
Primary	X	None
Secondary	X (partial)	None. However, spent sulfite liquor not discharged (collected for by-product manufacture: 15 TPD nutritional yeast and 70 TPD calcium lignosulfonate)
Tertiary	X	None
Other	—	12 savealls on paper machines
Air		
Particulate (fuel)	X	Fuel: 2.5% sulfur coal. No control equipment

Table 24.10
Rhinelander Mill: Pollution Control Record (continued)

Treatment	Overall Evaluation	Equipment
Particulate (production)	—	None needed
Gas and odor	—	None needed

Plant Emissions

To water:

| BOD | 53,000 lb/day (441 lb/ton of pulp; company figure) |
| Suspended solids | 17,000 lb/day (141 lb/ton of pulp; state figure) |

To air: N.A.

Pollution Control Expenditures:
$1,502,023 (primarily for savealls).

Legal Status:
(Water) Mill required to install water treatment facilities which will remove at least 80 percent of the BOD.
(Air) No abatement actions taken yet.

Future Plans:
The mill is planning to install a 30 million gallon capacity clarifier and an aeration system to satisfy the state water quality requirement. This $4 million water treatment facility will reduce the daily BOD to 11,000 pounds. In September, 1970, St. Regis announced that the sulfite operations at Rhinelander will be closing, probably in about two years. A company representative said, "Here's the classic case of a small old mill, where we just can't afford to make the required expenditures for pollution control under the new laws. It's sad. Rhinelander's not a wealthy area; 125 people are losing their jobs, but we have to close. The paper operations will continue and we will still be proceeding with water treatment plans."

Profile:
The Rhinelander mill has tried to utilize much of its waste material rather than discharging it. Bark is recovered and processed for use as fuel; the spent sulfite liquor and dissolved lignin are used to manufacture by-products; much of the potentially lost fiber from paper making is collected and reused; and water from the paper making operation is reused in the blow pits and bleaching processes.

However, the quality of the effluent is still very low, because there are no treatment facilities. In 1970, the Upper Wisconsin River Pollution Investigation Survey found that discharges from the mill "... are producing a great effect on biological as well as the physical characteristics of the river. This sampling site, which is approximately one-half mile below the Rhinelander Paper Division, reveals the bottom to be covered with fibers, wood chips, and slime growths heavy enough to make it difficult to sample the original bottom material ... The few organisms found were sludgeworms, sow bugs, scuds, and tolerant midges."

A student at Nicolet College in Rhinelander graphically described the effects of the mill on the local environment:

"The pollution situation of both the air and water is terrible. Sometimes the river turns weird yellowish colors and is full of foamy substance. But the worst thing is that you can't park your car out overnight—no one does it—or else by morning it is covered with a thick dirt scum which is hard to get off. And you can't hang your clothes out either. Some days the air is so bad that walking downtown is like walking in L.A. during one of those heavy smog attacks."

He added that the mill seems to have been responsible for almost all of the river's pollution, as it is the biggest industry in the area, and the water upstream of the mill is clear. As a result of the extensive pollution caused by this mill and St. Regis' conclusion that an adequate cleanup of existing operations would be economically unfeasible, the company made its announcement that the sulfite pulping mill would be closed down by the end of 1972.

CEP Estimate of Minimum Necessary Pollution Control Expenditure Not on Company Budget: $500,000 total, to improve power boiler particulate control.

References for Chapter 24

1. WJXT News Special Report, April 13, 1970. (Script sent to CEP by the Southeastern Environmental Council, P.O. Box 31278, Yukon Branch, Jacksonville, Florida 32230.)

2. Ibid.

3. Mr. Smith, quoted in the St. Regis News, reprinted as "Extension of Remarks" of Representative Charles E. Bennet of Florida, Congressional Record, May 22, 1970, p. E4644.

4. WJXT News Special Report.

5. Paper Age, February 17, 1970.

Chapter 25

SCOTT PAPER COMPANY
International Airport, Philadelphia, Pennsylvania 19113

Major Products:
Various paper products, plastics products (wrapping, cups, etc.), furniture.

Major Consumer Brands:
SCOTTIES, LADY SCOTT facial tissues, baby SCOTT disposable diapers,
SCOT TISSUE, WALDORF, SOFT WEAVE, LADY SCOT toilet tissues,
VIVA SCOT TOWEL paper towels, CUT-RITE wax paper, SCOTKINS napkins,
CONFIDETS sanitary napkins, FLOKOTE printing papers.

Financial Data ($ Millions):	1969	1968
Net Sales	731.5	684.6
Net Income	60.0	53.6
Capital Expenditures	71.2	69.4

Pollution Control Expenditures:
$ 15 million, 1960-1970; $97 million (estimated), 1970-1978.

Annual Meeting:
The annual meeting takes place during April in Philadelphia.

Officers:
President: Charles D. Dickey, Jr.
Chairman: Harrison F. Dunning, Moylan, Pennsylvania (Director: National
Biscuit Co., Bell Telephone Co. of Philadelphia)

Outside Directors:
Thomas S. Gates: Chairman of the Executive Committee, Morgan Guaranty
Trust Co., N.Y.
Charles B. Harding: Former Chairman of the Board, Smith, Barney and Co.,
Inc.
Curtis M. Hutchins: Chairman of the Board, Dead River Co., Bangor, Maine
W. M. Jenkins: Chairman of the Board, Seattle First National Bank
Charles H. Kellstadt: Retired Chairman of the Board, Sears, Roebuck and
Co.; Chairman of the Board and President, General Development Corp.
Ralph Lazarus: Chairman of the Board, Federated Department Stores, Inc.,
Cincinnati
William W. Scranton: Chairman of the Board, National Liberty Corp.,
former Governor of Pennsylvania
Samuel R. Sutphin: Chairman, Advisory Board, Plastic Coating Corp.
Jay L. Taylor: Owner, Jay Taylor Cattle Co., Chairman of the Board, Baker
and Taylor Drilling Co.

Breakdown of Total Company Sales:

Packaged products   63% of sales

Other paper products and pulp   28% "   "

Other products   9% "   "

Domestic Plants:
Packaged products   13 (including Brunswick, Ga.,
                                                                  50% owned by Mead)

S. D. Warren Division   6
Plastic Coating Corp.   4
Foam Division   2
Education Division   2
Disposables Division   2
Brown-Jordan Division   2
(indoor-outdoor furniture)

Timberholdings:
Scott owns 3.2 million acres, scattered about the world (an area about as
large as Connecticut), valued at $58,000,000.

Total Annual Pulp Production:
1,432,000 tons (1969)

Company Pollution Overview

Scott Paper Company operates pulp mills at nine locations scattered over the
United States, and is co-owner, with the Mead Corporation, of an 1,100 TPD
kraft mill in Brunswick, Georgia. Including half of Brunswick's production,
the company's mills produce 4084 tons of pulp and 3680 tons of paper daily,
using 259-264 million gallons of water.

Scott's pollution control record is generally unimpressive, although its
mills vary in age, size, and pollution control. Eight of Scott's mills have
adequate primary water treatment. Only the largest, the kraft mill at Mobile,
Alabama, has adequate secondary treatment, even though all but one small
groundwood mill produce bleached pulp, and thus discharge considerable
amounts of dissolved chemical and organic matter. Making matters worse,
five of the mills pulp by the sulfite process, the effluent of which has a very
high BOD. The Winslow, Maine, Anacortes, Washington, and Marinette,
Wisconsin mills discharge this spent sulfite liquor completely untreated;
Everett, Washington, provides only primary treatment, and Oconto Falls,
Wisconsin evaporates 50 percent of the liquor and discharges the rest after
primary and partial secondary treatment. All four of the company's kraft
mills have adequate primary treatment.

Scott's air pollution control is somewhat better. Three of the mills have
adequate gas and odor control, and six have adequate particulate control.

Only the Muskegon, Michigan, and Anacortes, Washington, mills have both.

Although the lack of pollution control at the company's two very small mills—in Menominee, Michigan, and Marinette, Wisconsin—should not be balanced as heavily as similar inadequacies at the larger mills, this overall pollution control record is far from satisfactory. It is also difficult to understand in light of the fact that Scott is definitely aware of pollution control, and in a 1970 brochure published the kind of specific information on pollution control equipment that CEP spent months trying to extract from other companies. Another contradiction is that Scott worked with Crown Zellerbach to develop the ammonia-base sulfite pulping process and long ago began converting its calcium-base sulfite mills. Yet only Oconto Falls has even partial recovery of its spent liquor.

As demonstrated by plans more impressive than past actions, and by its general policy of disclosure, Scott recognizes its pollution problems. It has indicated specific planned expenditures of $55 million for spent (ammonia) sulfite liquor recovery at Everett ($36 million) and Winslow ($19 million); and $2 million for added water treatment at Mobile and Oconto Falls. CEP estimates that an additional, unbudgeted $22.3 million—including half of the $10.25 million which should be shared with Mead to clean up the Brunswick mill—is required to install both planned and unplanned-but-needed pollution control equipment at all the Scott mills.

Table 25.1
Scott Paper Company: Pulp Production, Water Use, and Pollution Control

Mill Location	Pulp Prod. (TPD)	Other Prod. (TPD)	Water Use (MGPD)	Water Pollution Control			Air Pollution Control	
				Pri.	Sec.	Tert.	Part.	Gas, Odor
Mobile, Ala.	1,400	1,400	63	✓	✓	✗	✓	✗
Brunswick,* Ga.	550	125	37.5	✓	✗	✗	?	✗
Westbrook, Me.	300	515	25-30	✓	✗	✗	✓	✗
Winslow, Me.	430	550	21	✗	✗	✗	✗	✗
Menominee, Mich.	22	--	1	✓	—	—	—	—
Muskegon, Mich.	225	350	21	✓	✗	✗	✓	✓
Anacortes, Wash.	135		7	✓	✗	✗	✓	✓

Table 25.1
Scott Paper Company: Pulp Production, Water Use, and Pollution Control
(continued)

Mill Location	Pulp Prod. (TPD)	Other Prod. (TPD)	Water Use (MGPD)	Water Pollution Control			Air Pollution Control	
				Pri.	Sec.	Tert.	Part.	Gas, Odor
Everett, Wash.	850	450	65	✓	✗	✗	✗	✓
Marinette, Wisc.	57	175	6.7	✗	✗	✗	✓	✗
Oconto Falls, Wisc.	115	115	12	✓	✗	✗	✓	✗
Total	4,084	3,680	259.2-264.2					

*Half owned by Scott half owned by the Mead Corp. Production divided weekly
between Scott and Mead. (The water use and production figures for this plant
are divided between the two companies.)

Mobile Mill

Location:
Mobile, Alabama

Date Built:
1939, acquired by Scott in 1954

Process:
1,400 TPD un-, semi-, and fully-bleached kraft

Other Production:
1,400 TPD tissue, bottle cap, cup paper and stock, milk carton, facial tissue, etc.

Total Water Use:
63 MGPD

Source:
Wells, Mobile River

Discharge:
Mobile River (downstream of the International Paper Mill)

Table 25.2
Mobile Mill: Pollution Control Record

Treatment	Overall Evaluation	Equipment
Water		
Primary	✓	70 MGPD clarifier
Secondary	✓	Activated sludge system; 50 MGPD day capacity (built in 1961, enlarged 1965)
Tertiary	X	None
Air		
Particulate (fuel)	✓	Fuel: bark, gas, interruptible oil. 90% efficient mechanical collector
Particulate (production)	✓	Three recovery furnaces: 95% efficient precipitators. Four lime kiln scrubbers
Gas and odor	X	No odor control system
Plant Emissions		
To water:		
BOD	25,000 lb/day (18 lb/ton of pulp)	
Suspended solids	N.A.	
To air:	N.A.	

Pollution Control Expenditures:
$3 million for activated sludge plant.

Legal Status:
(Water) In compliance. (Mill was mentioned, but not criticized, at the federal water pollution abatement conference for Mobile Bay in January, 1970.)
(Air) State regulations not yet implemented.

Future Plans:
$1 million for separation and treatment of sanitary wastes from the main mill

sewege system to be completed by February, 1972; odor control facilities are
planned but no appropriation has yet been made.

Profile:
This mill provides good effluent treatment, and has the largest activated sludge
plant in the paper industry, with annual operating costs of over $500,000.
(Scott expects these to rise appreciably with the year-round operation, but
does not yet have a figure as to how much.) Until 1970, this secondary treatment system was operated only during the summer months; it is now operated
year-round.

In 1962, the mill received the Governor's Conservation Award from the
Alabama Wildlife Federation for "outstanding achievement in water conservation." This is noteworthy because even now—years later—the vast majority
of paper mills in the country do not yet have secondary treatment.

Air pollution control here, however, is more ordinary. There is no odor
control system; but particulate control is excellent, and Scott has done a good
job with this mill since acquiring it.

CEP Estimate of Minimum Necessary Pollution Control Expenditures Not
in Company Budget: $1 million total, for an odor control system.

Brunswick Mill

Location:
Brunswick, Georgia (jointly owned with Mead)

Date Built:
1938

Process:
1,100 TPD bleached kraft (half attributed to Mead)

Other Production:
250 TPD board

Total Water Use:
75 MGPD (half attributed to Mead)

Source:
Wells

Discharge:
Turtle River (tidal stream to Atlantic)

Table 25.3
Brunswick Mill: Pollution Control Record

Treatment	Overall Evaluation	Equipment
Water		
Primary	✓	12 million gallon capacity clarifier installed in 1970
Secondary	X	240 acre holding pond built in 1960 (inadequate)
Tertiary	X	None
Other	—	Outfall and diffuser
Air		
Particulate (fuel)	✓	Fuel: gas, oil, bark. Bark burners have 94% efficient dust collectors
Particulate (production)	?	Lime kiln: over 99% efficient venturi scrubbers, installed 1968
Gas and odor	X	No black liquor oxidation
Plant Emissions	N.A.	

Pollution Control Expenditures:
$4 million for the clarifier and deepening of the holding pond (in process).

Legal Status:
(Water) Secondary treatment required by the end of 1972.
(Air) Company program is under discussion with the state agency.

Future Plans:
The mill is planning a new recovery furnace with non-direct contact evaporation to control odor and more than 99 percent efficient precipitators. This unit will replace 2 of the 3 old units now operating.

Profile:
This mill, which is jointly owned by Scott and the Mead Corporation, is quite dirty and much neglected. Because it is such a large mill, with a significant impact on its environment, both companies are to be censured for waiting so long to take obvious measures to improve the situation. The planned improvements, which are being made in response to state requirements, are much

needed. However, they are not complete, and seem to be an attempt to clean up halfway. Even after they are completed, much will remain undone. Scott and Mead hope that the combination of a clarifier and deepened holding pond will provide satisfactory secondary treatment, but the state is dubious, and if a satisfactory BOD reduction is not achieved, further measures will be taken. Similarly, the new recovery furnace will be state-of-the-art technology, but the companies have not yet decided whether to add odor control equipment to the remaining old recovery unit. It is a little late for companies to be beginning halfway efforts.

CEP Estimate of Minimum Necessary Pollution Control Expenditures Not in Company Budget: $10.25 million total, of which $8 million is for a new recovery furnace with non-direct contact evaporators and precipitator; $500,000 is for odor control on an old recovery furnace; $500,000 is for a precipitator for an old recovery furnace; $250,000 is for improved bark boiler particulate control; and $1 million is for completion of the secondary treatment facilities.

Westbrook Mill (operated by S. D. Warren Division of Scott)

Location:
Westbrook, Maine

Date Built:
1854 (oldest mill in this study); acquired by Scott in 1967

Process:
300 TPD bleached kraft

Other Production:
515 TPD book, coated book, offset, and specialty papers

Total Water Use:
25-30 MGPD

Source:
Presumpscot River

Discharge:
Presumpscot River

Table 25.4
Westbrook Mill: Pollution Control Record

Treatment	Overall Evaluation	Equipment
Water		
Primary	✓	20 million gallon clarifier installed in 1966, "removes 60 tons of solid wastes daily" and reduces sludge to 30% solids
Secondary	X	None
Tertiary	X	None
Solid waste removal	—	On-site landfill
Air		
Particulate (fuel)	✓	Fuel: bark and oil. Bark burners have cyclone dust collector
Particulate (production)	✓	Recovery furnace: 99% efficient precipitator, venturi scrubber. Lime kiln: rotoclone dust collector "loses only about 4 lb/hour," and wet scrubber
Gas and odor	X	Weak black liquor oxidation, 90% efficient, installed in 1962 (inadequate); foam control unit-installed in 1969; chlorine scrubber for blow gases; digester relief gases burned in lime kiln

Plant Emissions

To water: N.A.

To air:

TRS   "Well below 120 ppm"

Particulates   N.A.

Pollution Control Expenditures:
$3.5 million: $1.8 million for air; $1.7 million for water.

Legal Status:
(Water) Compliance schedule requires secondary treatment in 1976.
(Air) State regulations not yet adopted.

Future Plans:
A vaporsphere to collect gases for incineration will be installed; "secondary treatment is under study to determine the feasibility of a proposed regional joint treatment system as an alternate to on-site treatment." The study will be completed in April, 1971.

Profile:
This mill is no exception to the rule that water treatment at any mill is bound to reflect the permissiveness of the state legislature. As has been noted, Maine's industries have about four years longer than those in any other state to install secondary treatment. Thus, secondary treatment is "under study" at Westbrook. In addition, the mill has a federal grant for a pilot demonstration of reclamation of usable solids from clarifier sludge. This is not directly related to water treatment but would provide a financial return on primary water treatment and alleviate sludge disposal space problems.

Air pollution control is exceptionally good for a Maine mill. Particulates are adequately controlled, and the odor control is nearly adequate. The company is not planning to add heavy black liquor oxidation at this time, but installation of a vaporsphere will reduce sulfurous emissions from other sources.

CEP Estimate of Minimum Necessary Pollution Control Expenditures Not in Company Budget: $2.8 million total, of which $500,000 is for an improved black liquor oxidation system; $200,000 is for improved bark boiler particulate control; $100,000 is for a vaporsphere (planned but no cost given); and $2 million is for secondary or joint water treatment.

Winslow Mill

Location:
Winslow, Maine

Date Built:
1900

Process:
430 TPD bleached calcium sulfite

Other Production:
550 TPD tissue, towels, wax base stock, tabulating card stock, and ledger grades

Total Water Use:
21 MGPD

Source:
Kennebec River

Discharge:
Kennebec River

Table 25.5
Winslow Mill: Pollution Control Record

Treatment	Overall Evaluation	Equipment
Water		
Primary	X	None. Spent sulfite liquor is discharged directly into the Kennebec River
Secondary	X	None
Tertiary	X	None
Other	—	Bark screening and filters; savealls on paper machines
Air		
Particulate (fuel)	X	Fuel: coal. 90% efficient cyclone dust collector (inadequate)
Particulate (production)	—	None needed
Gas and odor	X	None

Plant Emissions

To water:
BOD   N.A.
Suspended solids   38,000 lb/day (88 lb/ton of pulp)

To air: N.A.

Pollution Control Expenditures:
N.A.

Legal Status:
(Water) Compliance program approved by the state for secondary treatment
by 1976.
(Air) State regulations not yet approved.

Future Plans:
$19 million will be spent to reduce the pollution from this mill three years
before the state deadline. Two clarifiers with a total capacity of 25 million
gallons will be installed by October, 1973, at a cost of $3-4 million. A sewer
system to separate sanitary wastes for municipal treatment will be constructed.
Also in 1973, at an estimated cost of $15 million, the mill will convert to am-
monia base pulping and install a recovery system for the spent liquor and an
absorption system for sulfur dioxide emissions. These improvements are expected to reduce the mill's BOD discharge by 80 percent, or the equivalent of
secondary water treatment.

Profile:
From a pollution control point of view, this is the worst mill Scott operates.
It has never had any water treatment beyond saavalls and filters. All the
wastes—19 tons per day of fibers, chips, sulfur and bleaching chemicals, dissolved organic matter—are seemingly washed into the Kennebec through holes
in the floor.
     In light of Maine's lackadaisical attitude toward water pollution, it is not
surprising that improvements at this mill have not had as high a priority as
those at Scott's mills in other states. However, the company is planning a
much needed renovation of the mill, the first step of which will probably be
to find all the holes in the floor and plug them up. Then, two clarifiers will be
installed and the mill will be converted to the ammonia base sulfite pulping
process. These steps will greatly lessen the water pollution problem here,
reducing the BOD load by 80 percent.

     CEP Estimate of Minimum Necessary Pollution Control Expenditures Not
in Company Budget: $2.25 million total, of which $2 million is for secondary
water treatment and $250,000 is for improved power boiler particulate con-
trol.

Menominee Mill

Location:
Menominee, Michigan

Date Built:
1884, acquired in 1941

Process:
22 TPD groundwood

Other Production:
None

Total Water Use:
1 MGPD

Source:
Menominee River

Discharge:
Menominee River

Table 25.6
Menominee Mill: Pollution Control Record

Treatment	Overall Evaluation	Equipment
Water		
Primary	✓	Screening, savealls
Secondary	—	None needed
Tertiary	—	None needed
Air		
Particulate (fuel)	✓	Hydroelectric power. None needed
Particulate (production)	—	None needed
Gas and odor	—	None needed
Plant Emissions	N.A.	

Pollution Control Expenditures:
N.A.

Legal Status:
(Water) In compliance; has a commission order restricting waste discharges
but the state says control is not adequate. The mill was cited for suspended
solids and fibers emissions at the 1963 federal water pollution abatement
conference on the Menominee River.
(Air) Regulations brand new; no investigation yet.

Future Plans:
Additional fine screens.

Profile:
This mill is operated in conjunction with the Marinette mill across the river, and obviously, considering its very small size, minimal water use, and lack of air pollution potential, cannot be considered a major problem for the company or environment. In addition, as it has been a marginal operation for 20 years, CEP agrees that it would be economically unfeasible, and probably unnecessary, to make any major pollution control expenditures at this mill.

CEP Estimate of Minimum Necessary Pollution Control Expenditures Not in Company Budget: None.

Muskegon Mill (operated by S. D. Warren Division of Scott)

Location:
Muskegon, Michigan

Date Built:
1900, acquired by Scott in 1967

Process:
225 TPD bleached kraft

Other Production:
350 TPD machine coated book, lithographic, text papers; coated covers, cast coated papers; specialty papers

Total Water Use:
21 MGPD

Source:
Lake Muskegon

Discharge:
Lake Muskegon (to Lake Michigan)

Table 25.7
Muskegon Mill: Pollution Control Record

Treatment	Overall Evaluation	Equipment
Water		
Primary	✓	15 million gallon clarifier installed in 1954
Secondary	✕	None
Tertiary	✕	None
Other	—	Screening
Air		
Particulate (fuel)	✕	Fuel: coal, bark. 90% efficient flyash collector (inadequate)
Particulate (production)	✓	Recovery furnace precipitator, 98.5% efficient. Lime kiln scrubber, 99.6% efficient
Gas and odor	✓	Black liquor oxidation
Plant Emissions	N.A.	

Pollution Control Expenditures:
N.A.

Legal Status:
(Water) "Violated Michigan water quality standards for protection of recreational uses and aquatic life. Need additional treatment to reduce solids, BOD, turbidity, and color."[1]
(Air) Under abatement order for odor.

Future Plans:
Heavy black liquor oxidation to eliminate the foaming problem and a system to incinerate digester relief gases were installed in 1970. A preliminary water treatment engineering study has been completed and Scott is deciding on a secondary treatment system. It is hoping to join a proposed Muskegon county regional system.

Profile:
This mill has adequate primary water treatment and good particulate control

of the recovery furnace and lime kiln emissions. With regard to odor control, the construction of heavy black liquor oxidation and an incineration system for the digester blow gases should have brought it up to "sniff." Adequate water treatment may take a while because it is only at the planning and budgeting stage.

CEP Estimate of Minimum Necessary Pollution Control Expenditures Not in Company Budget: $2.25 million total, of which $250,000 is for improved coal boiler particulate control and $2 million is for secondary or joint water treatment.

Anacortes Mill

Location:
Anacortes, Washington

Date Built:
1920s (acquired by Scott in the 1940s)

Process:
135 TPD bleached ammonia base sulfite—converted in 1951

Other Production:
None

Total Water Use:
7 MGPD

Source:
Skagit River via shallow wells

Discharge:
Guemes Channel (Puget Sound)

Table 25.8
Anacortes Mill: Pollution Control Record

Treatment	Overall Evaluation	Equipment
Water		
Primary	✓	Settling basin for pulp mill rejects; no discharge

Table 25.8
Anacortes Mill: Pollution Control Record (continued)

Treatment	Overall Evaluation	Equipment
Secondary	✓	None, all spent liquor is discharged through a diffuser
Tertiary	✗	None
Solid waste removal	—	On-site landfill
Other	—	Screening, outfall pipe with diffuser
Air		
Particulate (fuel)	✓	Fuel: gas, bark. Fly ash collectors installed in 1968; "very good efficiency" according to the company
Particulate (production)	—	None needed
Gas and odor	✓	Absorption tower for acid-making relief gases
Plant Emissions	N.A.	

Pollution Control Expenditures:
$170,000.

Legal Status:
(Water) This mill was cited at the 1967 federal Puget Sound abatement conference which recommended primary treatment by October, 1970, and elimination of the solids discharge, extension of the outfall pipe, and improvement of the diffuser. The company wrote to CEP on August, 1970: "As to the legal status of the mill, we recommend that you omit the recommendations of the Puget Sound abatement conference in 1967 as these were guidelines for the state in developing its implementation program and the state has found it necessary to make some modifications or departure from them in its implementation program in the case of most mills. The legal status of the mill is better expressed, in our opinion, by stating that a Washington State Waste Discharge Permit No. T-28-66 was issued March 31, 1970. This requires, among other things that the outfall and diffuser improvements be completed by December 1931, 1971. The permit expires December 31, 1974."
(Air) State is in the process of establishing sulfite mill regulations.

Future Plans:
As Scott feels that the mill "is too small for a clarifier," it will install fine
screens and modify the outfall system.

Profile:
This is an old mill which Scott renovated and converted to ammonia base pulping. Unfortunately, no recovery system was built, and all spent liquor is still
discharged directly into Guemes Channel.
    Scott was the next-to-last of those companies cited at the Puget Sound
conference to agree with the state on a pollution abatement program (Rayonier,
a subsidiary of ITT, is the remaining hold-out), perhaps because it doubted
the conference findings. According to a Washington Post article, "Robert I.
Thieme, then general manager of the West Coast division of Scott Paper, tes-
tified at the hearing that 'we seriously question the validity and interpretation
of the scientific studies' in the report on pollution from mill wastes. 'As a
matter of fact,' Thieme said, 'the natural abundance of Puget Sound and its
recognized fertility are incompatible with the conclusion of the studies that
pollution exists in this area.' Thieme added that the nutrients in the waste
matter going into the Sound might even be good for the fish."[2] However, the
Council on Environmental Quality's Report on Ocean Dumping came to the opposite conclusion, reporting that "high priority should be given to protecting
those portions of the marine environment which are biologically most ac-
tive ... "[3]

    CEP Estimate of Minimum Necessary Pollution Control Expenditures Not
in Company Budget: $2.25 million total, of which $2 million is for complete
secondary water treatment and $250,000 is for improved power boiler partic-
ulate control.

Everett Mill

Location:
Everett, Washington

Date Built:
1931; acquired 1951; alterations in 1954, 1969

Process:
850-950 TPD bleached ammonia-base sulfite pulp

Other Production:
450 TPD toilet tissue in rolls, roll and folded towels, facial tissue, place
mats, napkins and sanitary napkins

Total Water Use:
65 MGPD

Source:
Upper Sultan River

Discharge:
Port Gardner Bay (Puget Sound)

Table 25.9
Everett Mill: Pollution Control Record

Treatment	Overall Evaluation	Equipment
Water		
Primary	✓	Two clarifiers installed in 1964, actual capacity, 20 million gallons, operated at 17 million gallons
Secondary	X	None-all spent sulfite liquor is discharged into Puget Sound
Tertiary	X	None
Other		Dock front diffusers system, outfall pipe and diffusers shared with the Weyerhaeuser sulfite mill in Everett
Air		
Particulate (fuel)	X	Fuel: gas, oil, hog fuel. Five 90% efficient cyclone dust collectors (inadequate)
Particulate (production)	—	None needed
Gas and odor	✓	Sulfur dioxide removed from the digester blow by absorption towers; 90% efficient
Plant Emissions		
To water:		
BOD		Approx. 840,000-940,000 lb/day (989 lb/ton of pulp)*
Suspended solids		N.A.
To air:		N.A.

Table 25.9
Everett Mill: Pollution Control Record (continued)

---

*Pollutional Effects of Pulp and Paper Mill Wastes in Puget Sound, U.S. Department of the Interior, Federal Water Quality Administration, and Washington State Pollution Control Commission, March, 1969, p. 253.

Pollution Control Expenditures:
$4 million, with annual operating costs of $150,000.

Legal Status:
(Water) The 1967 federal Puget Sound abatement conference recommended
that by 1972, the diffuser be enlarged, sludge deposits in Port Gardner Bay
be eliminated, chip spillage stopped, and waste sulfite liquor discharge reduced. After the conference, the company was given a permit [4] until 1978
to achieve an 80 percent reduction in spent sulfite liquor discharge.
(Air) Sulfite mill regulations now under development.

Future Plans:
An estimated $36 million, 2-stage program for spent sulfite liquor collection
and incineration systems (including two new recovery furnaces), reduction of
suspended solids discharge, and chip unloading controls will be completed in
August 1978. The first stage, a $19 million system to incinerate 40 percent
of the spent liquor, will be complete by April 1974. The second state, which
is expected to achieve the required 80 percent overall reduction in liquor discharge, is scheduled for completion by August 1978.

Profile:
This mill presents an odd picture: although it was converted from calcium-
base to ammonium-base sulfite pulping (half in 1954 and half in 1969) the company did not install a liquor incineration system. On the other hand, the company claims that it "was the only mill on Puget Sound to install clarifiers required of the mills in the 1962-1967 state permits."[5] Scott's policy has been
that additional water pollution prevention measures were not necessary because "... the mill is located on tide water with an excellent dispersion system and there is no interference with marine life because of oxygen deficiency."[6] As has been so often reiterated, large bodies of water are not immune
to pollution, it just takes longer.
    Most of the mill's water pollution problem will be solved when 80 percent
of the waste liquor is no longer "fertilizing" the bay. (Collection and incineration efficiencies of greater than 80 percent are almost impossible to obtain.)
However, this will be accomplished ten years after the Puget Sound conference
recommendations were issued. Again, much of the blame lies with the state,
which gave Scott at least five years more than are necessary to construct a
liquor recovery system.

CEP Estimate of Minimum Necessary Pollution Control Expenditure Not in Company Budget: $2.25 million total, of which $2 million is for secondary water treatment and $250,000 is for improved hog fuel boiler particulate control.

Marinette Mill

Location:
Marinette, Wisconsin (across the river from the Menominee mill)

Date Built:
1891, acquired in 1941

Process:
57 TPD bleached calcium sulfite

Other Production:
175 TPD toilet tissue in rolls, roll towels, facial tissues, and sanitary napkins

Total Water Use:
6.7 MGPD

Source:
Menominee River

Discharge:
Menominee River

Table 25.10
Marinette Mill: Pollution Control Record

Treatment	Overall Evaluation	Equipment
Water		
Primary	X	None
Secondary	X	None. 94% of the waste liquor is discharged into the Menominee, 6% is used for road binder (about 1.6 million gallons in 1967)
Tertiary	X	None
Other	—	Savealls, screening

Table 25.10
Marinette Mill: Pollution Control Record (continued)

Treatment	Overall Evaluation	Equipment
Air		
Particulate (fuel)	✓	Fuel: gas. No equipment needed
Particulate (production)	—	None needed
Gas and odor	✗	None

Plant Emissions

To water:

BOD	58,600 lb/day (1,000 lb/ton of pulp; state figure, 1969)
Suspended solids	9,980 lb/day (174 lb/ton of pulp; state figure, 1969)
To air:	N.A.

Pollution Control Expenditures:
N.A.

Legal Status:
(Water) In 1963, a federal abatement conference on the Menominee River said that for three miles below the mill there was pollution from wood chips. A state order was issued requiring submission of plans for reduction of pollution by July 1, 1964. Finally, a January 28, 1970, state abatement order said that the discharge of effluent from the pulping and papermaking operation constitutes a source of pollution of the Menominee River and Green Bay. It ordered a preliminary report by June, 1970, for construction of adequate waste treatment facilities to restrict daily BOD emissions to 6,000 pounds and suspended solids to 5,000 pounds to be in accord with recommendations of the federal Lake Michigan enforcement conference.
(Air) State regulations not yet implemented.

Future Plans:
This mill will possibly construct a clarifier.

Profile:
Scott has asked the Wisconsin Natural Resources Department to ease the pollution abatement order at this old mill which seems to have been awaiting

abatement action to close down.[7] The company explained that, "to meet the
suspended solids requirement, it is likely that an on-site clarifier will be required. As to meeting the BOD standard, a study of alternative procedures is
under way. It seems doubtful that this small pulp mill can sustain the burden
of costs for conventional evaporation and burning.[8]

It seems obvious that Scott will close the mill, rather than make the substantial investment to reduce the BOD discharge to the required 100 pounds per
ton of pulp.

CEP Estimate of Minimum Necessary Pollution Control Expenditures Not
in Company Budget: $4.25 million total, of which $3 million is for primary
and secondary water treatment and $1.25 million is for conversion to a re-
coverable base.

Oconto Falls Mill

Location:
Oconto Falls, Wisconsin

Date Built:
1890, acquired in 1951

Process:
115 TPD bleached sulfite; converted to ammonia base in 1959

Other Production:
115 TPD industrial roll and folded towels, household roll towels, and plastic
food wrap; 516 million gallons per year of road binder (1967 figure)

Total Water Use:
12 MGPD

Source:
Oconto River

Discharge:
Oconto River

Table 25.11
Oconto Falls Mill: Pollution Control Record

Treatment	Overall Evaluation	Equipment
Water		
Primary	✓	12 million gallon clarifier (removes 8,300 lbs/day of solids) installed in 1965
Secondary	X (partial)	50% of the spent sulfite liquor is not discharged, but evaporated and burned or sold; weak sulfite liquor goes to 22 million gallon holding lagoon and is released during the river's high flow; stream aeration at Stiles Pond Dam
Tertiary	X	None
Other	—	95% efficient savealls
Air		
Particulate (fuel)	✓	Fuel: gas. No equipment needed
Particulate (production)	—	None needed
Gas and odor	X	None

Plant Emissions

To water:

BOD   29,440 lb/day (255 lb/ton of pulp; state figure, 1969)

Suspended solids 5,200 lb/day (45 lb/ton of pulp; state figure, 1969)

To air: N.A.

Pollution Control Expenditures:
$2 million for water pollution control.

Legal Status:
(Water) State abatement order (dated December 30, 1969) called for a preliminary report in May, 1970 and a time schedule for reduction of BOD and

suspended solids to levels "no greater at any time than 10,000 pounds of BOD and 5,000 pounds per day of suspended solids . . . ." Scott expects to accomplish this by 1972.
(Air) State regulations not yet implemented.

Future Plans:
$1 million will be spent to enlarge the spent liquor treatment system to handle 90 percent of the liquor by 1973, to enlarge the clarifier, and to convert the holding pond to a secondary treatment aeration basin. (Cost is relatively low because it represents modification to existing equipment primarily).

Profile:
Considering the size and age of this mill, Scott has devoted a good deal of pollution control effort to it. Unfortunately, it is still a major source of air and water pollution. Conversion from a calcium base to an ammonia base sulfite pulping process did not effect as large an improvement in the quality of the effluent as possible because only 50 percent of the spent liquor is evaporated and the rest is discharged. Nonetheless, it is because the mill was converted that the planned pollution control improvements are feasible. Probably, if this were still a small, old, calcium base sulfite mill, it would be closed down rather than improved to meet the state requirements.

CEP Estimate of Minimum Necessary Pollution Control Expenditures Not in Company Budget: $200,000 total, for a sulfur dioxide absorption system.

References for Chapter 25

1. Letter to CEP from Michigan Water Resource Commission, May, 1970.

2. "Policing the Polluters: the Puget Sound Fight," Washington Post, February 16, 1970, pp. 44.

3. Report on Ocean Dumping, the Council on Environmental Quality, October 7, 1970, p. vi.

4. Permit No. T-3344.

5. Letter to CEP, August, 1970.

6. Ibid.

7. Pulp and Paper, September, 1970.

8. Letter to CEP, August, 1970.

Chapter 26

UNION CAMP CORPORATION
1600 Valley Road, Wayne, New Jersey 07470

Major Products:
Paper, lumber products, chemicals, hardware items.

Major Consumer Brand:
UNION CAMP grocery bags.

Financial Data ($ Millions):	1969	1968	1967
Net Sales	449.5	383.4	320.9
Net Income	30.4	26.4	24.8
Capital Expenditures	65.5	34.0	42.2

Pollution Control Expenditures:
$10 million, 1960-1970.

Annual Meeting:
The annual meeting is held during April in Newark, New Jersey.

Officers:
Chairman: Hugh D. Camp, New York City, New York (Director: First National State Bank of New Jersey)
President: Alexander Calder, Jr., Montclair, New Jersey (Director: Seaboard Coastline Railroad, Dillon-Beck Manufacturing Co.; Trustee: Bank of New York)

Outside Directors:
R. Manning Brown: Executive Vice President, New York Life Insurance Co.
Robert G. Calder
Kendall R. DeBevoise: Partner, Breed, Abbott and Morgan
Thomas T. Dunn: Retired Executive Vice President, Union Camp
Joseph H. King: Partner, Eastman, Dillon, Union Securities and Co.
Earl M. McGowan: Former Vice President, W. T. Smith Lumber Co.
Paul D. Camp, Jr.: Practicing Physician
W. Paul Stillman: Chairman of the Board, Mutual Beneficial Life Insurance Co.
Alfred S. Foote: Senior Vice President, Morgan Guaranty Trust Co.
Homer A. Vilas: Partner, Cyrus J. Lawrence and Sons

Breakdown of Total Company Production:

Paper and paperboard products   1,566,983 tons

Chemical products	142,417 tons
Lumber	85,369 thousands of board feet
Plywood	129,107 thousands of square feet

In 1969 paper products represented 24 percent of Union Camp's total sales.

Plants:
Union Camp has 50 domestic production facilities.

Paper and paperboard	4
Bags	6
Container and cartons	23
Chemicals	6
Lumber and plywood	6
Other	5

Timberholdings:
Union Camp owns almost 1,670,000 acres of timberlands valued at $27,886,000.

Annual Pulp Production:
1,662,000 tons (1969)

Company Pollution Overview

Union Camp, the fourteenth largest U.S. paper company, specializes in the production of heavy duty papers and brown paper bags. The company operates three large kraft mills in the South at Franklin, Virginia; Montgomery, Alabama; and Savannah, Georgia. The Savannah mill is the largest kraft mill in the world. The three mills produce 5,000 tons of pulp and 4,750 tons of paper using 93 million gallons of water each day.

In a July telephone interview, Union Camp's technical director, Malcolm L. Taylor, remarked: "What we think about pollution control isn't really relevant anymore. It has become academic. The federal government is going to enforce the laws anyway." Other statements made by company executives in the recent past have the same overtones of disinterest in the environment question and annoyance that it has been thrust on the company and the country. In a newspaper interview, Glenn Kimble, manager of air and water resources at the Savannah mill, stated:

"People are extremely emotional about losing a species ... but animals have been dying out every year clear back to the dinosaurs and in most cases man had nothing to do with it. For that matter, it probably won't hurt mankind a whole hell of a lot in the long run if the whooping crane doesn't quite make it."[1]

In a statement before the Georgia Water Quality Board, Savannah's manager, James Lientz, opposed required secondary treatment on the grounds that it "puts a heavy economic burden on existing industry by requiring treatment which produces no benefits and which is not needed for a waterway to serve its designated use."[2] Apparently he accepts the view that the Savannah harbor should serve only as an industrial and municipal sewer.

In the past ten years, the company has spent only $10 million on pollution control. Despite this limited expenditure, it managed to provide good water treatment at two mills by means of extensive lagooning systems. The conditions at Savannah, however, blacken the company's entire record. Until 1968, the Savannah mill discharged 40 million gallons of completely untreated effluent daily, accounting for 70 percent of the pollution in the lower Savannah River.[3] After it was cited at the 1965 federal pollution conference, only a clarifier was installed. Mill manager James Lientz conceded in a letter to the Georgia Water Quality Board, "any disturbance such as a high sludge load, need for maintenance, poor sludge characteristics, etc., now leaves us without enough capacity to handle all of the sludge."[4] There is no emergency holding pond, and only 55-65 percent of the mill's total effluent is treated.[5] Now, five years after the initial federal order to clean up, this enormous mill is installing an aerated lagoon for the secondary treatment required by 1972.

The air pollution control at two of three mills is inadequate. Despite a recently completed $57 million expansion and modernization program at the Franklin mill, there is no odor control and the coal and bark burning emissions continue to be "controlled" by inadequate mechanical dust collectors. The only pollution control equipment installed during this expansion seem to have been a precipitator and scrubber on the new recovery system. The Montgomery mill has quite good particulate and odor control equipment. The Savannah mill, which does have odor control, has totally inadequate particulate controls. The fact that this mill is operating in a major urban area should have insured a special effort to protect the city residents.

Union Camp was not particularly cooperative with CEP. The company's first several communiques reiterated the same message: "I regret to advise you that we must decline to furnish you the detailed numerical information you have requested in view of your inability to treat the information confidentially or to give us the opportunity to review comments you may wish to make on our operations."[6] The information finally sent was incomplete.

Union Camp has indicated budgeted expenditures of $15 million for water treatment at Savannah. CEP estimates that an additional $5-13 million is necessary to clean up all three mills.

Table 26.1
Union Camp Corporation: Pulp Production, Water Use, and Pollution Control

Mill Location	Pulp Prod. (TPD)	Other Prod. (TPD)	Water Use* (MGPD)	Water Pollution Control			Air Pollution Control	
				Pri.	Sec.	Tert.	Part.	Gas, Odor
Montgomery, Ala.	900	850	15	√	√	X	√	√
Savannah, Ga.	2,800	2,650	40	X	X	X	X	?
Franklin, Va.	1,300	1,250	38	√	√	X	X	X
Total	5,000	4,750	93					

*Process water only.

Montgomery Mill

Location:
Montgomery, Alabama (Prattville)

Date Built:
1966

Process:
900 TPD unbleached kraft (company figure)

Other Production:
850 TPD unbleached kraft linerboard

Total Water Use:
15 MGPD of process water; amount of cooling water unknown

Source:
Alabama River

Discharge:
Alabama River

Table 26.2
Montgomery Mill: Pollution Control Record

Treatment	Overall Evaluation	Equipment
Water		
Primary	✓	2.25 billion gallon settling pond, actually receives only 15 MGPD
Secondary	✓	1,500-acre holding lagoon with possible retention of 6-8 months
Tertiary	✗	None
Solid waste removal	—	"Sludges accumulated in the pond will ultimately be disposed of by land disposal"
Air		
Particulate (fuel)	✓	Fuel: gas, bark. Cyclone collector
Particulate (production)	✓	Recovery furnace: electrostatic precipitator, 99% efficient. Lime kiln: scrubber (no efficiency given)
Gas and odor	✓	Heavy black liquor oxidation
Plant Emissions		
To water:		
BOD	2,500 lb/day (2.8 lb/ton of pulp; company figure)	
Suspended solids	2,400 lb/day (2.7 lb/ton of pulp; company figure)	
To air:	N.A.	

Pollution Control Expenditures:
$1 million for storage and oxidation lagoons.

Legal Status:
(Water) In compliance.
(Air) No state program yet.

Future Plans:
There are no plans now for installing more facilities, because, as a company spokesman said, "It's a new mill and it's on a big river ... because this river is so big, we don't need as much water treatment as we do at Franklin. But," he added, "I'm quite sure we'll have to do more things as time goes on ... the state hasn't any air program yet."

Profile:
The Union Camp mill at Montgomery was built in 1966 at a capital cost of $55 million. The company obviously took account of a possible future tightening of state water and air quality standards, because, although there is no mechanical equipment, the settling pond and holding lagoon have a capacity far exceeding that currently required, and potential air pollutants are well controlled despite the lack of a state program.

     CEP Estimate of Minimum Necessary Pollution Control Expenditures Not in Company Budget: $250,000 total, to improve bark boiler particulate control.

Savannah Mill

Location:
Savannah, Georgia; the largest pulp and paper mill in the world

Date Built:
1936 (original capacity, 150 TPD)

Process:
2,500 TPD kraft (company figure)
300 TPD NSSC

Other Production:
2,650 TPD bag and wrapping papers, corrugated paper, linerboard; turpentine and tall oil

Total Water Use:
40 MGPD (36 MGPD discharged) (process water only)

Source:
Artesian well; local industrial water supply which is taken from a tributary of the Savannah River

Discharge:
Savannah River

Table 26.3
Savannah Mill: Pollution Control Record

Treatment	Overall Evaluation	Equipment
Water		
Primary	X	310-foot diameter mechanical clarifier (began operation 1968) removes 90% of solids from 65% of effluent
Secondary	X	None
Tertiary	X	None
Solid waste removal	—	Sludge dewatered in centrifuges, burned in bark boiler
Other	—	Two clarifiers: skimmers remove oil for mfg. of turpentine, tall oil
Air		
Particulate (fuel)	X	Fuel: bark, gas. Multiple cyclone collectors (inadequate)
Particulate (production)	X	Six recovery furnaces: precipitators, four 90% efficient, two 95% efficient. Three lime kilns: scrubbers, efficiency N.A.
Gas and odor	?	Heavy black liquor oxidation

Plant Emissions

To water:

BOD   98,000 lb/day (35 lb/ton of pulp; company figure). Or 118,422 lb/day (38 lb/ton of pulp; calculated from state equivalent population figure)

Settleable solids   9,000 lb/day (3.3 lb/ton of pulp; company figure). Or 11,323 lb/day (3.6 lb/ton of pulp; state figure)

To air:

TRS   N.A.

Particulates   2,100 lb/hr (19 lb/ton of pulp)

Pollution Control Expenditures:
$3.5 million for the clarifier and the pumping system to get the water to the clarifier.

Legal Status:
(Water) The mill was cited at a 1965 federal abatement conference and required to install primary treatment. The mill was also cited at the second session in 1969 of the conference and required to install secondary treatment by December, 1972, for 85 percent BOD reduction.
(Air) Not in compliance primarily because of excess particulate matter. The mill has until 1974 or 1975 to complete abatement facilities.

Future Plans:
As of May, 1970, the company was conducting engineering studies of how secondary treatment could be provided. Subsequent information indicated that the solution will be a $15 million program including a new evaporator and aerated lagoon.

"The air program hasn't gotten to a stage where we can discuss it much." The state says the solution will probably be a new recovery furnace.

Profile:
In the March 8, 1970 Atlanta Constitution, Rock Howard of the Georgia Water Quality Board stated:

"For 30 years, Union Camp has used billions and billions of gallons of fresh water, and they never spent a penny for pollution control until they were made to do it. They sat down there for thirty years and discharged billions of gallons of waste into the Savannah River—and that stuff's not feeding the fish, brother. After all this time, they ought to say, I'm happy to pay back nature for what I took . . . They ought to have sense enough to know that it's a different ball game today."[7]

In October, 1965, the federal pollution conference on the lower Savannah River pointed at this mill as "the largest kraft pulping operation in the world and certainly the greatest source of pollution to the lower Savannah." The conference cited the mill as the contributor of 70 percent of all the pollution in the lower Savannah River basin and noted that its daily BOD discharge of 135,000 pounds was eight times the discharge of Savannah's city sewer system. The mill's only pollution control at the time was screens and savealls. The conference required installation of primary treatment by 1969.

In response, Union Camp constructed a 310-foot diameter clarifier which went into operation in July, 1968, and reached sustained performance early in 1969. However, the treatment still left the mill in violation of the law. The clarifier reduced the BOD load by 15 percent and the suspended solids by 86 percent; the conference had ordered reductions of 25 percent and 90 percent, respectively.

A Ralph Nader task force recently prepared an in-depth study of this mill
and its history which claims that the clarifier has been inadequate from the
beginning, is not large enough to handle the mill effluent, has been consistently
overloaded, and leaks sludge. The report also cites data gathered by the
Georgia Water Quality Control Board, giving monthly average discharges and
maximum discharges of BOD and suspended solids. The average BOD discharge varied from 92,000 pounds per day (July, 1970) to 156,000 pounds per
day (February, 1970); the suspended solids discharge varied from 8,000 pounds
per day (October, 1969) to 41,000 pounds per day (December, 1969). More significant, however, are the daily maximum discharges because, "for many
pollutants and many kinds of aquatic life, the factor that determines how much
damage an industry will do to fish and microorganisms is the highest pollution
level, not the average."[8]

The gap between the monthly average and the maximum daily discharges
is huge. In July 1970, 2 years after the clarifier was installed, the maximum
was 1,600 percent of the average and 6,000 percent of the allowable discharge
as determined at the 1965 hearings. This discharge was accidental; the clarifier broke and the waste was sent directly to the river. Nevertheless, it is
inexcusable that the largest kraft mill in the world has no emergency holding
pond to prevent such occurrences.

The situation is even more disturbing because the company appears in no
hurry to take remedial action. On January 17, 1967, mill manager Lienz
spoke in opposition to mandatory secondary treatment at a Water Quality Control Board Meeting. He was against such a provision in the state law because
it

"puts a heavy economic burden on existing industry; by requiring treatment
which produces no benefits and which is not needed for a waterway to serve
its designated use."[9]

Three years later, in a letter to CEP, Union Camp reported that it was
still "conducting engineering studies of how secondary treatment can be provided at this very crowded location."[10]

The Savannah Mill's air pollution picture is (only in comparison to its
water pollution picture) not too bad. A heavy black liquor oxidation system
was installed for odor reduction. However, particulate controls are highly
inefficient, and the mill's soot and ash emissions reportedly "often blanket
the whole downtown Savannah business district." The state has most gener-
ously given Union Camp four years to correct this by adding better precipitators and scrubbers to its fuel and recovery boiler stacks.

CEP Estimate of Minimum Necessary Pollution Control Expenditures Not
in Company Budget: $1.25 million total, of which $250,000 is for improved
power boiler particulate control; $500,000 is for replacement of the 1940
precipitator; and $500,000 is for improved primary treatment. If, in addition,
a new recovery furnace is required, the total will be increased by $8 million.

Franklin Mill

Location:
Franklin, Virginia

Date Built:
1937; extensive expansion in 1970

Process:
1,300 TPD bleached and unbleached kraft pulp

Other Production:
1,250 TPD printing and fine paper; postcard, adding machine, and coating
base stock; plastic saturating and resin impregnating paper

Total Water Use:
38 MGPD

Source:
Artesian wells

Discharge:
Blackwater River (which forms the Chowan River one mile downstream and
flows into North Carolina)

Table 26.4
Franklin Mill: Pollution Control Record

Treatment	Overall Evaluation	Equipment
Water		
Primary	✓	12 billion gallon settling pond system retains water up to 280 days, discharges only during 120-day period in winter (has been in operation 6 years)
Secondary	✓	
Tertiary	✗	None
Solid waste removal	—	"Accumulated sludge will ultimately be disposed of by land disposal"
Air		
Particulate (fuel)	✗	Fuel: oil, bark, coal. Cyclone collectors (inadequate)

Table 26.4
Franklin Mill: Pollution Control Record (continued)

Treatment	Overall Evaluation	Equipment
Particulate (production)	X	Four recovery furnaces: precipitators, 92% to over 99% efficient (one furnace precipitator installed 1970). Three lime kilns: wet scrubbers, 95%-99.5% efficient. Demisters on two of four smelt dissolving vents.
Gas and odor	X	None

Plant Emissions

To water:

BOD	27,000 lb/day (20 lb/ton of pulp)
Suspended solids	3,000 lb/day (2 lb/ton of pulp)

To air:

TRS	N.A.
Particulates	44,000 lb/day (34 lb/ton of pulp)

Pollution Control Expenditures:
N.A.

Legal Status:
(Water) The mill is generally in compliance with state standards although a color problem exists. North Carolina is dissatisfied with the water quality. (Air) No information is available from the state agency.

Future Plans:
"We have no plans here for further water treatment, we feel we are complying with the requirements and see no ill effects on the river — we do have a few grumblings about the color, but..."

Profile:
Over the last two years, Union Camp spent $57 million on an expansion and modernization program for this mill which increased the production capacity by 33 percent. Yet where in this major company effort was the priority for pollution control? This very large kraft mill not only still operates without any odor control, but Union Camp has no plans to install it in the future. Par-

ticulate controls also remain inadequate, with only mechanical collectors to trap the soot and ash emissions from the coal and bark boilers.

In contrast, water pollution control is quite good. According to the state's regional water pollution control director, Gerold Yagel, the extensive state ponding system at this mill essentially removes the settleable and dissolved solids in the effluent, thus reducing the BOD by about 95 percent. In addition, all discharges from the pond are made during the river's high flow period in the winter and are stopped just prior to the annual herring run upstream. Monitoring of the effluent has never revealed any violations of the standards for dissolved oxygen, coliform bacteria, temperature, or pH.

This elaborate treatment is necessary because the Blackwater River is so small, and, in spite of its effectiveness, Mr. Yagel feels that the system can be improved somewhat, perhaps with aerators, to further improve the quality of the effluent.

There is also dissatisfaction with the color of the effluent. Although the darkness is not detrimental to the ecosystem of the river, it does violate general stream standards for color. The company has agreed to remove this discoloration as soon as there is an available technology.

CEP Estimate of Minimum Necessary Pollution Control Expenditures Not in Company Budget: $3.5 million total, of which $400,000 is for improved coal and bark boiler particulate control; $500,000 is for an odor control system; $600,000 is for upgrading three older precipitators or adding scrubbers to them; and $2 million is for a color removal system and clarifier.

References for Chapter 26

1. New York Times, May 4, 1970, p. 57.

2. From a draft version of The Water Lords, written by Ralph Nader's Study Group. Grossman Publishers, 1971.

3. Ibid.

4. Ibid.

5. Ibid.

6. Letter to CEP, April 27, 1970.

7. Draft of The Water Lords.

8. Ibid.

9. Ibid.

10. Letter to CEP, May 7, 1970.

Chapter 27

U.S. PLYWOOD-CHAMPION PAPERS INCORPORATED
777 Third Avenue, New York, New York 10017

Major Products:
Building materials, paper products, furniture and carpets.

Major Consumer Brands:
DREXEL, HERITAGE furniture, TREND, TEX-TUFT carpets, WELDWOOD
paneling, NOVOPLY particle board, PURE-PAC milk containers.

Financial Data ($ Millions):	1969	1968	1967
Net Sales	1,455.5	1,283.4	1,132.3
Net Income	64.4	54.7	41.4
Capital Expenditures	99.6	57.4	43.3

Pollution Control Expenditures:
$30.7 million, 1960-1969; $11.4 million planned.

Annual Meeting:
The annual meeting is held in mid-May in New York City.

Officers:
Chairman/President: Karl R. Bendetsen, New York City (Director: Westing-
house Electric Co.)

Outside Directors:
Dillon Anderson: Partner — Baker, Botts, Shepherd and Coate
S. W. Antoville: Retired Chairman of the Board, U.S. Plywood Corp.
John Bene: Special Advisor to the Government of Canada
Leonard G. Carpenter: President, Minneapolis Foundation
Francis E. Ferguson: President, Northwest Mutual Life Insurance Co.
Joseph B. Hall: Former Chairman of the Board, Kroger Co.
DeWitt Peterkin, Jr.: Executive Vice President, Morgan Guaranty Trust Co.
Herbert T. Randall: Retired Executive, Champion Papers, Inc.
Walter R. Severson: Partner, Severson, Werson, Berke and Bull
Dwight J. Thomson: Former Chairman of the Board, Champion Papers, Inc.

Breakdown of Total Company Production (1969 sales):

Building materials	$730,300,000
Paper, distribution, and converting	539,100,000
Interior furnishings	180,500,000
Other	5,600,000

Plants:
U.S. Plywood-Champion has domestic manufacturing plants at approximately
95 locations:

Building materials	41
Papers, envelopes, packaging and converting	30
Furniture and ornamental iron	24

Timberholdings:
U.S. Plywood-Champion owns 1,687,000 acres of timberlands. In addition the
company has cutting rights or other arrangements which give it effective use
of 5,000,000 acres in Canada and the United States. The total 1969 value of
U.S. Plywood-Champion's holdings: $144,003,000.

Annual Pulp Production:
768,575 tons (1969)

Company Pollution Overview

U.S. Plywood-Champion has two large kraft mills in Canton, North Carolina,
and Pasadena, Texas, which produce 2,665 tons of pulp daily using 129 million gallons of water. A third mill in Cortland, Alabama, began operation in
late 1970, and a fourth is being planned in Berners Bay, Alaska.

(Obviously, pulp production is not the only source of pollution in the manufacture of paper, and paper mills should be as vigilant as pulp mills in controlling their discharges. It has been reported that 616 pounds of a highly
toxic chemical accidentally leaked from a storage tank at U.S. Plywood-
Champion's Hamilton, Ohio, mill into the Great Miami River.[1] The company said that this leak "could have been the cause" of the deaths of 1.8 million fish. The mill does not have any antipollution devices to prevent a similar
occurrence; however, it will discontinue using the particular chemical until it
can provide adequate safeguards.)

The company has recently invested a great deal of money and effort in
pollution control and has been very open in discussing the problems involved.
The first step was hiring the former head of the Federal Water Pollution Control Administration, James Quigley, to be vice-president in charge of environment. Since then, the company has moved fast in water pollution. The
new Cortland mill was designed with the latest "in standard, established,
proven state-of-the-art"[2] equipment for which 6.2 percent of the capital
cost of the mill was allocated. At both Canton and Pasadena, elaborate $9
million water treatment systems have been recently installed, which achieve
90 percent reductions in BOD and settleable solids discharges.

From a conversation with Mr. Quigley it was learned that none of
Champion's plans or systems is experimental or innovative because several
years ago it invested in a new, revolutionary water treatment system which
collapsed after a few months of operation. This experience left the company
considerably chastened and quite conservative. As a consequence, any system

which it now installs has to have been tested and proven to work as designed. [3]

The company has been very negligent in cleaning the air; not until September, 1970, were extensive plans announced for the Canton mill, which has been particularly dirty for 60 years. However, by mid-1973, new equipment there should remove 92 percent of the particulate emissions and considerably reduce odorous gas emissions. The Pasadena mill is already equipped with black liquor oxidation and a 99 percent efficient precipitator on one recovery furnace, but the lime kiln scrubbers and the second recovery furnace precipitator are not adequate and no definite plans have been made to replace them.

It is clear that much of Champion's effort is being made in response to legal and public pressure. The company has spent $30.7 million since 1960 on its three mills and has planned expenditures of $11.4 million. CEP estimates that an additional $1.75-$9.75 million (the latter if a new recovery furnace is installed at Pasadena) will be required to fully control particulate emissions at the Pasadena and Canton mills, making a total of $13.2 or $21.2 million.

Table 27.1
U.S. Plywood-Champion Papers Incorporated: Pulp Production, Water Use, and Pollution Control

Mill Location	Pulp Prod. (TPD)	Other Prod. (TPD)	Water Use (MGPD)	Water Pollution Control			Air Pollution Control	
				Pri.	Sec.	Tert.	Part.	Gas, Odor
Cortland, Ala.	500	--	26	✓	✓	✗	✓	✓
Canton, N.C.	1,290	1,420	50	✓	✓	✗	✗	✗
Pasadena, Tex.	875	600	44	✓	✓	✗	✗	✓
Total	2,665	2,020	120					

Cortland Mill

Location:
Cortland, Alabama

Date Built:
1970

Process:
500 TPD bleached kraft

Other Production:
None

Total Water Use:
26 MGPD

Source:
Tennessee River

Discharge:
Tennessee River

Table 27.2
Cortland Mill: Pollution Control Record

Treatment	Overall Evaluation	Equipment
Water		
Primary	✓	30 million gallon clarifier
Secondary	✓	Aerated lagoons designed to remove 90% of the BOD and 100% of the settleable solids
Tertiary	✗	None
Solid waste removal	—	1,500-acre landfill
Air		
Particulate (fuel)	✓	Fuel: purchased power. No equipment needed
Particulate (production)	✓	Recovery furnace: precipitators over 99% efficient. Lime kiln: flooded disc scrubbers
Gas and odor	✓	Black liquor oxidation (design efficiency, 99.9%)
Plant Emissions	N.A.	

Pollution Control Expenditures:
$5 million, or 6.2 percent of the $80 million cost of the mill.

Legal Status:
This mill was built in accordance with all state requirements.

Future Plans:
None needed.

Profile:
U. S. Plywood-Champion hopes to have in Cortland, "a kraft mill that will be as good as any in the country, maybe in the world, from a pollution point of view."[4] The mill, which began full operation in early 1971, has the latest in "standard, established, proven state-of-the-art" equipment, and, according to Mr. Quigley, "the awareness of the environment and the importance of pollution loomed large from the beginning. ... We are reasonably optimistic that if things perform anywhere near standard in specification, we'll turn out both a water and air effluent that is going to be of pretty high quality."[5]

However, the mill will not have tertiary water treatment or thermal water pollution control. Nor will it have non-direct contact evaporators to cut down odor because, as Mr. Quigley explained, "our technical people have honest deep disagreements on their effectiveness as a production unit and as a pollution control mechanism; because in the process of reducing odors they turn out a great deal more sulfur dioxide and sulfur trioxide."[6]

Despite the lack of these measures, which may be added if they prove necessary, the Cortland mill has been built with much consideration given to pollution control.

CEP Estimate of Minimum Necessary Pollution Control Expenditures Not in Company Budget: None.

Canton Mill

Location:
Canton, North Carolina

Date Built:
1907; expanded several times since

Process:
1,290 TPD bleached kraft

Other Production:
1,420 TPD envelope, postal card, offset, and writing papers; board and mimeography papers. By-product manufacture: 1,400 gallons of turpentine, 42 tons of tall oil, and 487 tons of chemicals (caustic, chlorine, lime, chlorine dioxide)

Total Water Use:
50 MGPD

Source:
Pigeon River

Discharge:
Pigeon River

Table 27.3
Canton Mill: Pollution Control Record

Treatment	Overall Evaluation	Equipment
Water		
Primary	✓	Three clarifiers (installed 1960, 1965), capacity, 48.5 million gallons (one is 125-foot diameter, two are 200-foot diameter)
Secondary	✓	Two stabilization basins, two contact basins, two 200-foot diameter clarifiers; activated sludge, aeration system (installed July, 1970, reduces organic wastes 90%)
Tertiary	X	None
Solid waste disposal	—	Landfill
Air		
Particulate (fuel)	X	Fuel: soft coal, bark, oil. Three boilers with cyclone ash collectors
Particulate (production)	X	Four recovery furnaces: electrostatic precipitators. One is new, 99% efficient; others are not. Demister on recovery furnace. Scrubbers on lime kilns and evaporators
Gas and odor	X	No black liquor oxidation. Scrubber on one blow tower, other tower uncontrolled
Plant Emissions	N.A.	

An aerial view of U.S. Plywood-Champion's 18-acre waste water treatment facility at the Canton mill. The new secondary treatment plant (the large rectangular basin in the center and the two 200-foot clarifiers on the left), which cost $5 million, began operation in 1970. The primary plant (right) was completed in 1965 at a cost of $3.4 million. The combined facility processes over 47 million gallons of waste water daily from the mill, as well as the wastes from the town of Canton. The operating cost of the entire facility is estimated to be $4,000 daily. The secondary system removes oxygen-demanding organic matter from the effluent, while the function of the primary system is the removal of solid matter. Photo by Mike Zerby, Minneapolis Tribune; © The Minneapolis Star and Tribune Company.

The horizon is obscured by typical pulp mill emissions from U.S. Plywood-Champion's Canton mill. Photo by Karl A. Maier for CEP.

Pollution Control Expenditures:
U.S. Plywood-Champion has spent $10.7 million in the past ten years on pol-
lution control at this mill. Of this, $2.3 million has been spent since 1965 on
equipment to control air emissions including conversion of coal-fired lime
kilns to oil ($156,600); scrubbers on the evaporators, one of two blow towers,
condensate tank, lime kilns, and bleach plant ($476,900); and precipitators
and demisters on the new recovery furnace ($822,000). $8.4 million was invested in primary and secondary water treatment systems which cost an additional $4,000 per day to operate.

Legal Status:
(Water) In compliance but company is required to do studies to determine the
need for clorination facilities.
(Air) State planning to establish specific emissions standards for pulp mills
but this has not been done yet.

Future Plans:
U.S. Plywood-Champion is spending $11.4 million to bring this 62-year-old
mill up to modern air pollution control standards by mid-1973. Planned equipment is expected to reduce total particulate emissions by 92 percent and to
effect a considerable odor reduction.

The major part of the total expenditure will be for a new recovery furnace
to replace three old units. It will have three electrostatic precipitators, expected to remove 99 percent of the particulate emissions, and black liquor
oxidation for odor control. Since U.S. Plywood-Champion is planning no production expansion, the $8 million investment is being considered a pollution
control expenditure. However, it will greatly improve the mill's production
efficiency.

Another $2.5 million is budgeted for development and installation of elec-
trostatic precipitators for the coal burning boilers, and scrubbers for the
second blow tower.

Profile:
Canton is a typical mill town, and, as such, it is not a pleasant place to live.
The air is filled with smoke, ash, dust, and the characteristic kraft mill
smell. During several months these conditions are made even more severe
by almost constant air inversions which hold down the stagnant, humid smog
over hundreds of square miles—the entire valley. It is during these inversions that the highway signs are most needed. They read: "caution: dense fog
likely next 7 miles."

The water in the Pigeon River has been, until recently, as dirty as the
air in the valley. The Pigeon is much too small to absorb the effluent from a
1,300 ton per day mill. During the summer months its entire flow passes
through the mill; there is nothing left to dilute and disperse the wastes. Even
with the newly installed secondary water treatment system there will not be
many fish in the Pigeon River because the effluent will retain its 80-85 degree
temperature for at least several miles downstream.

These conditions exist because the Canton mill was "built without a thought of pollution control, and expanded many times on the basis of all sorts of considerations other than pollution control."[7] However, in September, 1970, U.S. Plywood-Champion took a major step. It held a press conference to discuss its problems at the mill and its plans, which exemplify the effort and expense necessary to surmount the consequences of years of neglect. The limited size of the mill site required installation of an expensive but compact water treatment system rather than less costly, extensive lagoons. Similarly, before scrubbers and precipitators can effectively control atmospheric emissions, much of the old, inefficient production equipment must be replaced. Champion is going to try to improve what Mr. Quigley has admitted is "the company's worst offender."[8]

CEP Estimate of Minimum Necessary Pollution Control Expenditures Not in Company Budget: $500,000 total, for improved bark burner particulate control.

Pasadena Mill

Location:
Pasadena, Texas

Date Built:
1937

Process:
800 TPD bleached kraft
75 TPD groundwood

Other Production:
600 TPD coated and uncoated letterpress and offset book papers; bond, duplicating, and tablet papers

Total Water Use:
44 MGPD

Source:
15 percent from wells; 85 percent from Lake Houston

Discharge:
Houston Ship Channel

Table 27.4
Pasadena Mill: Pollution Control Record

Treatment	Overall Evaluation	Equipment
Water		
Primary	✓	3-acre clarifier (44 million gallon, installed 1969) reduces settleable solids 90%
Secondary	✓	Activated sludge (installed Nov. 1970) reduces BOD 85%-90%
Tertiary	✗	None
Solid waste removal	—	Landfill
Air		
Particulate (fuel)	✓	Fuel: natural gas. None needed
Particulate (production)	✗	Two electrostatic precipitators on recovery furnaces: one 99% efficient, one old and inadequate. Gas scrubbers on all lime kilns
Gas and odor	✓	Black liquor oxidation. Demisters on vent tanks
Plant Emissions	N.A.	

Pollution Control Expenditures:
$15 million of which $9 million was for water pollution control. $12 million of this has been spent in the last three years.

Legal Status:
(Water) In compliance.
(Air) In compliance.

Future Plans:
There is no need to plan further expenditures on water treatment at this mill. According to the Texas Water Pollution Control Board, when the present facilities are completed, Pasadena will have the cleanest effluent of any mill in the state.

The company recognizes the need to clean air emissions at this mill, but no definite plans have been made.

Profile:
The Pasadena mill has had to expend a disproportionately large amount to upgrade its water treatment because of the limited size of the site. A clarifier and activated sludge facility were built to replace 27 acres of primary treatment holding ponds. The resultant water treatment system is considered to be the best in Texas and one of the best in the country.

The air pollution situation is not comparable. U.S. Plywood-Champion's vice-president, James Quigley, has admitted that the Pasadena mill has problems which must be solved in order to meet the high air pollution control standards in Texas; he conceded that local pressure by the media and politicians running for office was focusing attention on the mill and, in a sense, forcing it to take action. He added, "we have reached the point in Texas where we think we're in compliance, but barely... we recognize that this is a treadmill kind of thing and that we have to keep moving faster and faster."[9] Mr. Quigley defined the problems: "the lime kilns are not now adequately being scrubbed and we've got to come up with something better."[10] He added that he thought the recovery furnace particulate problems would ultimately be solved only by installing a new unit as will be done at the Canton mill.

CEP Estimate of Minimum Necessary Pollution Control Expenditures Not in Company Budget: $1.25 million total, of which $500,000 is for an electrostatic precipitator for the recovery furnace and $750,000 is for three improved lime kiln scrubbers. In addition, $8 million will be required if a new recovery furnace is built.

References for Chapter 27

1. Wall Street Journal, November 12, 1970.

2. Conversation with Mr. James Quigley, August 24, 1970.

3. Idem.

4. Idem.

5. Idem.

6. Idem.

7. Idem.

8. Idem.

9. Idem.

10. Idem.

Chapter 28

WESTVACO CORPORATION
299 Park Avenue, New York, New York 10017

Major Products:
Pulp, paper, paperboard, disposable party items, multiwall bags.

Financial Data ($ Millions):	1969	1968	1967
Net Sales	419.6	391.6	366.3
Net Income	21.9	20.7	21.4
Capital Expenditures	73.6	43.2	28.8

Pollution Control Expenditures:
$ 32 million (approximately), 1940-1970.

Annual Meeting:
The annual meeting is held in New York the fourth Tuesday in February.

Officers:
President: David L. Luke III, New York, New York (Director: B. F. Goodrich, Irving Trust Co., Clarkson Ind., U.S. Envelope Co., Clupak, Inc.)

Outside Directors:
Gene K. Beare: Executive Vice President, General Telephone and Electronics
Harold J. Berry: Chairman, Investment Banking Committee, Merrill, Lynch, Pierce, Fenner and Smith
William Elfers: General Partner, Greylock and Co.
David L. Hopkins, Jr.: Vice President, Morgan Guaranty Trust Co.
Barry T. Leithead: Chairman, Cluett, Peabody and Co.
John H. Mathis: Former Chairman, Lone Star Industries, Inc.
William Petersen: Retired Vice-Chairman, Irving Trust Co.

Breakdown of Total Company Production (tons, 1969):

Paper, paperboard and market pulp	1,666,000
Converted products	595,000
Chemical products	131,000
	2,392,000

Plants:
Westvaco has 31 domestic plants:

Paper and pulp	7
Corrugated box, folding carton, milk carton and multiwall bag	20

Building products   2
Chemicals   2

Timberholdings:
Westvaco owns a total of 1,160,000 acres of timberland mainly in North
Carolina, South Carolina, Virginia, West Virginia, and Kentucky.
The total 1969 value of Westvaco's holdings: $30,868,000 (at cost).

Annual Pulp Production:
1.5 million tons (1969)

Company Pollution Overview

Westvaco is in a difficult position. Apart from the brand new mill in Wickliffe,
Kentucky, all the company's mill have been expanded and rebuilt many times
since their construction in the early 1900s. In describing his company's pollution problems during a CEP interview, O. B. Burns, director of environmental controls, moaned, "We have so many problems with all these old mills —
I just hope you won't be too hard on us."

The Westvaco mills are strung along the East Coast: a small NSSC mill in
Mechanicville, New York (now shut down); a small kraft mill in Tyrone, Pennsylvania; and three large kraft mills in Luke, Maryland, Charleston, South
Carolina, and Covington, Virginia. The company's new Wickliffe mill is off
this main stream. These facilities produce 4696 tons of pulp and discharge
129 million gallons of water each day.

Not one of the Westvaco mills has both adequate water and air pollution
control. Covington and Tyrone have primary and secondary treatment. Luke
and Wickliffe have primary treatment, and Mechanicville and Charleston have
no treatment at all for their effluent. Only Wickliffe and Covington have the
necessary equipment to reduce soot and ash emissions, and only Charleston
and Wickliffe have odor control. Two of the mills have been primary targets
of dissatisfaction. At Covington, Virginia, where smoke and gas shrouded the
entire valley for decades, Westvaco was under heavy attack by a local citizens'
group for two years (and subsequently praised for its cleanup efforts). The
mill at Luke, Maryland, has been doubly offensive. Uncontrolled sulfur gas
emissions coming from a now closed down sulfite operation caused a complete defoliation of the surrounding mountainsides; in November, 1969, the
mill was charged by the FWQA as being one of the major polluters of the north
branch of the Potomac River.

Aside from those instances in which Westvaco has shown initiative — installing the industry's first activated sludge plant in 1955, the first electrostatic precipitator for kraft recovery furnace soot control in 1944, and the
first precipitator on a power boiler stack in 1966,* the company's overall pol-

----

*Most other companies have continued to call mechanical collectors adequate

lution control record is far from acceptable. Even at the new Wickliffe mill, Westvaco has taken the easy route, providing only primary water treatment before the effluent is sent through an outfall pipe into the Mississippi River. The state has agreed to this half-measure on the outdated assumption that a large body of water can absorb the pollution load.

The company is now at the point of deciding which mills are worth the large investments for a thorough cleanup. Some decisions have been made. Covington is now installing a $14.7 million system for air pollution control, which Westvaco claims will put the mill in excellent condition by the end of 1973. At Charleston, $5.3 million has been budgeted for primary and secondary water treatment and tighter control of particulate emissions. This mill should be quite clean by the end of 1972. And in December, 1970, Westvaco announced that pulping operations at Mechanicville and Tyrone would be shut down. Mechanicville had been operating under a water pollution abatement order since 1966, and Tyrone had a deadline for odor control.

However, one of the mills doesn't seem to be receiving much attention at all. Luke has no water control state deadline until 1974, and, despite its having been cited for discharging 16 million gallons of untreated effluent a day into the Potomac, has made no specific plans for cleaning up. CEP estimates that with all three large mills in operation, at least $2 million would have to be added to Westvaco's planned pollution control expenditure of $35 million. This would bring the company's minimum necessary investments to $37 million.

Table 28.1
Westvaco Corporation: Pulp Production, Water Use, and Pollution Control

Mill Location	Pulp Prod. (TPD)	Other Prod. (TPD)	Water Use (MGPD)	Water Pollution Control			Air Pollution Control	
				Pri.	Sec.	Tert.	Part.	Gas, Odor
Wickliffe, Ky.	600	350	28	✓	✗	✗	✓	✓
Luke, Md.	743	956	31	✓	✗	✗	✗	✗
Mechanicville,* N.Y.	158*	230*	2*	✗	✗	✗	✗	—
Tyrone,* Penn.	150*	177*	6*	✓	✓	✗	✗	✗

for controlling soot from coal and bark boilers —and have gone unchallenged. Westvaco took the first step in admitting that these collectors are not good enough.

Table 28.1
Westvaco Corporation: Pulp Production, Water Use, and Pollution Control
(continued)

Mill Location	Pulp Prod. (TPD)	Other Prod. (TPD)	Water Use (MGPD)	Water Pollution Control			Air Pollution Control	
				Pri.	Sec.	Tert.	Part.	Gas, Odor
Charleston, S.C.	2,000	1,813	45	X	X	X	X	✓
Covington, Va.	1,353	1,353	25	✓	✓	X	✓	X
Total	4,696	4,472	129					

*Not included in totals because the pulping operations have been shut down.

Wickliffe Mill

Location:
Wickliffe, Kentucky

Date Built:
Mill began operating in August, 1970

Process:
600 TPD bleached kraft pulp (planned capacity)

Other Production:
350 TPD uncoated bond and fine papers

Total Water Use:
28 MGPD

Source:
Mississippi River

Discharge:
Mississippi River

Table 28.2
Wickliffe Mill: Pollution Control Record

Treatment	Overall Evaluation	Equipment
Water		
Primary	✓	210-foot diameter clarifier
Secondary	X	None
Tertiary	X	None
Other	—	Outfall pipe with diffuser
Air		
Particulate (fuel)	✓	Fuel: purchased natural gas. No equipment needed
Particulate (production)	✓	Recovery furnace: 99.5% efficient precipitator and demister. Lime kiln: venturi scrubber
Gas and odor	✓	Black liquor oxidation
Plant Emissions	N.A.	

Pollution Control Expenditures:
$3.6 million of the total $80 million investment in the mill is ascribed to pollution control. (About $2.1 million for water clarifier and outfall pipe and about $1.5 million for air control—$800,000 for recovery furnace precipitator; $40,000 for demister; $220,000 for black liquor oxidation.

Legal Status:
(Water) In compliance.
(Air) In compliance.

Future Plans:
None

Profile:
Wickliffe, Westvaco's only new mill, began producing pulp in August, 1970. A major reason for choosing the mill's location may have been that the state pollution control board does not require secondary water treatment of mill effluent on the rather short-sighted assumption that the wide, fast-flowing Mississippi River can absorb the waste. Westvaco's expenditures for pollution

control equipment were thus only 4.6 percent of the total capital cost of the mill rather than the usual 10 percent which most new mills are expected to spend.

CEP Estimate of Minimum Necessary Pollution Control Expenditures Not on Company Budget: $1 million total, for an aerated lagoon or other secondary treatment.

Luke Mill

Location:
Luke, Maryland

Date Built:
1888; completely rebuilt 1955-1958 for $60 million

Process:
743 TPD bleached kraft

Other Production:
956 TPD coated offset, letterpress and gravure papers; coated cover, coated one-side label, uncoated printing and business papers

Total Water Use:
31 MGPD

Source:
Potomac River

Discharge:
Potomac River

Table 28.3
Luke Mill: Pollution Control Record

Treatment	Overall Evaluation	Equipment
Water		
Primary	√	On-site clarifier followed by joint treatment for 2/3 of the mill wastes at activated sludge plant, operated by Upper Potomac River Commission (However, the remaining 1/3 of efflu-

Table 28.3
Luke Mill: Pollution Control Record (continued)

Treatment	Overall Evaluation	Equipment
Secondary	X (partial)	ent discharged directly into the Potomac.)
Tertiary	X	None

Air		
Particulate (fuel)	X	Fuel: low sulfur oil, coal. Oil power boiler needs no controls. One coal boiler has 95% efficient precipitator. Other has no equipment.
Particulate (production)	X	Two recovery furnaces: 92%, 95% precipitators; demister on one furnace. Lime kiln: two 99.5% efficient scrubbers. Bag filter, three scrubbers in clay unloading area; dust collector in chipperhouse. Shroud on lime kiln (overloaded)
Gas and odor	X	Company designed black liquor oxidation system (installed 1967); overproduction reduces efficiency

Plant Emissions

To water:

BOD	21,050 lb/day (30 lb/ton of pulp; from treatment plant and direct discharge; FQWA figure, Nov. 1969)
Suspended solids	60,000 lb/day (from joint treatment plant and direct discharge)

To air: N.A.

Pollution Control Expenditures:
The mill has spent $5 million on capital investments for air and water pollution control equipment including: four electrostatic precipitators, five wet scrubbers, dust collectors, demisters and bag filter, black liquor oxidation, clarifier, and ash lagoon. In addition, Westvaco spends $750,000 annually as its share of the operating costs of the joint municipal water treatment plant.

Legal Status:
(Water) On February 13, 1970, the Maryland State Department of Water Resources issued an abatement order requiring that plant water treatment and sewer systems be tightened up by November 1974; and a plan for a reduction of mill effluent color be filed by December 1973 and implemented six months later.
(Air) In compliance.

Future Plans:
Westvaco is conducting a research project at its Charleston, South Carolina mill on a sulfur dioxide removal process using activated carbon. Installation of particulate and gas controls on the second power boiler at Luke is being delayed until this research is complete. A new $8.5 million, 1,000 TPD recovery furnace with a 99.9 percent efficient electrostatic precipitator, odor control, and 2 demisters due for August, 1972, start-up (2 furnaces now used will go on standby.)

Profile:
"This mill," reported Mr. Robert Dennis of the Potomac Basin Center, "has been one of the worst pollutors of the north branch of the Potomac for years." A federal water pollution control study [1] called the 45 mile stretch of river between the Luke mill and the town of Spring Gap, West Virginia one of the three most degraded parts of the north branch sub-basin; it has low dissolved oxygen counts and high waste loads. The study charged the Westvaco mill with contributing almost 40 percent of the total pollution load on this part of the river. Of the total load of 53,630 pounds of BOD coming from local towns and industries, Westvaco discharged about 21,050 pounds.

The mill uses 31 million gallons of water daily; of this, 11 million gallons are discharged directly into the Potomac with no treatment. This effluent has a BOD load of 16,000 pounds. The other 20 million gallons are piped to a joint municipal waste treatment plant at Westernport, run by the Upper Potomac River Commission, which has a capacity of 21 million gallons per day; one million gallons come from three nearby towns. The BOD load from the plant is 5,300 pounds per day.[2]

The FWPCA report further indicated that the pollution from the mill and joint treatment plant is particularly devastating to the 5-6 miles of river immediately downstream. The oxygen count there is often less than 1.0 milligrams per liter of water; the Maryland water standard for that location is 4.0 milligrams per liter.

Westvaco's mill has also been a prime air polluter in the Potomac River Valley. Mr. Dennis pointed out, "it is situated at the bottom of a steep canyon on the West Virginia bank of the river. The trees on the mountainside that rises opposite have been almost completely defoliated by the sulfur dioxide gas emissions from the old sulfite mill stacks. Although in the past several years a good deal of air control equipment has been installed, it may take years for the damage done to the trees to be repaired."

Odor control still presents a problem for this mill. Although a black liquor oxidation system was added in 1967, a company representative said "the system is not working as well as it should because the recovery furnace is being operated at overcapacity which makes gas control much more difficult." The company evidently intends to continue operating the furnace at overcapacity despite the odor problem, but claims that when a new, larger recovery furnace (1,000 tons) and accessory chemical recovery equipment go into operation in August 1972, the odor will be substantially reduced.

CEP Estimate of Minimum Necessary Pollution Control Expenditure Not on Company Budget: $500,000 total, for a new power boiler precipitator.

Mechanicville Mill

Location:
Mechanicville, New York

Date Built:
C. 1900

Process:
158 TPD bleached NSSC

Other Production:
230 TPD bond, writing, mimeo, ledger, opaque, papeterie, greeting card stock; music, map, envelope, place mat stock; soda straw stock; specialties

Total Water Use:
20 MGPD

Source:
Hudson River

Discharge:
Hudson River

Table 28.4
Mechanicville Mill: Pollution Control Record

Treatment	Overall Evaluation	Equipment
Water		
Primary	X	None

Table 28.4
Mechanicville Mill: Pollution Control Record (continued)

Treatment	Overall Evaluation	Equipment
Secondary	X	None. All spent pulping liquor and bleaching chemicals are discharged into Hudson
Tertiary	X	None

Air		
Particulate (fuel)	X	Fuel: No. 6 oil. None
Particulate (production)	—	None needed
Gas and odor	—	None needed

Plant Emissions

To water:

BOD	80,000 lb/day (company figure)
Suspended solids	N.A.

To air:	N.A.

Pollution Control Expenditures:
None.

Legal Status:
(Water) In September 1966 the mill was cited by the state for "foam, sludge, low dissolved oxygen, discoloration, floating and settleable solids in the effluent." New York gave Westvaco a deadline of March, 1971, to present an abatement plan.
(Air) The mill's boiler is in compliance.

Profile:
This mill, built around 1900, faced the problem of all obsolete, small pulping facilities whose marginal profitability would be completely eliminated by making the expenditures now being required for adequate pollution control. Since the mill was cited in 1966 for water pollution, no steps have been taken by Westvaco to alleviate the situation. Mechanicville has continued over the four

years to pour completely untreated effluent, including all the spent NSSC pulping liquor, into the Hudson.

A company representative concludes, "a full water treatment system
would cost about $3.5 million. That would be more than the book value of the
whole mill!" Pressure on the company resulted in the December announcement that the mill would be shut down.

CEP Estimate of Minimum Necessary Pollution Control Expenditure Not
on Company Budget: None, as the company has closed the pulping operation.

Tyrone Mill

Location:
Tyrone, Pennsylvania

Date Built:
C. 1900

Process:
150 TPD bleached kraft pulp

Other Production:
177 TPD papers

Total Water Use:
6 MGPD

Source:
Bald Eagle Creek, wells

Discharge:
Little Juniette River

Table 28.5
Tyrone Mill: Pollution Control Record

Treatment	Overall Evaluation	Equipment
Water		
Primary	✓	} Joint municipal treatment
Secondary	✓	
Tertiary	✗	None

Table 28.5
Tyrone Mill: Pollution Control Record (continued)

Treatment	Overall Evaluation	Equipment
Air		
Particulate (fuel)	X	Fuel: coal, oil. Three coal boilers: multiple cyclone collectors
Particulate (production)	X	No precipitator or scrubber on recovery furnace
Gas and odor	X	None
Plant Emissions		
To water:		
BOD	6,000 lb/day (40 lb/ton)	
Suspended solids	N.A.	
To air:	N.A.	

Pollution Control Expenditures:
The joint water treatment costs the company about $ 200,000 a year.

Legal Status:
(Water) In compliance, with joint system.
(Air) A state order was issued in March, 1970, for reduction of particulate emissions from the power boiler by January, 1971, and from the recovery furnace by January, 1972.

Future Plans:
The mill will be switching from coal to fuel oil to reduce particulate and soot emissions from the power boilers.

Profile:
The Tyrone mill, like the company's Mechanicville mill, was just too old and small to be worth a major investment in pollution control equipment. The recovery furnace had absolutely no control on it—for gas or soot. A company representative admitted this and added rather dejectedly, "it may be the only kraft recovery furnace in the industry with no precipitator or scrubber." Westvaco has known for several years that this plant was outdated and has considered moving it, but let it go on producing. Perhaps because of the re-

cent Pennsylvania State crackdown on air polluters, Westvaco made the decision to shut down pulping operations at Tyrone.

CEP Estimate of Minimum Necessary Pollution Control Expenditure Not on Company Budget: None, since the company has shut down pulping operations.

Charleston Mill

Location:
Charleston, South Carolina

Date Built:
1937—expanded and modernized since for $ 130 million

Process:
2,000 TPD unbleached kraft

Other Production:
1,813 TPD kraft linerboard, saturating papers, multiwall sack, and convertible papers

Total Water Use:
40-45 MGPD

Source:
Edisto River and Goose Creek Reservoir

Discharge:
Cooper River

Table 28.6
Charleston Mill: Pollution Control Record

Treatment	Overall Evaluation	Equipment
Water		
Primary	X	None
Secondary	X	None
Tertiary	X	None
Other	—	Screening. The caustic soda used in plant is filtered through activated carbon to remove mercury

Table 28.6
Charleston Mill: Pollution Control Record (continued)

Treatment	Overall Evaluation	Equipment
Air		
Particulate (fuel)	✓	Fuel: bark, oil. Bark boiler has 95% efficient scrubber installed in 1969
Particulate (production)	✗	Four recovery furnaces with electrostatic precipitators (90%, 93%, 97% and 97.5%). Four lime kilns, scrubbers
Gas and odor	✓	Company-designed black liquor oxidation system installed in 1966

Plant Emissions	
To water:	
BOD	65,000 lb/day (38 lb/ton; company estimate)
Suspended solids	N.A.
To air:	N.A.

Pollution Control Expenditures:
$3.5 million including $225,000 for bark boiler scrubber and $300,000 for black liquor oxidation.

Legal Status:
(Water) Under abatement order to correct water pollution.
(Air) Brand-new laws.

Future Plans:
The mill is constructing a $2.6 million clarifier to be completed by March, 1971, to provide adequate primary water treatment. (It will be the largest primary clarifier in the world.) The system will include facilities for collecting and incinerating the sludge. The state has approved Westvaco's construction schedule for this equipment; the company is studying a secondary water treatment plant, budgeted at $6 million.

The ash and particulates from the bark-burning power boiler are being well controlled by the newly installed wet scrubber, and the company claims that the new odor control system for the recovery operation is also "extremely good—better than most." However, $2.7 million is budgeted for the

next two years for more equipment and in-plant changes to increase particulate control in the chemical recovery operations. Plans for this include re-
routing some of the recovery furnace gases, adding a new precipitator, rebuilding the present precipitator to raise its effciency from 93 percent to 98
percent and improving one of the lime kiln scrubbers. The company says plans
will bring recovery unit precipitators up to 99.4, 99.8, 97 and 98 percent efficiencies.

Profile:
The Charleston mill is the largest and second newest of Westvaco's seven
mills. By August 1972 it will be equipped with both adequate air control and
primary water treatment at a total cost of $5.3 million. Secondary water treat-
ment will undoubtedly be required by the state. However, a state official said,
"the Cooper River, where the mill is located, is a big river. It's one of the
state's main waterways. It has lots of industry on it, but it also has a lot of
water and the pollution diffuses." Westvaco is now studying a $6 million sec-
ondary system.

For the past several years, Westvaco has been conducting a pilot project
for removing sulfur dioxide from stack emissions. The project, costing
$750,000 has been funded entirely by the company.

CEP Estimate of Minimum Necessary Pollution Control Expenditure Not
on Company Budget: None.

Covington Mill

Location:
Covington, Virginia

Date Built:
1899 (completely rebuilt 1947-1950 for $50 million)

Process:
1,048 TPD kraft of which 910 TPD is bleached
305 TPD NSSC

Other Production:
1,353 TPD envelope kraft, coded and uncoded carton board, bleached test
linerboard, semichemical corrugating medium, file folder tag and cup stock

Total Water Use:
25 MGPD

Source:
Jackson River

Discharge:
Jackson River to James River

Table 28.7
Covington Mill: Pollution Control Record

Treatment	Overall Evaluation	Equipment
Water		
Primary	✓	Clarifier
Secondary	✓	Activated sludge (first in the industry, installed 1955, upgraded 1959, 1962, 1964); waste water cooling
Tertiary	X	None
Other	—	Activated carbon towers for mercury contamination
Air		
Particulate (fuel)	✓	Fuel: coal, bark, gas. Gas boiler needs no controls. Two bark boilers have electrostatic precipitators, 98%, 99.5% efficient; no data on third bark boiler. Fifth boiler is standby, has 85% efficient dust collector
Particulate (production)	✓	13 small old rotary furnaces, 6 electrostatic precipitators (2, 94%; 1, 95%; 3, 99% efficient). Bag filter, 99% efficient. Lime kiln scrubber, 98% efficient
Gas and odor	X	No black liquor oxidation

Plant Emissions

To water:

BOD 10,000-12,000 lb/day (8 lb/ton of pulp; company figure) or 14,000 lb/day (10 lb/ton of pulp; state figure)

Suspended solids 21,000 lb/day (20 lb/ton of pulp)

To air: N.A.

Pollution Control Expenditures:
$4 million for construction and operating costs of the activated sludge facility
($1.5 million for construction) and $6 million for air pollution control. The 2
new power boiler precipitators cost $1.25 million.

Legal Status:
(Water) The mill is in compliance with the state law and has, according to
state officials, "one of the most sophisticated water treatment plants in Virginia." However, water pollution still presents a serious problem because of
the low flow of the Jackson River. More treatment will be required.
(Air) Completed first program with the state in 1970; now working on second.

Future Plans:
A $15 million program is now underway to overhaul and rebuild the mill's
kraft recovery operation. The 13 small rotary furnaces used to burn the spent
liquor will be replaced by a single 23-story furnace, equipped with black liquor
oxidation to reduce odor and a 99 percent efficient electrostatic precipitator.
When the new equipment goes into operation in late 1973, air pollution is expected to be reduced by 75 percent to only 200-400 pounds of emissions per
hour. In addition, a cooling tower is planned to reduce the heat of the waste
water before discharge.

Profile:
For 70 years the economic life of Covington has been sustained by Westvaco's
1,048-ton kraft mill. It pays a sizable share of the city's and Alleghany County's taxes and employs 2,000 of the town's 11,000 residents. In addition to being the most important industry in Covington, the Westvaco mill has long been
a serious polluter of the Covington Valley air and water.
        From an environmental point of view, said the Virginia State Pollution
Control Board, the mill should never have been built where it is. The Jackson
River is much too small and slow-flowing to absorb the mill's wastes, and
during drought periods (July through September) the entire river is chan-
nelled through mill pipes. Despite the extensive water treatment facilities
that have been built, mill effluent still renders that part of the river immediately downstream almost completely lifeless. A Washington Post article
reported: "The plant often stains the Jackson River for miles with a deep
brown, almost black substance, called Lignin (which)... interferes with photosynthesis, the process which in turn supports the fish population." Mill
manager Carl O. Skoggard did not discount this problem but called it "something facing us in the future."
        Noxious fumes and soot from the mill stacks have been the center of con-
troversy in Covington for years because the mill had very inadequate odor or
particulate control equipment. Dustfall around the mill was measured at 8 to
120 tons per square mile per month. It has now fallen to a level of about 2 to
15 tons maximum. The air pollution problem was aggravated by the fact that
the Covington region is subject to atmospheric inversions 150 to 200 days per
year.

Pressure for a mill cleanup began in the mid-1960s. The state air board was formed in 1966, and in 1968, a citizen's group formed, the Allegheny Crusade for Clean Air, Inc., to press for special local legislation to stop the mill from polluting. As a result of Westvaco's major improvements, it is now receiving commendations instead of complaints from the citizenry.

In mid-1970, Westvaco finally announced a $15 million program to rebuild recovery operations here including black liquor oxidation and a new electrostatic precipitator. By mid-1973, when the program is complete, the valley's air may be clear again.

CEP Estimate of Minimum Necessary Pollution Control Expenditure Not on Company Budget: $500,000 total, for tightening up the water system.

References for Chapter 28

1. Upper Potomac River Basin Water Quality Assessment, prepared by the Chesapeake Technical Support Laboratory (Middle Atlantic Region) for the Federal Water Pollution Control Administration, U.S. Department of the Interior, December, 1969.

2. Ibid. The study broke down other pollution sources on the 45-mile river section as follows: Celanese plant, discharging 30 million gallons per day, BOD 25,000 pounds; Kaiser, West Virginia, discharging 0.6 million gallons per day, BOD 400 pounds; town and industry of Cumberland, Maryland, discharging 5.2 million gallons per day, BOD 6,930 pounds.

Chapter 29

WEYERHAEUSER COMPANY
1021 South A Street, Tacoma, Washington 98401

Major Products:
Growing and harvesting timber, manufacturing and distributing all types of forest products.

Financial Data ($ Millions):	1969	1968
Net Sales	1,239.2	1,053.0
Net Income	131.4	106.4
Capital Expenditures	187.3	112.2

Pollution Control Expenditures:
$125 million to 1970, of which 20 percent went for personnel. In 1969, pollution control expenditures were $15.3 million.

Annual Meeting:
The annual meeting is held during April in Tacoma, Washington.

Officers:
Chairman: Norton Clapp, Medina, Washinton (Director: Seattle First National Bank, Metropolitan Building Corp.; Vice President, Boy Scouts of America; Trustee, University of Chicago)
President: George Weyerhaeuser, Tacoma, Washington (Director: Boeing Corp., Puget Sound National Bank)

Outside Directors:
William M. Allen: Chairman, Boeing Corp.
Carleton Blunt: Counsel, Bell, Boyd, Lloyd, Haddad and Burns
John H. Hauberg: President, Pacific Denkmann Co.
E. Bronson Ingram: President, Ingram Corp.
Robert H. Kieckhefer: Former President, Kieckhefer Container Corp.
John M. Musser: Former Vice President, Weyerhaeuser Co.
Robert D. O'Brien: Chairman, Pacific Car and Foundry Co.
General Edwin W. Rawlings: U.S.A.F. Retired, Retired Chairman, General Mills, Inc.
Thomas C. Taylor: Partner, Nixon, Hargrave, Devans and Dey
C. Davis Weyerhaeuser: Chairman, Dawson and Co.
Robert B. Wilson: Executive Vice President, National Bank of Oregon

Breakdown of Total Company Production:

Market Pulp   673,000 tons

Cartons	156,000 tons
Paperboard	1,054,000 "
Paper	299,000 "
Logs	3,695 million board feet
Lumber	1,724 "
Softwood plywood	856 million square feet
Shipping containers	12,743 "

Plants:
Weyerhaeuser has 101 plants, world wide:

Domestic facilities

Milk Carton plants	7
Folding carton plants	5
Shipping container plants	32
Printing and writing paper plants	5
Pulp and paper plants	23

Timberholdings:
Weyerhaeuser owns 5.6 million acres of forest land in the United States and
has management responsibility and harvest rights on 9.7 million acres more
abroad. The total value in 1969 was $395,885,000.

Annual Pulp Production (estimated on a basis of 355 production days per year):
2,183,250 tons (1969)

Company Pollution Overview

Weyerhaeuser has been a leader in pollution control in the pulp and paper industry. This is attributed to "farsighted" men in the company and "public
pressure."[1] Company engineers have participated in the most important
pollution control developments of the post-war period, and Weyerhaeuser has
led in application of the technology. Partly because of this early installation
of equipment, some mills now have out-of-date control facilities.
    The first experiments and installation of black liquor oxidation were car-
ried on by Weyerhaeuser, and Springfield and Everett had the first weak black
liquor oxidation systems in the industry. The vaporsphere, developed by
Weyerhaeuser to collect digester gases for lime kiln incineration, has been
widely used throughout the industry. Magnesium oxide sulfite pulping was
first developed in association with Babcock and Wilcox and used at Longview
in 1948; later, Weyerhaeuser was the first company to build a new mill of
that type.

Weyerhaeuser has installed state-of-the-art pollution control equipment at each of the nine mills planned, built, or expanded since World War II.

The company policy is described in a letter Mr. J. L. McClintock, director of environmental control, wrote to CEP July 13,1970. "It costs two or three times as much to modify an old mill as compared to building a modern new mill with the same environmental abatement devices. Many of the old mills cannot generate the capital required to do this job on a crash basis. Priorities and time are critical if the mills are to survive."

Weyerhaeuser operates 13 pulp mills at ten locations: three in the South, and two in Oklahoma and five sites in the Northwest. (Everett, Washington, is counted as two locations since the kraft and sulfite mills are a few miles apart and treat their wastes separately.) A large kraft mill at Valiant, Oklahoma, scheduled to start up in late 1971, will be the company's thirteenth pulp making facility, and will have state-of-the-art pollution control equipment. It was number 12, not 13, which was unlucky; at the new New Bern, North Carolina mill, the recovery furnace, a new low odor type of which the company was very proud, exploded soon after start-up. Weyerhaeuser's 12 currently operating mills produce 6,150 tons of pulp daily using 280 million gallons of water.

Of the six operating kraft mills, four have odor control systems, the largest (at Plymouth, North Carolina) and the smallest (at Pine Bluff, Arkansas) do not. At two locations (New Bern and Springfield) odor control is excellent. At Everett and Longview the systems are 15 years old and now out of date in terms of efficiency. However, even at 50 percent efficiency, they have kept tons of sulfurous gases from the atmosphere since the early 1950s. All three sulfite mills (Everett, Longview, and Cosmopolis, Washington) have good sulfur dioxide control systems.

Particulate emissions are adequately controlled at six of the nine pulping locations (9 of the 12 mills).

Weyerhaeuser won the Sports Foundations, Inc. Gold Medal in 1968 for excellence in water pollution control—the first time it was awarded to a corporation. (The following year, Owens-Illinois received it.) Of the ten Weyerhaeuser mill locations, two have virtually no water treatment. The remaining 8 all have adequate primary treatment and half also have secondary treatment now; of the remaining four that do not have adequate secondary treatment, two had what were once considered very fine ponding and lagooning systems but now need upgrading. This water control record is by no means "perfect." Still, Weyerhaeuser's efforts in this area and the extent of treatment provided are head and shoulders above the efforts of the largest pulp and paper companies.

The two prewar mills (Plymouth and Everett sulfite) are the exception to a generally responsible pollution control effort. One presents a sorry picture in air pollution control, one in water pollution control. Evidently they were both written off long ago, but not closed down. Everett is the only one of the company's three sulfite mills still using the nonrecoverable calcium base. It was one of those cited at the 1967 federal Puget Sound abatement conference because its discharge of untreated spent sulfite liquor was considered damaging to fish life. Weyerhaeuser has taken no visible pollution control action at Everett and will probably close it rather than convert to magnesium. However, the company is working on another new pulping technique, which, if it is developed in time, will be installed at Everett. Plymouth, on the other hand, shows neglect of air pollution control. Although good water pollution control equipment was installed there four years ago, this, the company's largest kraft mill, has no odor control equipment, and is only starting to install particulate control equipment. It is not likely that Weyerhaeuser is considering closing this mill.

As an innovative company, Weyerhaeuser seems relatively little influenced by state regulations. Mr. McClintock commented that state standards are "designed to bring laggards up to leaders ... As far as Weyerhaeuser is concerned, the real standard is neighbors in the community." He did, however, express doubt about whether Oregon and Washington's kraft mill particulate standards are realistic. Despite his doubts, two of the company's three kraft mills in those states (Springfield and Everett) are virtually in compliance now. Although pollution control expenditures are not the best gauge of cleanliness, they do indicate a priority level. Weyerhaeuser's pollution control expenditures for 1945 through 1969 were an astonishing $125 million. In 1969 they totalled $15.25 million: $11 million for capital equipment, $3 million for personnel, and $1.25 million for maintenance, research, and engineering. These totals do not include the cost of recovery furnaces (basic production equipment often included by company spokesman in "pollution control spending") and are thus even more impressive.

The company has budgeted $12 million for further pollution controls for the pulp mills. CEP estimates that additional necessary expenditures amount to $20.5 million. Half of this is for improving secondary treatment facilities and old, formerly adequate controls and cleaning up the air at the two neglected kraft mills (Pine Bluff, Arkansas and Plymouth, North Carolina) and half is for the conversion of the Everett sulfite mill. The total estimated Weyerhaeuser expenditure would then come to $32.5 million.

Table 29.1
Weyerhaeuser Company: Pulp Production, Water Use, and Pollution Control

Mill Location	Pulp Prod. (TPD)	Other Prod. (TPD)	Water Use (MGPD)	Water Pollution Control			Air Pollution Control	
				Pri.	Sec.	Tert.	Part.	Gas, Odor
Pine Bluff, Ark.	200	200	3.9	✓	✗	✗	✗	✗
New Bern, N.C.	600	--	27	✓	✓	✗	✓	✓
Plymouth, N.C.	1,550	1,475	42	✓	✓	✗	✗	✗

Table 29.1
Weyerhaeuser Company: Pulp Production, Water Use, and Pollution Control
(continued)

Mill Location	Pulp Prod. (TPD)	Other Prod. (TPD)	Water Use (MGPD)	Water Pollution Control			Air Pollution Control	
				Pri.	Sec.	Tert.	Part.	Gas, Odor
Craig, Okla.	450	450	14.5	✓	✗	✗	✓	—
Valiant, Okla.	1,600*		20*	✓	✓	✗	✓	✓
Springfield, Ore.	1,150	1,150	20	✓	✓	✗	✓	✓
Cosmopolis, Wash.	400	--	28	✓	✗	✗	✓	✓
Everett (sulfite), Wash.	300	--	28	✗	✗	✗	✗	✓
Everett (kraft), Wash.	360	--	27	✓	✗	✗	✓	✗
Longview, Wash.	1,140	1,400	90	✓	✗	✗	✓	✗
Total	6,150		280.4					

*The Valiant mill is under construction and is not included in the totals.

Pine Bluff Mill

Location:
Pine Bluff, Arkansas (Dierks Division)

Date Built:
1957, acquired 1969

Process:
200 TPD kraft

Other Production:
200 TPD convertible kraft paper, multiwall sack, bag and sack, asphalting, creping, kraft specialities, kraft board

Total Water Use:
3.9 MGPD

Source:
N.A.

Discharge:
Arkansas River

Table 29.2
Pine Bluff Mill: Pollution Control Record

Treatment	Overall Evaluation	Equipment
Water		
Primary	✓	Clarifier
Secondary	✗	None
Tertiary	✗	None
Air		
Particulate (fuel)	✓	Fuel: gas and purchased power. None needed
Particulate (production)	✗	Recovery furnace: precipitator 85% efficient (inadequate). Lime kiln scrubber
Gas and odor	✗	None
Plant Emissions		
To water:		
BOD	5,000 lb/day (25 lb/ton of pulp)	
Suspended solids	N.A.	
To air:	N.A.	

Pollution Control Expenditures:
N.A.

Legal Status:
(Water) Secondary treatment required by end of 1972.
(Air) In compliance.

Future Plans:
120 acre lagoon for secondary treatment scheduled to start up in March 1971.

Profile:
This very small kraft mill, which was recently acquired by Weyerhaeuser, lacks adequate air and water pollution control. There is no odor control equipment, nor plans to install it; the recovery furnace precipitator is out of date. Secondary treatment, as planned, should provide adequate water pollution control, especially since water use at the mill is very low. This mill is not old enough to be abandoned in the area of air pollution control, even though the previous owners were lax; state requirements are not very demanding.

CEP Estimate of Minimum Necessary Pollution Control Expenditures Not on Company Budget: $1 million total, of which $500,000 is for a black liquor oxidation odor control system and $500,000 is for a scrubber on recovery furnace to back up the precipitator or new precipitator.

New Bern Mill

Location:
New Bern, North Carolina

Date Built:
1969

Process:
600 TPD bleached kraft

Other Production:
None

Total Water Use:
27 MGPD

Source:
N.A.

Discharge:
Neuse River

Table 29.3
New Bern Mill: Pollution Control Record

Treatment	Overall Evaluation	Equipment
Water		
Primary	✓	27 million gallon clarifier
Secondary	✓	70-acre holding pond; 98 acre aeration pond (10 day retention)
Tertiary	✗	None
Air		
Particulate (fuel)	?	Fuel: N.A.
Particulate (production)	✓	Recovery furnace: precipitator, 99.6% rated efficiency. Lime kiln: precipitator, 99.6% rated efficiency
Gas and odor	✓	Non-direct contact evaporation system; vaporsphere for collection of gases from digester and evaporation for incineration in lime kiln. Heat exchangers to prevent odor during recovery process
Plant Emissions		
To water:		
BOD		No data yet because of start-up conditions
Suspended solids		N.A.
To air:		
TRS		Hydrogen sulfide approx. 1 ppm (company estimate)
Particulates		N.A.

Pollution Control Expenditures:
$6 million of the total $50 million capital cost of the mill; $3 million for air and $3 million for water.

Legal Status:
(Water) In compliance.
(Air) In compliance.

Future Plans:
None (new mill).

Profile:
The mill is not quite yet up to capacity, so the final results of the odor and
particulate control systems cannot be given. However, the company has installed state-of-the-art air pollution control equipment at a substantial investment, showing a great concern for pollution control. The company calls New
Bern "the cleanest mill ever," and the water treatment system is certainly
adequate. However, start-up problems and the unfortunate accidental explosion of the recovery furnace in June 1970 make it unlikely that the odor control
efficiency is as good as was expected.

CEP Estimate of Minimum Necessary Pollution Control Expenditures Not
on Company Budget: None.

Plymouth Mill

Location:
Plymouth, North Carolina

Date Built:
1937

Process:
1,250 TPD kraft, of which 450 tons are bleached
300 TPD NSSC

Other Production:
1,275 TPD kraft linerboard, NSSC corrugating medium, milk carton board,
polyethlene coated board
200 TPD fine papers

Total Water Use:
42 MGPD

Source:
Roanoke River

Discharge:
Roanoke River

Table 29.4
Plymouth Mill: Pollution Control Record

Treatment	Overall Evaluation	Equipment
Water		
Primary	✓	300-acre holding pond (14 day retention, installed 1966)
Secondary	✓	Aeration of effluent before discharge
Tertiary	X	None
Air		
Particulate (fuel)	X	Fuel: bark, oil. Multiple cyclone separators (inadequate)
Particulate (production)	X	Lime kiln: scrubbers installed 1970; recovery furnace: controls N.A.
Gas and odor	X	None
Plant Emissions	N.A.	

Pollution Control Expenditures:
$2 million for water equipment; $200,000 for lime kiln scrubbers. (Total mill capital investment estimated at $100 million).

Legal Status:
(Water) In compliance, secondary treatment provided before North Carolina required it; the mill has permit which expires in December, 1974.
(Air) Presently in compliance.

Future Plans:
None.

Profile:
This mill presents considerable contrast. While water pollution control was undertaken relatively early, there has never been any air pollution control to speak of. The lime kiln scrubbers were just installed, and there is no odor control equipment.

The company commented on water control by saying that since the installation of the holding pond, mill effluent has only 13 percent of the effect on dissolved oxygen in the river that it did before. This is a rather odd way of describing treatment efficiency. Nevertheless, someone likes the system; the North Carolina Wildlife Federation gave it an award in 1968.

CEP Estimate of Minimum Necessary Pollution Control Expenditures Not on Company Budget: $2.25 million total, of which $1 million is for an odor control system (high because mill is so large); $250,000 is for a power boiler scrubber; and $1 million is for a recovery furnace precipitator.

Craig Mill

Location:
Craig, Oklahoma (Dierks Division)

Date Built:
1961, acquired in 1969

Process:
450 TPD groundwood

Other Production:
Wallboard, medium density board

Total Water Use:
14.5 MGPD

Source:
N.A.

Discharge:
Mountain Fork River

Table 29.5
Craig Mill: Pollution Control Record

Treatment	Overall Evaluation	Equipment
Water		
Primary	√	40-foot clarifier
Secondary	X	120-acre settling pond; 160-acre aeration lagoon (installed 1967; inadequate)
Tertiary	X	None

Table 29.5
Craig Mill: Pollution Control Record (continued)

Treatment	Overall Evaluation	Equipment
Solid waste removal	—	Sludge from clarifier is recycled and added to the fiberboard
Air		
Particulate (fuel)	√	Fuel: natural gas, bark. Bark burners have cyclones and scrubbers
Particulate (production)	—	None needed
Gas and odor	—	None needed

Plant Emissions

To water:

BOD	Approx. 22 lb/ton of pulp*
Suspended solids	N.A.

*BOD is reduced from 3,100 ppm to 83 ppm in the water treatment system, a "reduction of 1,460 pounds per day," according to J. L. McClintock, the company's director of environmental control.

Pollution Control Expenditures:
Total of $600,000, of which $311,000 was spent for the water treatment system.

Legal Status:
(Water) In compliance.
(Air) In compliance.

Future Plans:
None.

Profile:
This mill was acquired by Weyerhaeuser in 1969, and although it has considerable equipment, the reported BOD is still very high for a groundwood mill, and improved secondary treatment is needed. Given the fact that the state provides a substantial tax incentive for pollution control equipment, the

company should be able to manage better water treatment. Air pollution control measures seem adequate.

CEP Estimate of Minimum Necessary Pollution Control Expenditures Not on Company Budget: $1 million total, for improved secondary treatment.

Valiant Mill

Location:
Valiant, Oklahoma

Date Built:
Construction began 1971

Process:
1,600 TPD kraft

Other Production:
None

Total Water Use:
20 MGPD

Source:
N.A.

Discharge:
Red River

Table 29.6
Valiant Mill: Pollution Control Record

Treatment	Overall Evaluation	Equipment
Water		
Primary	√	200-foot clarifier
Secondary	√	60-acre aeration lagoon; separation of sanitary wastes for lagoon oxidation and chlorination
Tertiary	X	None

Table 29.6
Valiant Mill: Pollution Control Record (continued)

Treatment	Overall Evaluation	Equipment
Air		
Particulate (fuel)	✓	Fuel: oil, gas. Centrifugal dust collectors, 96.8% rated efficiency, on bark burners
Particulate (production)	✓	Recovery furnace precipitators, 99.8% rated efficiency; lime kiln wet scrubber, 99.7% rated efficiency
Gas and odor	✓	Non-direct contact evaporation; vapor-sphere for gas collection

Plant Emissions	
To water:	
BOD	Expect 85% reduction
Suspended solids	Data not yet available
To air:	
TRS	Data not yet available
Particulates	Data not yet available

Pollution Control Expenditures:
$6.8 million of the mill's total $100 million capital cost.

Legal Status:
Expected to be in compliance when operating.

Future Plans:
New mill; none.

Profile:
This mill will have state-of-the-art air pollution control equipment. The company says that the "total particulate discharge will be less than from a large urban apartment house with a central coal burning heating system." Water pollution control plans are also good. No complaints —a very fine effort.

CEP Estimate of Minimum Necessary Pollution Control Expenditures Not on Company Budget: None.

Springfield Mill

Location:
Springfield, Oregon

Date Built:
1949

Process:
1,150 TPD unbleached kraft

Other Production:
1,150 TPD kraft linerboard

Total Water Use:
20 MGPD

Source:
N.A.

Discharge:
McKensie River

Table 29.7
Springfield Mill: Pollution Control Record

Treatment	Overall Evaluation	Equipment
Water		
Primary	✓	} Aeration lagoons
Secondary	✓	
Tertiary	✗	None
Solid waste removal	—	Company landfill
Other	—	Spray irrigation of hot process water containing evaporator condensate (since 1961)

Table 29.7
Springfield Mill: Pollution Control Record (continued)

Treatment	Overall Evaluation	Equipment
Air		
Particulate (fuel)	√	Fuel: oil, gas. Multiple cyclones on stack; collected fly ash is sold for manufacture of briquets
Particulate (production)	√	Three recovery boilers: 89% efficient precipitators
Gas and odor	√	Vaporsphere to collect digester blow gas for lime kiln incineration; black liquor oxidation, 95% efficient (installed 1953, and later improved)

Plant Emissions	
To water:	
BOD	Summer, less than 3,000 lb/day (3 lb/ton of pulp) Winter, less than 4,000 lb/day (4 lb/ton of pulp)
Suspended solids	N.A.
To air:	
TRS	240 lb/day (0.2 lb/ton of pulp)
Particulates	4,000 lb/day (3.5 lb/ton of pulp)

Pollution Control Expenditures:
Total $8 million, of which $6 million was for water and $2 million was for air.

Legal Status:
(Water) In compliance; BOD as reported satisfies permit.
(Air) On state schedule for kraft mills; one recovery furnace presently complies; a new recovery furnace will replace the others by mid-1971.

Future Plans:
A new $8.5 million state-of-the-art recovery furnace will replace the two non-complying units in January 1971. With this new furnace, particulate control will be over 90 percent.

Profile:

This is not a new mill, but Weyerhaeuser has provided excellent pollution control here from the start, and continues to do so. This was the first mill at which the company installed a vaporsphere to cut odorous emissions about 90 percent, and black liquor oxidation. Emissions figures indicate that the mill provides adequate pollution control in all areas, and the reported TRS is fine. Particulate controls, which were state-of-the-art at installation, now need the planned upgrading. The sale of fly ash to a manufacturer of charcoal briquets makes boiler controls a break-even operation.

Water pollution control has not been neglected here either. The mill won awards in 1964 and 1968 (as part of the company Gold Medal) for innovation in water pollution control, and is one of very few mills in the Northwest with secondary treatment. The spray irrigation system was also an innovation. In 1967, the FWPCA said, "The huge Weyerhaeuser pulp and paper plant at Springfield is one of the most efficient mills in the industry in terms of the ratio of discharge wastes to production."[2] The mill is continuing to act as a guinea pig for pollution control experimentation, and presently has a $170,000 FWQA grant to develop an improved method of removing BOD and recovering valuable by-products from kraft mill effluent. The grant was for a pilot program at Longview but Weyerhaeuser decided to spend more of its own money and run a full-scale program here.

All in all, Springfield has shown a continuing excellent effort. Weyerhaeuser's pollution control director, comparing it to other mills in 1969, named it and Kamloops, British Columbia as the company's best mills, and said, "It has gone through three expansions and has been modernized. Public pressure has been greater in Springfield, so there has been more of a need to try to correct problems there."[3]

CEP Estimate of Minimum Necessary Pollution Control Expenditures Not on Company Budget: None.

Cosmopolis Mill

Location:
Cosmopolis, Washington

Date Built:
1957

Process:
400 TPD bleached magnesium oxide base sulfite

Other Production:
None

Total Water Use:
28 MGPD (of which 6 MGPD are for the bleach plant)

Source:
Wynoochee River and Lake Aberdeen

Discharge:
Chehalis River (Gray's Harbor Estuary, Puget Sound)

Table 29.8
Cosmopolis Mill: Pollution Control Record

Treatment	Overall Evaluation	Equipment
Water		
Primary	✓	60 million gallon settling ponds
Secondary	✗ (partial)	80% of the spent sulfite liquor is recovered; 11 million gallon aeration lagoon used for 5 summer months; four holding ponds, with 4-day capacity discharge on outgoing tide only
Tertiary	✗	None
Solid waste removal	—	Company-owned landfill adjacent to settling ponds
Other	—	Outfall with diffuser
Air		
Particulate (fuel)	✗	Fuel: oil, bark. Cyclone separators (inadequate)
Particulate (production)	✗	Recovery furnace: precipitator and scrubber, overall efficiency 85%; cyclone separator to remove magnesium oxide ash
Gas and odor	✓	Four absorption towers to remove sulfur dioxide and particulate matter

Table 29.8
Cosmopolis Mill: Pollution Control Record (continued)

Treatment	Overall Evaluation	Equipment
Plant Emissions		
To water:		
BOD	60,000 lb/day (123 lb/ton of pulp)*	
Suspended solids	N.A.	
To air:		
TRS	Sulfur dioxide estimated 2,750 lb/day (70 lb/ton of pulp)	
Particulates	N.A.	

*This represents about 50% BOD reduction. River is monitored for dissolved oxygen, temperature, and salinity.

Pollution Control Expenditures:
Total of $9.27 million, of which $4.27 million was spent at the time of construction and $5 million has been spent since. $144,788 of this was spent between 1967 and 1970.

Legal Status:
(Water) Permit issued in June 1970 requires further BOD reduction to a maximum of 37,300 pounds per day during 5 months of low river flow.
(Air) In compliance; state sulfite regulations being developed.

Future Plans:
Additional aeration lagoons (cost $1.5 million) to achieve required 43 percent BOD reduction are tentatively planned. A new $400,000 diffuser outfall is planned.

Profile:
This was the first U.S. mill built to pulp by the magnesium-base sulfite process in which evaporation and burning of the spent sulfite liquor to recover magnesium oxide and sulfur dioxide are part of the production process. Weyerhaeuser made water pollution control efforts when the mill was constructed, but more equipment is needed to provide complete secondary treatment in lieu of the current stop-gap holding and aeration systems. However, gas and particulate emission controls are good. Some slight BOD reduction is probably

achieved in the holding ponds, but not enough to handle bleaching wastes and the recovery plant effluent. Although the company discharges only on the outgoing tide, this does not serve to flush the effluent from the harbor. In fact, waste materials can take as long as 40 days to move through the harbor to the ocean, because of currents and tidal action. The inadequacy of the waste water treatment is reflected by the fact that the dissolved oxygen content in Gray's Harbor is lower than that in a nearby waterway.

The company's last water permit expired in October 1968, and negotiations for a new one wereonly concluded in June 1970. Still, company plans for achieving the additional BOD reductions are not set.

CEP Estimate of Minimum Necessary Pollution Control Expenditures Not on Company Budget: $2.5 million, of which $250,000 is for a scrubber for power boiler; $250,000 is for improvement of particulate control on recovery furnace; and $2 million is for secondary treatment (maximum probably less).

Everett Sulfite Mill

Location:
Everett, Washington

Date Built:
1936

Process:
300 TPD bleached calcium sulfite

Other Production:
None

Total Water Use:
28 MGPD

Source:
Sultan River Basin (same as city water supply)

Discharge:
Port Gardner Bay

Table 29.9
Everett Sulfite Mill: Pollution Control Record

Treatment	Overall Evaluation	Equipment
Water		
Primary	X	Clarifier for barking effluent only (inadequate)
Secondary	X	None. Spent sulfite liquor discharged without treatment through outfall pipe shared with Scott's sulfite mill in Everett
Tertiary	X	None
Solid waste removal	—	Sludge from bark clarifier sold for use as soil mulch
Other	—	Outfall and three deepwater diffusers into Port Gardner Bay
Air		
Particulate (fuel)	X	Fuel: bark, oil. Multiple cyclone collectors (inadequate)
Particulate (production)	—	None needed
Gas and odor	√	Absorption towers for sulfur dioxide gas

Plant Emissions

To water:

BOD   300,000 lb/day (1,000 lb/ton of pulp)

Suspended solids   N.A.

To air: N.A.

Pollution Control Expenditures:
N.A.

Legal Status:
(Water) This mill was cited in the 1967 Puget Sound abatement conference [4] for discharges of sulfite waste liquor which caused harm to fish. The present permit requires a plan for removal of 80 percent of the spent liquor or a maximum discharge of 2.2 million pounds per day (based on 10 percent solids by weight) by March 31, 1972.
(Air) Complies with sulfur dioxide limits.

Future Plans:
The mill is studying how it will comply with the state water requirements. It is planning a scrubber for bleach plant chlorine emissions.

Profile:
The company has made little attempt to provide pollution control equipment at this mill. Only a fraction of the effluent, and none of the pulping effluent, receives even primary treatment. Bleached calcium sulfite effluent has the highest BOD potential of any type of mill waste. Because Everett's discharge is combined with Scott's, it is difficult to ascribe BOD to the two companies; but this does not improve the water quality in Port Gardner Bay. Company comments that fishermen have not been discouraged but rather flourish in the discharge area, must be considered in light of the Puget Sound study findings that "the prevailing water quality of Everett Harbor, the lower reach of the Snohomish River, and the surface waters of the broad reaches of the study area has been found to be injurious, or less than satisfactory for many marine forms—[including] juvenile salmon ... Such damages derive, almost wholly, from the weak pulping wastes and paper mill wastes discharged by the Scott and Weyerhaeuser mills."[5]

In response, Weyerhaeuser is ready to quote a three-year company-supported study by the University of Washington Fisheries Research Institute which reported that no adverse effects from liquid mill wastes were observed on fish in the harbor. Quoting studies is scarcely the point anymore; discharge of untreated waste so rich in BOD is bound, in time, to harm biota. Although fish and fishermen are attracted to the warmth of the effluent stream, this is no more proof of its nontoxicity than the attraction of fish to power plant effluent until the hot water kills them. Here its not the heat but the chemicals.

Since this plant was built, its waste liquors have been discharged without treatment into the waters of Puget Sound. It is wrong to assume Puget Sound has an infinite capacity for the assimilation of wastes. The Washington State agency has tolerated this situation for a long time. Five years after the federal study, after probable harm to biota was documented, the mill is only required to provide a plan for alleviating the situation. At present, the company reports that it sees three possible alternatives: (1) convert the mill to a recoverable base—but the estimated $15 million cost is prohibitive; (2) use an entirely new pulping method which has less effluent and is faster—over four years, $2.5 million is being spent on research for this aim; (3) abandon the

mill. At present, the third alternative seems likely. The lack of any investment at this plant indicates that Weyerhaeuser has long planned this action.
Everett is one of the larger paper-producing cities in the West, with four
mills—Scott, Weyerhaeuser (two sites), Simpson Lee—in the city limits.
Everett, with a population of 60,000 together with its pulp mills, consumes
more water than the city of Seattle, with a population of 600,000 and no mills.

CEP Estimate of Minimum Necessary Pollution Control Expenditures Not
on Company Budget: $11.5 million total, of which $4 million is for primary
and secondary water treatment; $250,000 is for a power plant scrubber; $7
million is for conversion to a recoverable pulping process; and $250,000 is
for a scrubber for bleach plant emissions.

Everett Kraft Mill

Location:
Everett, Washington

Date Built:
1952

Process:
360 TPD bleached kraft

Other Production:
None

Total Water Use:
27 MGPD

Source:
Snohomish River

Discharge:
Snohomish River

Table 29.10
Everett Kraft Mill: Pollution Control Record

Treatment	Overall Evaluation	Equipment
Water		
Primary	√	Retention ponds hold all effluent and discharge it on outgoing tide; also provide some partial secondary treatment as well
Secondary	X	None
Tertiary	X	None
Other	—	Outfall pipe
Air		
Particulate (fuel)	√	Fuel: gas. None needed
Particulate (production)	√	Recovery furnace: precipitator and scrubber, 98% efficient. Lime kiln: precipitator and scrubber, 85% efficient
Gas and odor	X	Vaporsphere to collect digester gases for lime kiln incineration (removes 50% of sulfurous gases). Weak black liquor oxidation (installed, 1953; second system in the industry, now inadequate)

Plant Emissions	
To water:	N.A.
To air:	
TRS	650 lb/day (less than 2 lb/ton of pulp)
Particulates	650 lb/day (less than 2 lb/ton of pulp)

Pollution Control Expenditures:
$1 million for air between 1955 and 1970.

Legal Status:
(Water) Permit requires 85 percent BOD reduction via secondary treatment.
(Air) Presently complies with 1975 particulate standards; expect to meet odor
deadline on schedule—May 1974.

Future Plans:
The company is planning an aeration system for secondary treatment by 1972
and either strong black liquor oxidation or a new system which is being developed under a pilot program with Nalco Chemical Company for catalytic oxidation of odorous gases.

Profile:
Weyerhaeuser took very good, very early action for air pollution control here,
but neglected water pollution, as it did at the other Everett mill. This mill
presents a contrast of pioneering work in air pollution control—both the scrubber and oxidation system were innovations—but minimal water pollution control. Finally, state pressure is bringing water treatment and the already good
odor control will improve. Percentage efficiency implies that the lime kiln
needs improvement; emission figures indicate that it does not.

   CEP Estimate of Minimum Necessary Pollution Control Expenditures Not
on Company Budget: $1.2 million total, of which $200,000 is for improvements to the odor control system and $1 million is for a secondary water
treatment system (planned but no costs given).

Longview Mills

Location:
Longview, Washington (three mills)

Date Built:
1931, 1948, 1954

Process:
300 TPD bleached magnesium oxide base sulfite
640 TPD bleached kraft
200 TPD NSSC

Other Production:
220 TPD unbleached corrugating medium
260 TPD bleached kraft paperboard
120 TPD coated and uncoated bleached boards
400 TPD polyethylene coated board
350 TPD bleached sulfite pulp
50 TPD Mitscherlich pulp

Total Water Use:
90 MGPD

Source:
N.A.

Discharge:
Columbia River

Table 29.11
Longview Mill: Pollution Control Record

Treatment	Overall Evaluation	Equipment
Water		
Primary	√	295-foot clarifier (installed, 1968)
Secondary	X (partial)	None, but 80% of the sulfite liquor is recovered
Tertiary	X	None
Solid waste removal	—	Landfill for clarifier sludge
Other	—	Outfall pipe
Air		
Particulate (fuel)	√	Fuel: oil, gas. None needed
Particulate (production)	√	Recovery furnace precipitator. Lime kiln scrubber
Gas and odor	X	Magnesium oxide absorption towers for sulfur dioxide control; weak black liquor oxidation, 45% efficient (installed, 1954, 1956; inadequate)
Plant Emissions	N.A.	

Pollution Control Expenditures:
Total of $8 million, of which $1.7 million was for the clarifiers and $203,000 was for black liquor oxidation.

Legal Status:
(Water) The only pulp mill in Washington in compliance — 80 percent spent sulfite liquor removal required.
(Air) Power boiler in compliance; company says the kraft mill will meet state time schedule.

Future Plans:
The mill plans to increase the efficiency of its black liquor oxidation to 95 percent at a cost of $160,000.

Profile:
This large production complex, with a variety of pulp and paper output and a huge water use, presents substantial pollution control problems. As elsewhere, Weyerhaeuser long ago took pollution control measures but needs to update them now that technology has improved. This was the first sulfite mill in the world to convert to magnesium; one of the first kraft mills to have black liquor oxidation. The mill needs secondary water treatment, even though the state does not yet require it. Primary treatment is not adequate for the production and bleaching effluent here.

  This mill was one of those cited by the Justice Department in July, 1970, for discharging mercury. Here as elsewhere (Georgia-Pacific, Bellingham) the mercury came from the plant which provides bleaching chemicals for this and Weyerhaeuser's other West Coast mills. The company has taken steps to control mercury emissions.

  CEP Estimate of Minimum Necessary Pollution Control Expenditures Not on Company Budget: $2.1 million total, of which $2 million is for secondary water treatment and $100,000 is for a vaporsphere.

References for Chapter 29

1. Director of Environmental Control J. M. McClintock, in a conversation with CEP, March 1970.

2. Federal Water Pollution Control Administration, quoted in the Weyerhaeuser booklet Our Environment, 1970, p. 16.

3. J. M. McClintock, "The New Ball Game in Environmental Control," Weyerhaeuser Management, November, 1969.

4. Pollutional Effects of Pulp and Paper Wastes in Puget Sound, U.S. Department of the Interior, Federal Water Quality Administration, and Washington State Pollution Control Commission, March, 1967, p. 258.

5. Ibid., pp. 301-302.

FEDERAL CRITERIA FROM THE "GREEN BOOK"

Water Quality Criteria [1], often called the "Green Book," presents guidelines for state water pollution control laws in terms of recommended standards of water quality. These standards are determined by what amount of pollution various bodies of water and their inhabitants can absorb. The study recommends that secondary treatment be generally required of all waste sources. Another recommendation is the inclusion of an antidegradation clause, guaranteeing at least the present quality of a stream.

The study makes specific recommendations for water quality standards in terms of what minimum protection is needed for growth, reproduction, and survival of various species of fish. A good state program should comply with the guidelines. A summary of the recommendations is presented here. [2] Following it is a glossary of the terms used therein and in state air and water quality criteria.

1. Dissolved materials. These should not be increased by more than one-third of the concentration that is characteristic of the natural condition of the subject water.

2. pH should be in the range 6.0-9.0.

3. Temperature. For warm water biota, there should be no increase greater than five degrees; in lakes no more than three degrees. The following maximum temperatures for various species must be adhered to:

a. 93 deg: growth of catfish, gar, white or yellow bass, spotted bass, buffalo, carpsucker, threadfin shad, and gizzard shad.

b. 90 deg: growth of largemouth bass, drum, bluegill, and crappie.

c. 84 deg: growth of pike, perch, walleye, smallmouth bass and sauger.

d. 80 deg: spawning and egg development of catfish, buffalo, threadfin shad, and gizzard shad.

e. 75 deg: spawning and egg development of largemouth bass, white and yellow bass, and spotted bass.

f. 68 deg: growth or migration routes of salmonids and for egg development of perch and smallmouth bass.

g. 55 deg: spawning and egg development of salmon and trout (other than lake trout).

h. 48 deg: spawning and egg development of lake trout, walleye, northern pike, sauger, and Atlantic salmon.

4. Dissolved oxygen.

a. For diversified warm water biota, including game fish, dissolved oxygen concentration should be above 5 mg/l (milligrams per liter). Under extreme conditions they may range between 5 and 4 mg/l for short periods, provided the water quality is favorable in all other respects. These requirements "should apply to all waters except administratively established mixing zones. In lakes, such zones must be restricted so as to limit the effect on the biota. In streams, there must be adequate and safe passageways for migrating forms." b. For cold water biota, it is desirable that dissolved oxygen concentrations be at or near saturation. This is especially important in spawning areas where dissolved oxygen levels must not be below 7 mg/l at any time. For good growth and the general well-being of trout, salmon, and their associated biota, dissolved oxygen concentrations should not be below 6 mg/l. Under extreme conditions, they may range between 6 and 5 mg/l for short periods provided the water quality is favorable in all other respects and normal daily and seasonal fluctuations occur. In large streams that have some stratification or that serve principally as migratory routes, dissolved oxygen levels may range between 5 and 4 mg/l for periods up to six hours, but should never be below 4 mg/l at any time or place.

5. Turbidity. This should not exceed 50 Jackson units in warm water streams, or 10 Jackson units in cold water streams.

6. Settleable materials. Since it is known that even minor deposits of settleable materials inhibit the growth of normal stream and lake flora, no such materials should be added to these waters in quantity that adversely affect the natural biota.

7. Color and transparency. For the effective photosynthetic production of oxygen, it is required that 10 percent of the incident light reach the bottom of any photosynthetic zone in which adequate dissolved oxygen concentrations are to be maintained. Therefore, all floating materials of foreign origin should be excluded from streams and lakes.

8. Odor and taste. All materials that will impart odor or taste to fish or edible invertebrates should be excluded from receiving waters at levels that produce tainting.

9. Marine and estuarine organisms. For their protection, there should be:

a. no changes in channels, basin geometry, or freshwater influx so as to change salinity more than 10 percent.
b. no change in currents.
c. a pH level between 6.7 and 8.5.
d. no greater temperature increase than 4 degrees in September through May, and no more than 1.5 degree in June through August.
e. a dissolved oxygen concentration greater than 5 mg/l, and never less than 4 mg/l in estuaries and tidal tributaries.

Glossary

The terms used to describe air and water standards tend to be similar. "Ambient Air/Water Criteria" describe maximum or minimum allowable concentrations of pollutants. Gases in ambient air standards, or solids or turbidity in water have maxima indicated by parts per million (ppm) of air or water. The allowable concentrations of pollutants vary depending upon the type of area. Industrial areas are permitted higher soot levels than are residential areas. For water standards, a higher bacteria concentration is allowed in waters used primarily for navigation than in waters used for swimming and fishing. Once air or water standards are established, effluent requirements are set up to ensure that safe concentrations are not exceeded. That is, each source of pollution in an area can put a certain amount of matter into the air or water without the total exceeding the natural capability of the resource to cleanse itself. Thus, a sort of rationing system can be established. For example, Wisconsin has established limits on effluents into the Fox River, having calculated the total effluent sources that can be handled by the river. Or, a city, having decided upon maximum acceptable concentrations of sulfur dioxide, can place limits on emissions of sulfur, so that the criteria can be met.

Dissolved oxygen is measured either by weight in milligrams per liter (mg/l) or by volume in parts per million. The terms are almost equivalent since there is little difference between a concentration of 5 milligrams per liter of dissolved oxygen and one of 5 parts per million. One ppm is equal to one glassful of water in a regular Olympic swimming pool.

A third way of describing the amount of oxygen contained in water is by percent of saturation. This involves a complicated measurement, taking into account the weather and the season; it appears preferable to use one of the absolute standards. Dissolved oxygen is important in water for the same reason oxygen is important in air. Fish and plants demand oxygen in the water to live. Dead plants or waste matter in a body of water use oxygen in decomposition. Certain amounts of oxygen must be left over after decomposition to sustain the water's biota. If there is an inadequate amount of oxygen in a stream, fish will die or be stunted in growth, and the water will lose its capacity to cleanse itself. Lake Erie is an extreme case of a body of water without sufficient oxygen to maintain the life cycle. The waste discharged into the lake uses the oxygen in decomposition, and insufficient oxygen is left for maintenance of aquatic life. Fish vary considerably in their oxygen requirements. Trout are the most sensitive, and require 7 parts per million of dissolved oxygen.

Biochemical oxygen demand (BOD) is a measure of the oxygen requirements (for decomposition) of wastes emitted into a river. The higher the BOD the more oxygen in the water will be needed to handle the effluent. A pound of BOD can also be translated to "population equivalents," that is, to a measure of how many people's sewage is equivalent in oxygen use to the amount of industrial or other waste.

The coliform count is an indirect measure indicating the probability of the presence of pathogenic (disease-causing) organisms in waste put into the water. Coliform-count limits are essential for bathing waters, drinking water, and shellfish waters, since these waters lead to the direct human intake of microorganisms. The coliform count can be measured two ways: total and fecal. It is measured in relation to 100 liters of water and also by weight.

Fecal coliform is the most precise bacteria measure. Although less than the "total," it varies with total coliform count, and the total is easier to measure. The U.S. Public Health Service has defined an allowable maximum coliform count for waters in which shellfish are harvested. When shellfish live in dirty water and move to clean water, they rid themselves of the objectionable compounds before harvesting. Coliform measurements are in terms of the "most probable number" (MPN) per 100 liters of water.

The pH measures the acidity (lower than 7) or alkalinity (above 7) of water. When fish ingest highly acidic or alkaline water the imbalance can destroy their body chemistry and can kill the fish. A pH of under 2 represents an acidity comparable to that of sulfuric acid; pH above 9 is comparable to caustic soda. The names of the pulping processes demonstrate the need to control pH in mill emissions: kraft or alkaline sulfate, sulfite or acid sulfite. The NSSC process is not neutral but slightly acidic, and soda pulping uses caustic soda. It is obvious that uncontrolled emissions of extreme pH can harm humans who come into contact with the water. In some cases natural conditions of the water are outside the 6-8 neutral range. For example, marshes and swamps are alkaline.

Temperature standards are based upon the following considerations: In summer, under natural conditions, many fish live within a few degrees of a fatal temperature. Certain temperature spans are better for spawning and growth than others. Temperature changes may trigger activity inappropriate to natural conditions so that fish may hatch too early, before food is available. Fish tolerance for sudden temperature changes is much lower than for gradual temperature changes. Warm water can hold less dissolved oxygen than cool water. Thus the addition of warm water to a stream will result in lower concentrations of oxygen.

The Ringelman chart is a piece of cellophane or plastic with gradations of shading. When held up to a smoke plume a number on the chart will designate the point where the plume is as dark as the shading. Ringelman # 1 indicates the lightest smoke; higher numbers indicate darker smoke. This measures, indirectly, the particulate content of smoke from a stack. Many cities and states have regulations prohibiting smoke emissions of a density higher than # 2 or # 3 on the Ringelman chart. The chart was developed when coal was the principal fuel burned for power and heat and the main known component of air pollution was soot. There are other measures, such as a soiling index, which measures by weight the particulate matter that lands in a certain area per month. But the Ringelman chart is the most common effluent standard in air pollution. It is not sophisticated and measures only one component of air pollution.

Total reduced sulfur (TRS). Since the odor of a kraft mill is mainly caused by gases containing sulfur in its unoxidized form, the terminology "total reduced sulfur" (TRS) has been used to collectively group these odorous compounds. Four chief sulfur-containing gases are included within this TRS grouping: Hydrogen sulfide (H_2S), methyl mercaptan (MM), dimethyl sulfide (DMS), and dimethyl disulfide (DMDS). All of these gases are known to be extremely odorous, with concentrations of less than 10 parts per billion parts of air often detectable by the human nose. Individual odor threshold sensitivities may range from 0.2 to 35 ppb. The concentration at which a faint odor is detected is called the "odor threshold" (see Table A1.1).

Table A1.1
Odor Thresholds

Gas	Mean Odor Threshold (ppb)	Odor Threshold Range (ppb)
H_2S	4.5	1-22
MM	1.0	2-3.4
DMS	2.5	0.2-9.8
DMDS	7.6	0.9-35

All state water quality standards include the "four free-froms" as minimum classifications. These specify that sludge, odor, color, and wastes in general may not be discharged so as to cause objectionable or observable harm to water quality. They are spelled out only for Florida, as the "minimum conditions for all waters in the state," and for Georgia as "general criteria."

References for Appendix 1

1. Water Quality Criteria, report of the National Technical Advisory Committee to the Secretary of the Interior, Washington, D.C.: U.S. Government Printing Office, 1968.

2. Ibid., pp. 32-34.

3. The Louisiana Air Control Program, Louisiana State Department of Health, March 2, 1970.

THE MARLIN-BRAGDON REPORT

"... The pressure on corporations to meet social goals often is seen as a negative influence on their earnings ... Not necessarily so, argue two researchers ... the companies in this [the paper] industry that get the highest grades from environmentalists also are the most profitable ... The report will provide some ammunition for analysts and portfolio managers who may be looking for some economic encouragement for backing "good guy" companies."[1]

"If the study's findings hold up when applied to other industries, there is a very reassuring moral for businessmen to learn from it. There is not necessarily a fundamental conflict between profitability and pollution control. The manager who does good may also do well."[2]

On October 8, 1971, a paper prepared by Joseph H. Bragdon, Jr. (Account Executive, H. C. Wainwright & Company) and John A. Marlin (Assistant Professor of Economics and Finance, Baruch College, City University of New York), was presented at the annual meeting of the Financial Management Association at Denver, Colorado.

Entitled "Pollution and Profitability: The Case of the Pulp and Paper Industry"[3] the paper was based on the first version of Paper Profits released by CEP in 1971. It compared pollution control information from Paper Profits with corporate financial data, and determined that the cleaner pulp and paper companies show higher profits than those with worse pollution control records.

The authors analyzed 17 of the 24 companies reviewed by CEP, excluding those that were either very small or for which paper products represented a disproportionately small percentage of total sales. Of the 17, five were heavily involved in mergers effecting temporary abnormalities in either pollution control or earnings data, and thus were not included in the final correlations.

The companies were ranked according to five measures of profitability and three measures of pollution control performance, and were correlated according to the Spearman Rank System. (Table A2.1) The profitability measures were: earnings per share growth, 1965-1970 and 1970-1971; average return on equity, 1965-1970 and 1970; and average return on capital 1965-1970. The pollution control measures are weighted averages of the percentage of the companies' plants with adequate controls. Index A gives equal weight to water treatment, particulate control and gas and odor control; Index B gives water the same weight as both types of air pollution control; and Index C rates the companies overall environmental record on the basis of CEP data as good (1), average (2), or poor (3). This gross measure indicates a close relationship between pollution control and profitability, contradicting the traditional view that the two are not compatible.

Table A2.1
Spearman Rank Correlations for Twelve Pulp and Paper Companies

	Earnings per Share Growth		Return on Equity		Return on Capital
	1965-1970	1970-1971 (est.)	1965-1970	1970	1965-1970
Index A	0.514	0.223	0.682	0.546	0.329
Index B	0.560	0.227	0.623	0.434	0.259
Environmental Rating	0.628	0.557	0.691	0.502	0.544

The authors rejected the view that pollution control is a function primarily of prior profitability, noting that Weyerhaeuser was a leader in pollution control before it attained its highly profitable current position, while International Paper was profitable for many years without bringing its pollution control equipment up to the current state of the art.

They proposed instead four possible explanations for the fact that the earnings performers during the 1965-1970 period seem to include a disproportionate number of companies with strong records in the areas of pollution control. The primary explanation was that a good pollution control record reflects good management, and that good management leads to higher profits. Another was that a well-designed long term pollution control program (as opposed to spot efforts to comply with laws, regulations and abatement orders) reduces costs in such areas as labor, absenteeism, plant and equipment maintenance, legal fees and fines, and taxes. A third possible explanation was that responsible companies obtain readier access to capital markets, at a lower cost; and a fourth, which the authors feel will become increasingly important, that clients are more willing to do business with a responsible firm.

References for Appendix 2

1. "Heard on the Street," Wall Street Journal, August 4, 1971.

2. "Bet on the White Hats," editorial, Business Week, August 7, 1971.

3. To be published in the April, 1972, issue of Risk Managment (American Society of Insurance Managers, New York), and in a forthcoming issue of Management Sciences: Publications.

A FOLLOW-UP STUDY: DECEMBER 1970 TO JUNE 1972

Since December 1970, the companies surveyed by the Council on Economic Priorities have made measurable but by no means dramatic improvement in controlling their pollution. Three companies — Crown Zellerbach, Georgia Pacific, and International Paper — are on their way toward joining Owens-Illinois and Weyerhaeuser as industry environmental leaders. Those at the opposite end of the spectrum in 1970 — Diamond International, Potlatch Forests, and St. Regis — are in the process of solving some of their problems. Yet there is a long way to go before the industry reaches state-of-the-art air and water pollution control at every mill. (The table in this appendix indicate completed projects with a ⓥ .)

Long ago the federal government suggested 1972 as the deadline for primary and secondary treatment of industrial effluent. Yet by the end of this year, the percentage of pulp mill effluent (in this study) receiving primary and secondary treatment will have increased from 31% to only 49%.* That receiving primary treatment only will have increased from 34% to 37.5%; and that receiving no treatment will have decreased from 33% to 13.5%. Thus, despite the improvement, half of the two billion gallons of waste water discharged daily by the 101 mills in this update survey will still be inadequately treated.**

Moreover, fewer than one-third of the mills for which effluent quality data is available are reducing their pollution to levels that the Environmental Protection Agency considers achievable with "reasonable control practice."*** 14 of 51 mills have reached "achievable" BOD levels; 10 of 35 have reached "achievable" suspended solids levels.

----

*This includes the 2 mills (Hodge, Louisiana and Crossett, Arkansas) which have primary, secondary, and tertiary treatment. These mills discharge 65 million gallons of effluent daily, representing 2% of the total effluent).
**This update survey does not include American Can, Scott, Mead, or Riegel. The figures for total water use and percentages receiving treatment do not include International Paper.
***These EPA Advisory Federal Effluent Guidelines were developed for guidance in considering discharge permit applications. The achievable levels per ton of pulp are:
Sulfate (kraft) pulp BOD: 6-12 pounds; Suspended solids: 5-10 pounds
Sulfite pulp   BOD: 40-80 pounds; Suspended solids: 20 pounds
NSSC pulp   BOD: 25 pounds; Suspended solids: 15 pounds
Groundwood pulp   BOD: 5-6 pounds; Suspended solids: 9-10 pounds
(Air and Water News, January 24, 1972, p. 5.)

The companies' progress in controlling air pollution has been comparable to that of water pollution. Between December, 1970, and December, 1972, the percentage of pulp produced with adequate gas and odor control will have increased from 44% to 58%; that produced with adequate particulate control will have increased from 48% to 59%. But again, only half of the 101 mills will be controlling both air pollutants.

The pulp and paper industry's environmental changes coincided with economic and attitudinal ones. The industry continued its economic downturn during 1971, showing a trend toward closing down older, unprofitable mills, expanding capacity at a slower rate, and increasing its expenditures for pollution control. At the end of 1971, profits were down 20%, 1.5 million tons of paper and paperboard and 0.5 million tons of pulp capacity had been shut down, and the American Paper Institute reported that planned pulp capacity would increase only 2.3% annually between 1972 and 1974, a 50% drop from the 1956-1971 growth rate. Concurrently, according to the API, pollution control expenditures increased from $113 million in 1970 to $234 million in 1971. (Of course, these figures often include expenditures for such basic production equipment as recovery furnaces, lime kilns, etc., which improve capacity and efficiency while reducing pollution.) These industry-wide economic conditions are clearly visible among the 20 companies surveyed in this update study. While only one new U.S. mill went into operation during 1971,* many planned production increases were cut back, and eight marginally profitable mills, owned by four companies, shut down their pulping operations.

The changes in the industry's approach and attitude toward pollution control have been more positive than its economics. Although there have been no drastic changes in technology, companies, perhaps anticipating increasingly stringent legal requirements, are installing and operating equipment for maximum possible emissions control. All recovery furnaces now under construction include 99+% efficient precipitators and either non-direct contact evaporation or black liquor oxidation. And scrubbers on smelt dissolving tank vents and systems for collecting and incinerating the odorous, noncondensible gases emitted from digesters, blow tanks, and evaporators are becoming standard equipment.

Even more encouraging is the fact that companies are talking about their environmental progress, plans, and problems in greater detail. All except American Can, Scott, and Mead cooperated with CEP's quest for updated information. Potlatch, for example, which last year withheld all but the most general public relations type information, this year supplied detailed data on equipment, operations, and emissions. Crown Zellerbach displayed a similar openness, and even International Paper, although still reticent about details, was willing to discuss its projects. For the first time, a trade publication, Pulp and Paper, published a complete survey of the pollution control plans and

---

*Weyerhaeuser's Valiant, Oklahoma mill.

expenditures of each U.S. paper company (in the January, 1972, issue). If the actual progress has not been as great as possible, at least the communications gap seems to be closing.

American Can Company

The company released no updated information on its pollution control activities.

Boise Cascade Corporation

Despite an $85 million loss in 1971, Boise Cascade is continuing its planned pollution control projects. By the end of 1972, primary and secondary treatment will be operating at four of the company's six mills. These include Wallula, Washington, which discontinued its project to use effluent for spray irrigation; and Salem, Oregon, where a recently installed ammonia sulfite recovery system replaced the impoundment basins. However, secondary treatment at International Falls, Minnesota, is still at the planning stage with no scheduled completion date; and the Steilacoom, Washington, mill, which now discharges its effluent (after primary treatment) through a new deepwater diffuser outfall pipe, has no plans for further treatment.

Boise Cascade's projects to meet state air pollution compliance schedules include: power boiler particulate controls at Steilacoom; an expensive change to pump rather than "blow" the pulp from the digester to prevent gas escape at Salem, Oregon; and scrubbers on the smelt tank vents and noncondensible gas collection and incineration systems at three of its four kraft mills (International Falls, Wallula, and St. Helen's, Oregon). However, Wallula and International Falls will continue operating with no recovery furnace control system until 1975, and no improvements have been defined for the mills at St. Helen's and De Ridder, Louisiana.

Consolidated Papers Incorporated

Consolidated exemplifies the potentially adverse effects of increasingly stringent pollution control regulations on companies with small, marginal mills. Unable to finance a thorough and affirmative water pollution abatement program, it can only take stop-gap measures as crises arise. At Appleton, Wisconsin, the company requested and was granted an extension of its water pollution abatement order because "economic and technological facts are such that we cannot meet present state orders and deadline other than by closure of the operation." At Wisconsin Rapids, Consolidated shut down the bleached groundwood pulping operation. At Whiting, Wisconsin the company did succeed in completing a $600,000 primary treatment plant, just one week before the December, 1971, deadline. The only pollution control plan on the company schedule is a water treatment facility which will handle the wastes from the separate Wisconsin Rapids paper and kraft pulp mills—but it will not be begun until 1973.

Unfortunately, even if Consolidated's measures are effective in holding off state action, they will not put the mills in adequate condition. There will still be no secondary water treatment anywhere; and air pollution problems will remain unsolved at two of its four mills: Wisconsin Rapids kraft and Biron.

Continental Can Company

Consistent with its past performance, Continental Can continues to make patchwork efforts at pollution control. The Hodge, Louisiana, tertiary water treatment began operating in the fall of 1971, and when the Port Wentworth, Georgia, secondary treatment is completed in December, 1972, three of the company's four mills will be discharging adequately treated effluent. However Hopewell, Virginia, is still four years away from having any treatment at all.

Less progress has been made toward reducing air pollution. The company continues to regard odor control as a technological mystery. Aside from a liquor concentrator installed for partial odor control at Augusta, Georgia, it has scheduled only a low odor recovery furnace as part of the $100 million Hodge expansion. There are no specific plans for odor abatement at the other two mills; nor have any of the improvements in particulate control been large enough to change CEP's overall mill evaluations. There have been no changes at Hopewell. At Port Wentworth power boiler and lime kiln emissions are now reduced, but the recovery furnace precipitators remain inadequate. At Augusta, scrubbers have been added to the lime kiln and one recovery furnace, but the other furnace will lack a precipitator until December, 1973. Only at Hodge will Continental Can's cleanup be uniform and total.

Crown Zellerbach Corporation

In 1975, when Crown Zellerbach's $100 million pollution control program is complete, the company anticipates having state-of-the-art particulate and odor control equipment at each of its nine mills and primary and secondary water treatment at six.

The air program includes equipping every kraft mill with black liquor oxidation, systems for collecting and incinerating noncondensible (odorous) gases, scrubbers on smelt dissolving tank vents and lime kilns, and 98-99+% efficient recovery furnace precipitators (often followed by secondary scrubbers). This equipment, which will be installed in stages during the next two years, will reduce total reduced sulfur emissions to between one and two pounds per ton of kraft pulp, and particulate emissions to between four and six pounds per ton of pulp — nearly low enough to meet Oregon's standards, the strictest in the nation.

Unfortunately few of these improvements, with the significant exception of those at Camas, Washington, will be completed before 1974. In 1971, both Camas mills were completely revamped. Scrubbers and an odor control system were installed in the kraft mill preparatory to the 1974 construction of a new recovery furnace; and the sulfite mill was converted to the recoverable magnefite process. The major reductions in air and water pollution effected by these changes will be augmented in 1974 by the addition of the new recovery furnace and secondary treatment.

Primary and secondary water treatment systems are under construction at the two Louisiana and the West Linn, Oregon, mills. The Camas and Wauna, Oregon, mills will not be similarly equipped until 1975 and unless plans are revised the three mills which discharge into the Pacific Ocean will go on operating indefinitely without secondary treatment.

Diamond International

Diamond International is slowly making progress in abating the pollution from its four mills. The coal burning power boilers at Plattsburg, New York, have been converted to oil; but two of the small mills continue to operate with inadequate power boiler particulate control. By the end of 1972, the mills at Red Bluff, California, and Ogdensburg, New York, will have primary and secondary water treatment; and Plattsburg is participating in a joint industrial-municipal treatment project which will be completed in October, 1973.

At the company's fourth and largest mill, Old Town, Maine, a new recovery furnace designed for 99.5% particulate control and hydrogen sulfide reduction to one part per million began operating in 1971. However, construction of the mill's effluent treatment plant will not begin until June, 1973. While this schedule is in compliance with state law, it does little to enhance Diamond's environmental image. Old Town, the company's only bleached sulfite pulp mill, is by far the worst polluter, discharging nearly three times as much untreated waste water as the other three mills combined. Unfortunately, these factors have not prompted Diamond to overlook Maine's lenient 1976 deadline and push for the earliest possible completion date.

Fibreboard Corporation

Pursuant to a $200,000 study of its effluent, Fibreboard is finally installing water treatment at its only pulping facility, the San Joaquin mill in Antioch, California. The $1.5 million system now under construction will provide primary but no secondary treatment for both pulp and recently expanded paper mill wastes. The company's other environmental expenditure was $172,000 to convert the lime kiln and power boilers to natural gas.

Georgia-Pacific Corporation

By 1974, Georgia-Pacific will have completed its extensive $60 million pollution control program and all further projects will be to upgrade equipment as new technology develops. Unfortunately, like Crown Zellerbach, Georgia-Pacific continues to consider diffuser outfall pipes adequate for ocean discharge and thus has no plans for installing secondary treatment at three west coast mills.

Water pollution control facilities are now under construction at six of the company's nine mills. Those for primary treatment are scheduled for late 1972 completion; those for secondary will be completed in 1973. Instead of installing standardized clarifying and lagooning systems, Georgia-Pacific is experimenting with several different techniques. At the Lyons Falls, New York, mill, the settleable solids will be removed by surface skimming rather than sedimentation. At Port Hudson, Louisiana, an elaborate system of vertical organ-like pipes set 15 to 30 feet below the water surface of a reservoir will blow over one ton of air per hour into the effluent. This aeration system will be the first of its kind in the industry. At Woodland, Maine, and Crossett, Arkansas, the very successful company-designed color-removal system is in operation.

Despite its efforts, Georgia-Pacific seems to have become a target for the federal Environmental Protection Agency. In 1970, the company was erroneously indicted by the Justice Department for mercury discharge; the charge was dropped when the evidence showed the government's error. In 1971, the company became so impatient with the continual delays caused by changing federal requirements that it dropped out of the much vaunted Bellingham, Washington, joint municipal-industrial treatment plant and began constructing its own facility. Then, in 1972, the Justice Department filed suit, under the 1899 Rivers and Harbors Act, accusing Georgia-Pacific of polluting the St. Croix River in Maine.

Again the charges seem hard to understand. The Woodland mill has primary treatment, sanitary sewage treatment, and a color-removal system for bleaching wastes in operation, and full secondary treatment ready for construction. In a caustic refutation of the charges, company Environmental Controls Director Matthew Gould called the Justice Department action "an obvious political hoax," and a local newspaper opined that the suit had more to do with administration attempts to discredit presidential aspirants in their own states than with environmental protection. Whatever the legal outcome, the action may in fact slow down rather than accelerate Woodland's water treatment construction schedule.

Georgia-Pacific's air pollution control program is less extensive than that for water pollution because most of the work has already been done. It includes augmenting black liquor oxidation with noncondensible gas collection and incineration at four mills, installing new precipitators at two mills, and converting the Toledo, Oregon, power boilers to natural gas. These improvements will be completed by the end of 1972, leaving only the problem of three mills with inadequate power boiler particulate control.

Great Northern Nekoosa Corporation

Although Great Northern Nekoosa expects to spend approximately $30 million on pollution abatement projects between 1972 and 1976, little improvement will be visible for several years. The water treatment systems at Ashdown, Arkansas, and Cedar Springs, Georgia, have been upgraded, and construction of a primary system at Millinocket, Maine, will be completed by November, 1972. However, the other three mills will have no effluent treatment facilities for two more years, although the installation of a recovery system at Port Edwards, Wisconsin will reduce the mill's incredibly high BOD discharge (650 pounds per ton of pulp) by 80%.

Air pollution control plans are much vaguer. Money has been budgeted, and the states have approved abatement schedules for three mills. But the company has done little beyond recognizing its problems and promising to solve them. It has outlined no specific programs for Ashdown, Millinocket, or Nekoosa, Wisconsin; has undertaken no new construction except for the conversion of Port Edwards' coal boilers to gas/oil; and will continue to operate all three of its kraft mills without any odor control for several years.

Hammermill Paper Company

Like many companies with below average pollution control records, Hammermill is on the treadmill of solving each problem only as it becomes critical. The problem-by-problem approach has been successful in controlling water pollution at three of its four mills. The company is improving the lagoon system at Selma, Alabama, and installing primary and partial secondary treatment facilities at Kaukana, Wisconsin. It has achieved the required BOD reduction at Lockhaven, Pennsylvania, by shutting down the pulp mill and reducing paper production. These changes have reduced Lockhaven's water intake by 50%, eliminated the color problem, and permitted longer effluent retention in the clarifier.

However, in the area of air pollution control, progress is slow. Studies are underway to determine how Selma and Kaukana can meet kraft mill particulate and odor emissions standards, but no action has yet been taken. The soft coal power boilers at Erie, Pennsylvania, do not meet new requirements. And the new power boilers at Lockhaven, installed in 1971 and equipped with two $400,000, 93% efficient dust collectors, will not meet new federal standards.

Hammermill's one important achievement is the conversion of the Erie NSSC mill to the recoverable Neutracell II process developed by the company. The $40 million project, which included the installation of 99.5% efficient precipitators and scrubbers and the shutdown of three of the seven old coal boilers, was completed in 1971. The controversial technique of injecting spent liquor into deep wells was simultaneously discontinued; and in 1974 all effluent will be treated in a joint industrial-municipal plant. The final improvement made at Erie was a new 3,600-foot thermoplastic outfall pipe to diffuse bleach plant effluent. Although company officials have long asserted that the direct discharge of these wastes had no detrimental effect on the ecology of Lake Erie, they have proudly admitted that since the installation of the diffuser pipeline "lake conditions and fishing are far better than they have ever been . . . and there is really a noticeable difference."

Hoerner Waldorf Corporation

With the exception of primary effluent treatment planned for 1973 at Ontonagon, Michigan, all of Hoerner Waldorf's pollution control efforts are focused on the recovery system at its Missoula, Montana, mill. Although an Environmental Defense Fund suit against the mill was dismissed by the U.S. District Court, public pressure and the state board of health insured that the planned improvements would be carried out.

Phase I has been completed: two scrubbers were installed on the power boiler, and in April 1971 one recovery furnace was converted to a low odor mode of operation and equipped with a precipitator. Phase II, a second low odor recovery furnace with precipitator, is now under construction. When completed, the Missoula improvements will reduce odorous emissions by 90% and particulate emissions by 99%, and the mill will stand as an example of the potency of public pressure.

International Paper

Since the 1970 announcement of its $101 million expenditure for pollution control,* International Paper has defined its terms, begun releasing information about the specific program being undertaken at each mill, and made considerable progress. From available information, it appears that the goals of 99% particulate removal, "substantial control over kraft mill odors," and primary and secondary treatment at all 15 mills may be achieved within three years.

In the past 18 months, the company has added clarifiers to three mills and aerated lagoons or trickling filters to six. Similar equipment is scheduled for completion by late 1972 at another two locations (Mobile, Alabama, and Moss Point, Mississippi), and by 1974 at two more (Natchez, Mississippi, and Gardiner, Oregon). The natural oxidation lagoons at Camden, Arkansas, and Bastrop and Springhill, Louisiana, will be upgraded with floating aerators.

The company's major remaining unsolved water pollution problem is the 300-acre bed of sludge in Lake Champlain, alledgedly created by effluent discharges from the old Ticonderoga pulp mill. In an attempt to assess responsibility for past pollution, the state of Vermont has sued to force New York State and International Paper to remove the sludge bed. The U.S. Supreme Court will hear the case sometime this year (1972).

International Paper's air pollution control program is as extensive as that for water treatment. During 1972 and 1973, new low odor recovery furnaces with high efficiency precipitators will start up at Panama City, Florida, and Mobile, Alabama; new black liquor oxidation systems will be operating at five other mills; new precipitators will be installed at six mills; and scrubbers will be replaced where needed. The last stage of the program, a new low-odor

---

*The expenditure has been revised to $125 million.

recovery system designed to meet state emissions standards at Gardiner, Oregon, will be completed in 1975.

Despite its positive program, International Paper still seems to have conflicting internal philosophies about the question of public disclosure. It remains unwilling to reveal many important details about its pollution control projects, yet its generally cooperative response to this update was a refreshing change from 1970's closed, defiant attitude.

Kimberly-Clark Corporation

Kimberly-Clark effectively eliminated a good part of its pollution problem by selling two of its five mills: Anderson, California, and Niagara Falls, New York; and negotiating to sell a third: Niagara, Wisconsin. It is consolidating all coated printing paper operations at Kimberly, Wisconsin, which, unfortunately, is the company's worst polluter. The consolidation has necessitated construction of a $2.8 million effluent collection and primary treatment system designed to reduce suspended solids to 20 pounds per ton (or a maximum of 12,250 pounds per day). The mill's only air pollution problem, excessive particulate emissions from the power boilers, was partially solved in 1971 when a new gas boiler replaced some of the coal burners. The company has indicated no plans for installing secondary treatment at Kimberly or improving the inadequate emissions control at its other remaining pulp mill in Coosa Pines, Alabama.

Marcor, Incorporated (Container Corporation of America)

Marcor made some inroads on the water pollution problems it was facing in 1970 at two of its three mills. In the past year, it installed a $2.5 million effluent collection and primary treatment system at Fernandina Beach, Florida, and secondary aeration and settling ponds at Circleville, Ohio. At the Brewton, Alabama, mill new oxidation ponds, installed in late 1970, provide the required 90% BOD reduction; however, in response to a recommendation made at the third federal conference on Escambia Bay, the company now plans an additional $2 million secondary treatment facility for bleach plant wastes.

Fernandina Beach is now the only company mill with no secondary water treatment; its air pollution control inadequacies similarly darken the company's record. There are no plans to upgrade the old equipment to equal the efficiency levels of the new precipitator, or to improve power boiler particulate control. In contrast, Circleville now burns low sulfur oil and has applied for conversion to gas as soon as allotments are unfrozen (which may not be for three or four years). And at Brewton, the old recovery furnace precipitator has been rebuilt and the previously planned improvements will be completed by mid-1972.

The Mead Corporation

The company released no updated information on its pollution control activities.

Owens-Illinois

Owens-Illinois has maintained its progressive and responsible environmental posture by elaborating on already adequate air and water treatment facilities at all four mills. Improvements in water pollution control range from an extensive program at Valdosta, Georgia (175-foot diameter clarifier, sludge pond, increased cooling tower capacity, ponds to contain spills) to the installation of additional aeration and defoaming equipment on the secondary systems at Big Island, Virginia.

Similarly, with few exceptions, the company's air pollution control program has been directed at raising efficiencies from 92-94% to 99%. Two of the three problems — inadequate power boiler particulate control at Big Island, Virginia, and inadequate particulate control from power boilers and recovery furnace at Tomahawk, Wisconsin — have been corrected. The third is on a state compliance schedule.

The program to upgrade equipment efficiencies includes additional particulate control at Valdosta and Orange, Texas; modified and expanded gas and odor control at Tomahawk and Orange; and an experimental process change at Big Island, Virginia, to determine the technological feasibility of producing pulp without using sulfur.

These and other programs insure Owens-Illinois' status as a paper industry environmental leader.

Potlatch Forests, Inc.

Outside of the addition of primary treatment at Cloquet, Minnesota, Potlatch made no major pollution control improvements at its two mills. However, it has changed its plans and its attitude toward public disclosure. The 1971 publication of a shareholder's brochure detailing the improvement plans and schedule is encouraging.

Like all companies which waited for pollution control to be required by law, Potlatch is now hampered by bureaucratic uncertainties. At Lewiston, Idaho, the plans for a secondary aeration system are still stalled pending completion of an Army Corps of Engineers dam project. The system is not scheduled for completion until 1974 and the company was apparently unable to devise any interim treatment.

At Cloquet, Minnesota, the situation is more complex. The mill operated with no effluent treatment until the 1968-1969 St. Louis River Standards set a December 1973 pollution abatement deadline. Potlatch and other Cloquet industries decided, in 1970, to construct a joint industrial-municipal treatment plant. Subsequently, a regional project to include Duluth, Minnesota, and possibly Superior, Wisconsin was proposed. Even if the plans for this regional system are accepted by the federal EPA, the St. Louis River will not be receiving adequately treated effluent for at least another four years.

To satisfy interim requirements, Potlatch has installed a clarifier and chemical additive system to reduce suspended solids, and an aerator in the

river to "help prevent fish kills." It will achieve the (amended) December 1973
BOD and suspended solids reductions by closing down the sulfite mill one
month before the deadline. All further abatement action will await the eventual
completion of the regional system.

Not unexpectedly the installation of modern air pollution control equipment,
which is closely related to improved production efficiency, is on a more immediate timetable. At both Lewiston and Cloquet, low odor recovery systems,
precipitators, and miscellaneous scrubbers are under construction. By mid-
1973, both mills will have effected considerable reductions in gas and particulate emissions but neither will have state-of-the-art water pollution control.

Riegel Paper Corporation

Riegel merged with Federal Paperboard Company and is not included in this
update.

St. Regis Paper Company

St. Regis has made few significant air pollution control improvements over
the last year. A secondary scrubber was added to the precipitator at Monticello, Mississippi, a third lime kiln scrubber was installed at Jacksonville,
Florida, and the Deferiet, New York, boilers were converted to oil. However,
nearly two-thirds of the company's total planned expenditure of $90 million
will go toward rebuilding the recovery systems at Tacoma, Washington and
Pensacola, Florida—programs which will improve production efficiency as
well as air pollution control at these two mills. Neither of these two recovery
systems will be completed for two to three years; and plans for the badly
needed "emissions control" at Jacksonville have not yet even been defined.

Several of the company's major water pollution problems are being solved
with pulp mill shutdowns. Cornell, Wisconsin, will be closed completely in
July, 1972; and at Rhinelander, Wisconsin, and Deferiet, New York, the sulfite mills will close and only the paper mills will remain in operation. Three
other mills have installed abatement facilities: Tacoma (primary treatment);
Jacksonville, Florida (primary and secondary); and Pensacola (secondary).
And at Sartell, Minnesota, primary and secondary treatment plants are under
construction.

St. Regis's most interesting project is the development of an experimental
offset paper made from unsegregated solid waste. The raw material, "urban
fibre," was pulped by a machine developed by the Environmental Protection
Agency and the Black and Clawson Company to separate metals, glass, and
paper for reclamation. St. Regis made the paper, thus demonstrating the technological feasibility of the project. However, commercial utilization depends
on the economics of waste collection and transport. According to the company,
paper made from "urban fibre" will be economically feasible only in those
cities with municipal recycling plants and paper mills, and right now, Franklin, Ohio, has the only such recycling plant.

Scott Paper Company

The company released no updated information on its pollution control activities.

Union Camp Corporation

Union Camp ranks high in its sudden and startling committment to pollution control. Its attitude toward public disclosure has improved as much as its abatement plans.

The biggest changes are underway at the Savannah, Georgia, mill, where an extensive $17 million water treatment system is under construction and a complete $34 million revamping of the recovery system is planned. By the end of 1972, the company anticipates that the mill's BOD discharge will be reduced from 35 to 7 pounds per ton of pulp; and by 1975, that particulate and odorous emissions from the recovery operation will be reduced by two-thirds.

Plans for the other mills are less extraordinary, but so is the magnitude of their problems. $2.5 million is being spent to improve effluent collection and treatment at Montgomery, Alabama, and Franklin, Virginia; and $400,000 is budgeted for air pollution control at these mills. Unfortunately, the company's program will not be completed until 1975, and even then, the power boiler particulate control and recovery furnace odor control at Franklin will remain inadequate.

U.S. Plywood-Champion Papers Incorporated

U.S. Plywood-Champion continues to be honest in discussing its pollution problems and constructive in solving them. Its program for providing primary and secondary water treatment at every mill was completed in March, 1971, but an unexpected excess solids and color problem has developed at Pasadena, Texas. The activated sludge treatment is not working as designed. Moreover, the mill is faced with very stringent effluent regulations proposed for the Houston Ship Channel which would reduce the permissable BOD discharge from 50 ppm to 13 ppm. Temporary and long range solutions to the Pasadena effluent problem are now being investigated.

The company's air pollution control program has been considerably expanded. In addition to the multimillion dollar improvements under construction at Canton, North Carolina, systems for collection and incineration of noncondensible gases will be installed at Pasadena and the brand new Courtland, Alabama, mill; and Pasadena mill will have a new recovery furnace, and, for the first time in the paper industry, a precipitator on the lime kiln.

Westvaco Corporation

Equipment to solve all of Westvaco's air pollution problems as much as is technologically possible has been installed or is under construction. At Luke,

Maryland, a low sulfur oil boiler was completed in late 1971; a new recovery furnace is scheduled for August, 1972, completion; and a system for noncondensible gas collection and new precipitators are planned for 1973.

At Charleston, South Carolina, the three precipitators were upgraded and a noncondensible gas collection system was added; and at Covington, Virginia, the new recovery system designed to replace the 13 small rotary furnaces will be completed in mid-1973.

Most of the needed effluent treatment facilities are further from completion. While primary treatment was installed at Charleston in 1971, the remaining effluent from the Luke mill (mainly condenser wastes) will not be directed to the secondary plant until 1973; and secondary treatment at Charleston and Wickliffe, Kentucky, is not scheduled for completion for three more years.

Westvaco is continuing development of its two pilot projects utilizing activated carbon to remove pollutants from the air and water. At Covington, an experimental effluent color-removal system is projected, but will not be operational for another five years. At Charleston and Luke the project to recover the sulfur dioxide in the flue gases as saleable sulfur has been expanded with an EPA grant. If successful, it will provide an alternative to the use of low sulfur fuels in many industries, including electric power. Unfortunately, the company feels that commercial application of the process is still three to four years away.

Weyerhaeuser Company

With a 1971-1972 pollution control expenditure of nearly $39 million,* by the end of this year seven of the company's ten mills will be operating with primary and secondary water treatment and six will have fully adequate gas or particulate controls.

Among the changes which have produced improvements in air and water quality are: secondary treatment facilities at the Pine Bluff, Arkansas, Craig, Oklahoma, and Everett, Washington, kraft mills; expanded primary treatment at Cosmopolis, Washington; a new recovery furnace at Springfield, Oregon; and odor control at Longview, Washington. Upon the 1974 replacement of the kraft recovery furnaces at Plymouth, North Carolina, and Everett and Longview, Washington, there will be significantly improved emissions control at all but the company's smallest mill at Pine Bluff, Arkansas.

Weyerhaeuser's biggest pollution problem, the Everett, Washington, sulfite mill, is still unresolved. On January 11, 1972, the company announced that it would cease operations as of the May 31, 1973, state water pollution abatement deadline rather than spend $10-15 million to convert the mill to a less polluting process. This was the first announcement of a major industrial shutdown in Washington based on pollution regulations. Accusations and con-

----
*Includes $10 million for the Kamloops, British Columbia, mill.

fusion followed. State ecology officials first blamed the uncertainties of federal laws and later concluded that perhaps corporate economic considerations were a larger factor in the decision than environmental protection. Although company spokesmen repeatedly denied that Everett is either uneconomical or nearing obsolescence, many observers pointed out that it is a small, marginally-efficient mill out of step with Weyerhaeuser's large and highly automated, integrated production complexes.

Subsequent events lend credence to this view. In May, the New York Times reported that the company had applied to the Washington State Department of Ecology for a permit to develop a $200-$300 million pulp, paper, and forestry complex over the next ten years at Longview. The application, similar to those filed for electric power plant siting, examines the total environmental impact of the proposed project. It also includes a request to continue operating the Everett sulfite mill for 40 months beyond its scheduled closing date.

Unfortunately, no mention of the Longview project was made during the furor surrounding the company's January closure announcement. It was this silence that led critics to propose that Weyerhaeuser fomented a jobs vs. environment conflict in Everett to serve its own political interests, ease itself out of a financial liability, and attain a 40-month permit extension.

Table A3.1
Results of the Follow-Up Study: Company Pulp Production, Water Use, and Pollution Control

Location	Pulp Prod. (TPD)	Other Prod. (TPD)	Water Use (MGPD)	Water Pollution Control			Air Pollution Control	
				Primary	Secondary	Tertiary	Particulate	Gas and Odor
American Can Company	N.A.							
Boise Cascade Corporation								
DeRidder, La.	1,200	1,050	27	✓	✓	✗	✓	✗
International Falls, Minn.	750	380	35	✓	✗	✗	✗	✗
St. Helen's, Ore.	825	540	40	✓	(✓)	✗	✓	✗
Salem, Ore.	250	230	18	✓	(✓)	✗	✓	✓
Steilacoom, Wash.	350	350	6	(✓)	✗	✗	(✓)	—
Wallula, Wash.	640	610	7	Under constr. Due end 1972	✗	✗	✗	✗
Consolidated Papers Incorporated								
Appleton, Wisc.	150	--	12	✗	✗	✗	(✓)	✗
Whiting, Wisc.	100	265	4	(✓)	✗	✗	✓	—
Wisconsin Rapids, Wisc.	Pulp mill closed in March, 1971. Paper mill continues to operate							

Location				Status marks				
Wisconsin Rapids, Wisc.	360	--	10	X	X	X	✓	✓
Biron, Wisc.	200	560	10	✓	X	X	X	—
Continental Can Company								
Augusta, Ga.	800	675	38	✓	✓	X	X	X
Port Wentworth, Ga.	625	625	18	✓	Under constr. Due Dec. 1972	X	X	X
Hodge, La.	820	700	14-16	✓	✓	⊘	New recovery furnace under constr.	
Hopewell, Va.	1,000	1,100	18.2	X	X	X	✓	X
Crown Zellerbach Corporation								
Fairhaven, Calif.	500	--	30	X	X	X	✓	✓
Bogalusa, La.	1,500	1,750	30	✓	Under constr. Due 1972	X	✓	X
St. Francisville, La.	720	770	35	Under constr. Due 1972	✓ / Under constr. Due 1973	X	✓	✓
Lebanon, Ore.	105	100	4	✓	✓	X	⊘	✓
Wauna, Ore.	1,200	760	50	✓	X	X	✓	✓
West Linn, Ore.	340	500	15	✓	Under constr. Due 1972	X	?	—
Camas, Wash.	1,225	1,050	90	✓ ⊘	X	X	⊘	⊘
Port Angeles, Wash.	195	400	14	✓	X	X	✓	—
Port Townsend, Wash.	420	400	12	✓	X	X	✓	✓

Table A3.1
Results of the Follow-Up Study: Company Pulp Production, Water Use, and Pollution Control (continued)

Location	Pulp Prod. (TPD)	Other Prod. (TPD)	Water Use (MGPD)	Water Pollution Control Primary	Secondary	Tertiary	Air Pollution Control Particulate	Gas and Odor
Diamond International								
Red Bluff, Calif.	80	--	2.25	√	⊘	X	X	—
Old Town, Me.	550	100	20.5	X	X	X	?	?
Ogdensburg, N.Y.	N.A.	N.A.	4.4	Under constr. Due Oct. 1972	X	X	X	—
Plattsburgh, N.Y.	N.A.	N.A.	1.5	Joint treatment plant under constr. Due Oct. 1973		X	X	—
Fibreboard Corporation								
San Joaquin Mill, Antioch, Calif.	750	?	20	Under constr. Due 1972	X	X	√	√
Georgia-Pacific Corporation								
Ketchican, Ala.	630		42	Under constr. Due 1972	X	X	X	√
Crossett, Ark.	1,250	760	50	√	√	√	√	√
Samoa, Calif.	550		25	X	X	X	√	√
Port Hudson, La.	600		20	Under constr. Due Oct. 1972	Under constr. Due Dec. 1973	X	New precip. due end 1972	√
Woodland, Me.	800	500	30	√	Under constr. Due Dec. 1973	Partial	√	√

Location								
Lyons Falls, N.Y.	70	160	4		×	×	×	—
Plattsburgh, N.Y.	70	155	2.5	Under constr. Due Sept. 1973	×	×	✓	—
Toledo, Ore.	1,000	880	13	Joint treatment plant due Oct. 1973	×	×	(✓)	✓
Bellingham, Wash.	500	535	39	Under constr. 85% liquor recovery Due end 1972	×	×	×	✓
Great Northern Nekoosa Corporation								
Ashdown, Ark.	550	230	20	(✓)	×	×	✓	×
Cedar Springs, Ga.	2,000	2,000	26	✓	×	×	(?)	×
E. Millinocket, Me.	800	1,100	20	×	×	×	—	—
Millinocket, Me.	1,100	900	30	Under constr. Due Nov. 1972	×	×	×	—
Nekoosa, Wisc.	340	400	30	×	×	×	×	×
Pt. Edwards, Wisc.	230	360	12	×	×	×	(✓)	—
Hammermill Paper Company								
Selma, Ala.	500		20	✓	×	×	×	×
Erie, Pa.	600	425	32	×	×	×	(✓)*	✓

*Production equipment in compliance, power boilers equipped, but will not be able to meet issued requirements.

Table A3.1
Results of the Follow-Up Study: Company Pulp Production, Water Use, and Pollution Control (continued)

Location	Pulp Prod. (TPD)	Other Prod. (TPD)	Water Use (MGPD)	Water Pollution Control			Air Pollution Control	
				Primary	Secondary	Tertiary	Particulate	Gas and Odor
Lockhaven, Pa.	Pulp mill shut down. Paper mill continues to operate							
Kaukauna, Wisc.	380	500	28	Under constr. Due end 1972	X	X	X	X
Hoerner Waldorf Corporation								
Ontonagon, Mich.	200	200	5	X	X	X	X	—
St. Paul, Minn.	300	340	8.5	✓	✓	X	✓	—
Missoula, Mont.	1,250	950	16	✓	(✓)	X	New recovery furnace under constr. Due end 1972	
International Paper								
Mobile, Ala.	1,500	1,250	34.4	✓	Under constr. Due end 1972	X	New recovery furnace under constr. Due 1973	✓
Camden, Ark.	750	665		✓	✓	X	✓	X
Pine Bluff, Ark.	1,550	1,600		✓	(✓)	X	✓	X
Panama City, Fla.	2,050	700		✓	(✓)	X	Two recovery furnaces under constr. Due late 1972	
Bastrop, La.	1,700	1,530		✓	✓	X	✓	✓

Springhill, La.	1,725	1,300	✓	X	✓	X
Jay, Me.	675	525	⊘	X	✓	⊘
Moss Point, Miss.	700	700	✓	X Under constr. Due end 1972	✓	X
Natchez, Miss.	950	--	Under constr. Due end 1972	X Under constr. Due end 1974	New precip. due Spring 1973	X
Vicksburg, Miss.	1,200	1,000	✓	X	✓	✓
Corinth, N.Y.	255	500	⊘	X	✓	—
North Tonawanda, N.Y.	140	250	⊘	X	⊘	?
Ticonderoga,* N.Y.	550	--	✓	X	✓	✓
Gardiner, Ore.	545	545	Under constr. Due July 1973 X	X	X	✓
Georgetown, S. C.	2,230	1,660	⊘	X	New precip. due 1973	⊘

Kimberly-Clark Corporation

Coosa Pines, Ala.	1,520	1,200	40	✓	X	X	X
Anderson, Calif.	Mill sold to Simpson Lee Paper Company						

* New mill opened on April 24, 1971.

Table A3.1
Results of the Follow-Up Study: Company Pulp Production, Water Use, and Pollution Control (continued)

Location	Pulp Prod. (TPD)	Other Prod. (TPD)	Water Use (MGPD)	Water Pollution Control			Air Pollution Control	
				Primary	Secondary	Tertiary	Particulate	Gas and Odor
Niagara Falls, N.Y.	Mill sold to Cellu Products Co.							
Kimberly, Wisc.	80	440	15	Under constr. Due end 1972	×	×	×	—
Niagara, Wisc. Mill being sold								
Marcor, Incorporated (Container Corporation of America)								
Brewton, Ala.	900	150	38	✓	✓	×	(✓)	New recovery furnace under constr. Due mid 1972
Fernandina Beach, Fla.	1,700	0	30	(✓)	×	×	×	✓
Circleville, O.	180	120	2	✓	(✓)	×	(✓) *	—
The Mead Corporation N.A.								
Owens-Illinois								
Valdosta, Ga.	870	825	12	✓	✓	×	✓	✓
Orange, Tex.	900	900	10	✓	✓	×	✓	✓
Big Island, Va.	510	510	12	✓	✓	×	✓	—

*Power boilers converted to oil; application pending for natural gas.

							Improvements under constr.
Tomahawk, Wisc.	550	620	7	✓	✓	X	✓
Potlatch Forests, Inc.							
Lewiston, Id.	830	685	32	✓ Under constr. Due 1974	X	New recovery furnace due end 1972	
Cloquet, Minn.	440	330	20	Ⓥ	X	X	New recovery furnace due mid-1973
Riegel Paper Corporation — Merged with Federal Paperboard Company, and not included in update							
St. Regis Paper Company							
Jacksonville, Fla.	1,370	1,300	78	Under constr. Due July 1972 X	X	✓	
Pensacola, Fla.	900	910	28	Ⓥ	✓	X	X
Bucksport, Me.	270	600	12	X	X	X	—
Sartell, Minn.	125	200	5.8	Under constr. Due 1973	X	✓	—
Monticello, Miss.	1,620	1,500	20	✓	X	✓	
Deferiet, N.Y.	Pulp mill shut down June 1971; only paper mill remains						
Tacoma, Wash.	900	610	30	Ⓥ	X	X	New recovery system under constr. Due early 1974
Cornell, Wisc.	Entire mill shut down June 1972						
Rhinelander, Wisc.	Pulp mill will shut down; only paper mill will remain						
Scott Paper Company — No information available							

Table A3.1
Results of the Follow-Up Study: Company Pulp Production, Water Use, and Pollution Control (continued)

Location	Pulp Prod. (TPD)	Other Prod. (TPD)	Water Use (MGPD)	Water Pollution Control			Air Pollution Control	
				Primary	Secondary	Tertiary	Particulate	Gas and Odor
Union Camp Corporation								
Montgomery, Ala.	900	850	15	✓	✓	✗	✓	✓
Savannah, Ga.	2,800	2,650	40	(✓)	Under constr. Due 1972	✗	New recovery system under constr. Due 1974	
Franklin, Va.	1,400	1,300	38	✓	✓	✗	✗	✗
U.S. Plywood-Champion Papers Incorporated								
Cortland, Ala.	500	--	26	✓	✓	✗	✓	✓
Canton, N.C.	1,290	1,420	50	✓	✓	✗	Improvements under constr. Due July 1973	
Pasadena, Tex.	900		44	✓	✓ *	✗	Under constr. Due 1973	
Westvaco Corporation								
Wickliffe, Ky.	600	350	28	✓	✗	✗	✓	✓
Luke, Md.	745	900	31	✓	Partial	✗	New recovery system under constr. Due Aug. 1972	
Charleston, S.C.	2,000	1,850	43	(✓)	✗	✗	(✓)	

*Secondary treatment system not working as designed.

Location								
Covington, Va.	1,350	1,350	25	√		X	X	New recovery system under constr. Due mid-1973
Weyerhaeuser Company								
Pine Bluff, Ark.	200	200	3.9	√	⊘	X	X	X
New Bern, N.C.	800		30	√	√	X	√	√
Plymouth, N.C.	1,550	1,475	42	√	√	X		New recovery furnace under constr. Due 1974
Craig, Okla.	450	450	14.5	√	⊘	X	√	—
Valiant, Okla.	1,600		21	√	√	X	√	√
Springfield, Ore.	1,150	1,150	15	√	√	X	√	√
Cosmopolis, Wash.	400		28	√	X	X	√	√
Everett* (sulfite), Wash.	300		28	X	X	X	X	√
Everett (kraft), Wash.	360		27	√	Under constr. Due end 1972	X	√	X
Longview, Wash.	1,140	1,400	90	√	X	X	√	√

---

*Mill's future unresolved. Scheduled for June 1973 shutdown, but application for extension pending.

INDEX OF CONSUMER BRANDS

American Can Company

AURORA toilet and facial tissues
BUTTERICK patterns and home catalogue
DIXIE paper cups and plates
GALA toilet and facial tissues
NORTHERN toilet and facial tissues
VOGUE patterns and pattern book
WAXTEX waxed paper
Others:
PRINTING SERVICES OPERATIONS CO., a nonpublishing printer of business and trade periodicals
Grocer Mfr. and Progressive Grocer grocery industry magazines
UNIQUE zippers, pinking shears and sewing accessories
WORLD PUBLISHING CO. bibles and dictionaries

Boise Cascade Corporation

ARISTOCRAT travel trailers
BOISE CASCADE prefabricated homes
BEHAVIOR TODAY magazine
CRM books and records
CITATION travel trailers
CORSAIR travel trailers
DIVCO WAYNE travel trailers
GOLDEN HILLS real estate
HARBOR COVE real estate
INCLINE VILLAGE real estate
KINGSBERRY homes
LAKE ARROWHEAD real estate
LAKE DON PEDRO real estate
LIFE TIME mobile homes
OCEAN PINES real estate
PINE MOUNTAIN LAKE real estate
PRINCESS cruises
PSYCHOLOGY TODAY magazine
SKI INCLINE real estate
U.S. LAND INC. real estate
WOODBINE LAKE real estate

Consolidated Papers Incorporated

CAPRI book publishing grades

CONSOBARR paperboard
CONSO GLOSS web offset
CONSO SEMI GLOSS web offset
CONSOLIDATED GLOSS letterpress
CONSOLIDATED SEMI GLOSS letterpress
CONSOLITH GLOSS lithography
CONSOLITH OPAQUE lithography
CONSOWEB BRILLIANT web offset
CONSOWEB MODERN web offset
CONSOWEB VELVET web offset
CONSOWELD 6 laminated plastic
CONSOWELD 10 laminated plastic
CONTEX book publishing grades
DURA BEAUTY VERTICAL SURFACING laminated plastic
MODERN GLOSS letterpress
MODERNROTO rotogravure
PALOMA COATED EMBOSSED lithography
PALOMA COATED MATTE lithography
PALOMA MATTE web offset
PRODUCTION GLOSS letterpress
PRODUCTOROTO rotogravure
SIROCCO EMBOSSED COVER lithography
SIROCCO EMBOSSED ENAMEL lithography
STAPEL pulp and chemical products
MINNESOTA MINING & MANUFACTURING markets coated carbonless paper
produced by Consolidated

Continental Can Company

ADVANCE bags
ADVANCE COTE paperboard
BENCOSEAL foil laminations
BONDTWIST closures
BONDWARE paper plates and cups
CANTRAK multi packs
CLUPAK extensible paper
CONABLAN colored paper bags
CONALOY plastic bottles
CONELAST multiwall bags
CONOCLEER multipacks
EGG SAFETY cartons
EMERALD paper bags
FLAV-O-TAINER bags, bakery cartons
HAZELWARE glassware
JAK-ET-PAK cartons, partitions, multi packs
LEVERPAK fibre drums
MOLESKIN paper bags

OHIO paper bags
PALEFACE paper bags
PAY-OFF-PAK fiber drums
PERF-O-ROLL polyethylene sheets
RAIN-IN-THE-FACE paper bags
RING PULL cans
SAFETY FIRST paper bags
SAMPSON paper bags
SAV-A-PAN baking cartons
SCOTT paper bags
SMARTOP closures
STAPAK fiber drums
SUPERSAKS paper bags
TEE PAK synthetic cellulose tubing used in meat processing
THERMOSEAL BAGS paper bags
TWINSAKS paper bags
TWIST-OFF closures
UPAK fibre drums
WA-HA paper bags

Crown Zellerbach Corporation

CHIFFON facial tissues, napkins, paper towels, lunch bags
COMFORT toilet tissue
CROWN printing papers
CROWN SEAL polyethylene film-packaging for bakery products
FAMILY PAK paper napkins
FIXTURE FOLD paper napkins
SILK napkins, facial tissue
ZEE napkins, paper towels, toilet and facial tissues, napkin dispensers,
wax paper, sandwich bags, lunch bags, garbage bags
Other:
Crush International Ltd. manufactures CRUSH beverages, MASONS root
beer, HIRES root beer, ORANGE CRUSH beverage, SUN DROP diet cola

Diamond International

BEE playing cards
BERIGARD trays
BLUE RIBBON household paper products
BICYCLE playing cards
CEL PAK trays
CHICTAINER trays
CONGRESS playing cards
FRUIT-SHEL trays
FOODTAINER trays
LITHWHITE coated paperboard

POCKETBOOK matches
POULTRYTAINER trays
PULPEX pie plates
ROUNDTAINER trays
TALLY HO playing cards
VANITY FAIR bathroom tissue, facial tissue, paper towels, paper products
VERSATIL trays
U.S. playing cards

Fibreboard Corporation

BARRIERMATIC cartoning system
FIBRESIX bottle system
JUNIOR PAK MASTER corrugated containers
PABCO building materials, insulation materials, folding cartons, corrugated containers, lumber and plywood
PAK MASTER wraparound corrugated containers

Georgia-Pacific Corporation

CORONET facial tissue, toilet tissue, paper towels and napkins
G.P. paneling for homes
KUSHION-AIRE doors and windows for homes
M.D. facial and toilet tissue
NU-SASH replacement window
ROYAL OAK charcoal briquettes

Great Northern Nekoosa Corporation

NEKOOSA Offset
NEKOOSA Opaque

Hammermill Paper Company

BECKETT paper products
BUCKEYE-BECKETT offset, cover, opaque, text, and specialty papers
BURGESS envelopes
COAST envelopes
DUPLEX envelopes
HAMMERMILL BOND business papers, fine writing papers
OLD COLONY envelopes
STRATHMORE technical papers, book publishing, and coated printing papers
THILMANY specialized papers and bags, industrial packaging
WATERVLIET coated book, offset, cover, postcard, and other printing papers

Hoerner Waldorf Corporation

International Paper

CONFIL fabric
DAVOL health and medical products
FACELLE ROYALE facial and boathroom tissue, paper towels and napkins
FRESHABYES disposable diapers
GARBAX disposal system, disposable paper bags
ICE-PAK containers
SANEEN disposable diapers, folded towels, bibs, washcloths, examination
gowns
SPACEMAKERS homes
SUPERCEL bathroom tissue, dispenser napkins, disposable wipers

Kimberly-Clark Corporation

DELSEY toilet tissue, automatic bowl cleaner
FEMS sanitary napkins
KAYDRY towels
KIMBIES disposable diapers
KIMLON disposable garments and sheets
KIMPAK interior packaging materials
KIMTOWELS towels
KIMWIPE disposable towels
KLEENEX tissues, paper towels, napkins
KLEENEX BOUTIQUE tissues
KOTEX sanitary napkins and tampons
KOTEX KOTIQUE after shower mist, beauty bath, douche powder and liquid,
feminine deodorant spray, etc.
LITHOWIPES towels
MARVALON adhesive coverings
TERI-TOWEL towels

Marcor (Container Corporation of America)

CONCORATEX RC100 paperboard
MONTGOMERY WARD products

The Mead Corporation

CLUSTER-CASE single ply shipper
CLUSTER-VUE packaging for beverage cans
GILBERT bond paper
MONTAG stationary
OCEANEYE underwater camera housing
POLAR-CORR paperboard
SARGENTS plastic coloring pencils, clay, chalk

STANLEY furniture
WESTAB crayons, notebooks, bookcovers, stationary

Owens-Illinois

ASPHALT-PAK
COMBO-BIN
DURACAN
DURALON
FANFOLD
FLARETOP
HICONE glass bottles
KONTUR-A-PAK
LIBBEY glasses and tableware
LILY paper cups, plates, bowls
LILYFOME plastic plates and dishes
OWENS-ILLINOIS SCIENTIFIC MARTINI MAKER
PORTA-BAG
VABAR

Potlatch Forests, Inc.

CARLTON business and printing papers
DOVERPRINT paper towelling, facial and toilet tissue
KARMA business and printing papers
MOUNTIE business and printing papers
MUSTANG business and printing papers
NORTHWEST business and printing papers
RANGER business and printing papers
SPA paper towelling, facial and toilet tissue
SWANEE paper towelling, facial and toilet tissue
VELOPAQUE business and printing papers
VELURE paper towelling, facial and toilet tissue
VINTAGE business and printing papers

Riegel Paper Corporation

CAROLINA printing papers
DECOLAR laminates
FAIRTEX color coordinated metalic yarns
FOLDCAN packaging system
JERSEY printing papers
MAKEREADY papers
MONOTHERM electrical laminates
RIEGELPAK to collate and carton flexible pouches of food
RIEGELTHERM pharmaceutical packaging system

St. Regis Paper Company

CROSS plastic bags
MURRAY bags
NIFTY school supplies
SUPERIOR paper plates and cups
WHIRLPOOL Trash Master bags

Scott Paper Company

BABYSCOT panties and disposable diapers
BROWN JORDAN furniture
CONFIDETS sanitary napkins
CUT-RITE waxed paper
FLOKOTE printing papers
LADY SCOTT facial tissues, toilet tissues
SCOTKINS napkins
SCOTT baby diapers, educational learning aids, electrostatic copiers
SCOTTIES tissue
SCOTTISSUE toilet tissue
SCOTTOWELS paper towels
SOFT-WEVE toilet tissue
VIVA paper toilet tissue, napkins, paper towels
WALDORF toilet tissue
WARRENS printing papers

Union Camp Corporation

CABINET CUBES storage units
HONEYCOMB products
STRETCH-PAX cartons
UNION CAMP packaging and plastics

U.S. Plywood-Champion Papers Incorporated

DREXEL furniture
HERITAGE furniture
JEAN-ALAN carpets
MEADOWCRAFT wrought iron furniture
NOVOPLY particle board
PURE-PAC milk cartons
RUB N GLUE
TREND carpets
TEX-TUFT carpets
U.S. Plywood contract service systems
WELDWOOD paneling

Westvaco Corporation

Weyerhaeuser Company

JAY PEAK ski area
STEAKHOUSE charcoal briquettes
WEYERHAEUSER garden products
WEYERHAEUSER WATER MATE

U.S. DEPARTMENT OF THE INTERIOR CONFERENCES ON POLLUTION
OF INTERSTATE WATERS

Coosa River System, August 27, 1963.

Coosa River System, (Georgia-Alabama) April 11, 1968.

Escambia River Basin (Alabama-Florida) January 20, 1970.

Lake Champlain and its Tributary Basin (New York-Vermont) November 13
and December 19-20, 1968.

Lake Michigan and its Tributary Basin (Wisconsin-Illinois Indiana-Michigan)
January 31, 1968.

Lower Savannah River and Its Esturaries, Tributaries, and Connecting
Waters (Georgia-South Carolina) Feb. 2, 1965.

Pearl River (Mississippi-Louisiana) October 22, 1963.

Penobscot River and Upper Penobscot Bay and their Tributaries (Maine)
April 20, 1967.

Perdido Bay and its Tributaries (Florida-Alabama) January 22, 1970.

BIBLIOGRAPHY

I. Books

American Public Health Association, American Society of Civil Engineers, American Water Works Association, Water Pollution Control Federation. Glossary of Water and Wastewater Control Engineering, 1969.

Britt, Kenneth W., ed. Handbook of Pulp and Paper Technology. New York: Reinhold Publishing Corporation, 1964.

Casey, James P. Pulp and Paper Chemistry and Chemical Technology, Vol. I. New York: Interscience Publishers, Inc., 1952.

Lockwood Publishing Co., Inc. Lockwood's Directory of the Paper and Allied Trades. New York: 1969.

National Wildlife Federation. Conservation Directory 1970. Washington, D.C.: National Wildlife Federation, 1970.

Stern, Arthur C., ed. Sources of Air Pollution and Their Control (Vol. III of Air Pollution). New York: Academic Press, 1968.

Udall, Stewart L. The Quiet Crisis. New York: Avon Books, 1964.

II. Magazine Articles

Berland, Theodore. "Our Dirty Sky," Today's Health, March, 1966.

Collins, T. T., Jr., "New Systems Proposed for Kraft Mill Odor Control and Heat Recovery," Paper Trade Journal, May 31, 1965.

Collins, T. T., Jr., "The Venturi-Scrubber on Lime Kiln Stack Gases," TAPPI, January 1959.

Faltermayer, Edmund K. "We can Afford Clean Air," Fortune, November, 1965.

Klein, David H. and Edward D. Goldberg. "Mercury in the Marine Environment" Environmental Science & Technology; n.d.

Lando, Barry. "Save Our Air," The New Republic, October 11, 1969.

Nalesnik, Richard P. "A 20th Century Program for Water Pollution Control," Water and Wastes Engineering, February, 1968.

Olin, Joe H. "API Hears Prospects in Industry, Government and the Economy," Paper Trade Journal, March 10, 1969.

Olin, Joe H. "NCASI Meeting Covers a Wide Range of Technical Topics," Paper Trade Journal, March 10, 1969.

Pearl, Irwin A. "Waste Product Use Helps Paper Industry Control Pollution," Environmental Science & Technology, September, 1968.

Peet, Creighton. "The Effluent of the Affluent," American Forests, May, 1969.

Phinney, Robert E. "Air and Water Management in the Paper and Pulp industry, As Seen from Manufacturing," Southern Pulp and Paper Manufacturer, September 10, 1969.

Ross, Edward N. "Pulp-Paper Mills are Depolluters! Ross Declares," Paper Age, May, 1970.

. . . "Boise Southern: the instant giant," Chem 26, July, 1969.

. . . "Galveston Bay: Test Case of an Estuary in Crisis," Science, February, 1970.

. . . "Top 50," Chem 26, June, 1970.

. . . Chemical Engineering, October 14, 1968.

III. Trade Journal Articles

1. Combustion Engineering

Fernades, J. H. "Industrial and Municipal Air Pollution Control Techniques and Practices in the United States," March 5, 1970.

Gommi, J. V. "Air Contact Evaporator System for Kraft Recovery Furnace Odor Abatement," February 15, 1970.

Hochmuth, F. W. "An Odor Control System for Chemical Recovery Units," September 16-18, 1968.

Murphy, V. P. "Industrial Water Treatment," March 11, 1966.

Owens, V. P. "Odor Abatement from Recovery Units," May 28, 1968.

Owens, V. P. "Recent Developments in Chemical and Heat Recovery in Pulp Mills," May 23-27, 1967.

Owens, V. P. "Trends in Design of Kraft Pulping Chemical Recovery Units," May 27-29, 1969.

Smith, Edmond L. "Kraft Mill Chemical Recovery Units: The Third Generation," November 2, 1964.

Smith, Edmond L. "Pulp Mill Pollution Control with Two-Stage Recovery Cycles," September 24-27, 1967.

Smith, Edmond L. "Sulfite Pulping and Pollution Control," June 1967.

Wangerin, D. D. "Are We Getting the Most Out Of By-Product Fuels?" February 11-13, 1969.

2. Pulp and Paper

Brooks, Sheldon "Are Our Mills Really Doing an Effective job on fiber and white water losses?" December, 1969.

Brooks, Sheldon. "The '70s Beckon: Will We Enter Them With a Half-hearted Anti-pollution Plan?" January, 1970.

Drummond, Robert M. "How International Paper Controlled Its Kraft Mill Odor at Ticonderoga, N.Y.," May, 1969.

Farin, William G. "Want a More Direct Approach to Spent Liquor Processing?" August, 1969.

Gantzhorn, Eugene. "It Was Almost All Waste Treatment at Secondary Fiber Pulping Conference," November 4, 1968.

Guerrier, James J. "Bleaching: Is Oxygen Coming to the Fore?" August, 1970.

Guerrier, James J. "NCASI budgeted at $1 million-plus," May, 1970.

Heckroth, Charles W. "Soap Box Oratory Not Needed for Air/Water Control: Time, Money and Facts are Essential," September 30, 1968.

Julson, J. O. "Environmental Protection: How Much Will it Cost Your Mill?" April, 1969.

Lardieri, Nicholas J. "Saving Secret: Air for Stream Standards," October 21, 1968.

Martin, Fred. "Secondary Oxidation Overcomes Odor from Kraft Recovery," June, 1969.

Palladino, Anthony J. "Pulp Mill Effluents: What They Are, and How They are Handled and Treated," February, 1970.

Shaw, Charles L. "Kraft Mill Emission: Pro and Con," April, 1969.

Sullivan, Michael D. "Toughest Water/Air Rules Met by New Mill," May, 1970.

Van Derveer, Paul D. "Odor! Yield! Pollution! Pulpmen on Target with New Approaches," December, 1969.

Wahlstrom, Jack. "Beware of Hydrogen Sulfide: How Protective Measures Save Employees," May, 1969.

Williams, Harris K. "H_2S: Not Enough is Known about It," May, 1969.

. . . "Closed-circuit $C10_2$ Generator Eliminates H_2SO_4 Effluent," May, 1970.

. . . "New Waste Water Treatment System Shows High Fiber Recovery Rates," March, 1970.

. . . "Three-Stage Program at Ontario Paper to Eliminate Pulping Waste Organics," January, 1970.

IV. Reports

American Paper Institute. How the American Paper Industry Performed in 1969. A Report on Its Progress, Problems and Prospects.

. . . The Paper Industry's Part in Protecting the Environment. November, 1970.

Bolton, Roger E. The Corporation and the Environment. A Report on the October 1968 Conference. Williamstown, Massachusetts. June, 1969.

Chapman, William C. Long-Term Storage as an Effective Method of Treating Pulp and Paper Mill Wastes. National Council of the Paper Industry for Air and Stream Improvement, Inc. Technical Bulletin No. 218. 1968.

Courtright, Robert C., and Bond, Carl E. Potential Toxicity of Kraft Mill Effluent After Oceanic Discharge. Engineering Experiment Station and Department of Fisheries and Wildlife. Oregon State University.

Harding, C. I., et al. Clearing the Air in Jacksonville. Air Improvement Committee of Greater Jacksonville, Inc. October, 1967.

National Goals Research Staff. Toward Balanced Growth: Quantity with Quality. Report to the President of the United States. July 4, 1970.

Gordon, C. C. Hoerner Waldorf Pre-Trial Report to Garlington. September, 1969.

Green, W. G. Industrial Waste Water Purification. The Bionomic Systems Corporation. 1966.

Hall, J. Alfred. The Pulp and Paper Industry in the Northwest. U.S. Department of Agriculture, Forest Service. 1969.

. . . Wood, Pulp and Paper, and People in the Northwest.

Louis Harris and Associates, Inc. The Public's View of Environmental Problems in the State of Oregon. Study No. 1990. March, 1970.

Hendrickson, E. R., et al. Systems Analysis Study of Emissions Control in the Wood Pulping Industry, n.p., 1970.

Kleppe, Peder J. and Rogers, Charles N. Survey of Water Utilization and Waste Control Practices in the Southern Pulp and Paper Industry, n.p., June, 1970.

Locke, Edwin A. Emerging Managerial Attitudes in the U.S. Paper Industry. Address Given at the Annual Meeting, Canadian Pulp and Paper Association. Montreal, Canada. January 30, 1969.

. . . Potentials and Problems of Paper Packaging in the 1970s. Remarks before the American Management Association. 39th National Packaging Conference and Exposition, New York City. April 22, 1970.

. . . Wanted: A Unified Strategy for Environmental Protection. Address Given at the Second Annual Meeting, American Forest Institute. Washington, D.C. October 27, 1969.

Michelson, Irving, and Touris, Boris. The Costs of Living in Polluted Air
Versus the Costs of Controlling Air Pollution in the New York-New Jersey
Metropolitan Area, n.p., March, 1968.

National Council of the Paper Industry for Air and Stream Improvement, Inc.
(NCASI). Atmospheric Pollution Literature Review 1969. Atmospheric Pollution Technical Bulletin No. 45. 1970.

. . . A Guide to the Determination of Biochemical Oxygen Demand, Dissolved
Oxygen, Suspended and Settleable Solids, and Turbidity of Pulp and Paper
Mill Effluents. Technical Bulletin No. 230. 1969.

. . . National Council Regulatory Review. Bulletin No. 58. 1969.

Owen, G. L. Air Pollution in Missoula. Published by the Author. Missoula,
Montana. 1968.

Sheridan, Dr. R. Summary Report Concerning Alteration of the Missoula
Valley Ecosystem, n.p., summer, 1969.

. . . A Report on the Kraft Pulping Industry's Progress in Complying with
the Emission Regulations of April, 1969, n.p.

V. Proceedings

Hendrickson, Dr. E. R. Proceedings of the International Conference on Atmospheric Emissions from Sulfate Pulping. Sanibel Island. April 28, 1966.
n.p.

Institute of Paper Chemistry. Proceedings of the Thirty-Fourth Executives'
Conference, May 7-8, 1970. n.p.

Technical Association of the Pulp and Paper Industry. 7th Water and Air Conference. Minneapolis, Minnesota. June 7-10, 1970.

VI. U.S. Government Documents

U.S. Congress. Senate. National Emission Standards Study, Report of the
Secretary of Health, Education and Welfare, S. Doc. No. 91-63, 91st Congress, 2nd Session, 1970.

. . . Testimony of Edwin A. Locke, Jr., President, American Paper Institute, on S. 2005, Resource Recovery Act of 1969, before the Subcommittee
on Air and Water Pollution of the Committee on Public Works, February 26,
1970.

U.S. Congress. House. Message from the President of the United States
transmitting A Report of the Council on Environmental Quality on Ocean
Dumping. 91st Congress, 2nd Session. House Document No. 91-399. October,
1970.

U.S. Department of Health, Education and Welfare. Control of Atmospheric Emissions in the Wood Pulping Industry. E. R. Hendrickson et al. 1970. Vols. I, II, and III.

. . . Summary of Conference in the Matter of Pollution of Interstate Waters of the Snake River and its Tributaries. Lewiston, Idaho. January 15, 1964.

U.S. Department of the Interior. Federal Water Pollution Control Administration. The Cost of Clean Water, Vol. III of Industrial Waste Profiles, No. 3, Paper Mills. Washington, D.C.: Government Printing Office, 1967.

. . . The Economics of Clean Water, Vol. I, II, III, and summary. Washington, D.C.: Government Printing Office, 1970.

. . . Effects of Pollution on Water Quality, Escambia River and Bay, Florida. Washington, D.C.: U.S. Government Printing Office, 1970.

. . . Effects of Pollution on Water Quality, Perdido River and Bay, Alabama and Florida, Washington, D.C.: U.S. Government Printing Office, 1970.

. . . Pollution Affecting Shellfish Harvesting in Mobile Bay, Alabama, Washington, D.C.: U.S. Government Printing Office, 1969.

. . . Pollutional Effects of Pulp and Paper Mill Wastes in Puget Sound, A Report on Studies Conducted by the Washington State Enforcement Project. Washington, D.C.: U.S. Government Printing Office, 1967.

. . . Pollution Caused Fish Kills: 1968, Ninth Annual Report. Washington, D.C.: U.S. Government Printing Office, 1969.

. . . A Primer on Waste Water Treatment, Washington, D.C.: U.S. Government Printing Office, 1969.

. . . Proceedings of the Conference in the matter of Pollution of the Interstate Waters of the Lower Savannah River and its Estuaries, Tributaries and Connecting Waters: Georgia-South Carolina. And Summary. Second Session. Savannah, Georgia. October 29, 1969.

. . . Proceedings of the Conference on the matter of Pollution of the Navigable Waters of Puget Sound, the Strait of Juan de Fuca and their Tributaries and Estuaries (Washington). Second Session. Seattle, Washington. September 6-7, 1967 and October 6, 1967. Vols. I, II, and III.

. . . Proceedings of the Second Session of the Conference in the matter of Pollution of Lake Michigan and its Tributary Basin. Chicago, Illinois. February 25, 1969. Vols. I, II, and Summary.

. . . Conference Proceedings in the matter of Pollution of the Interstate Waters of the Pearl River (Mississippi-Louisiana). Second Session. Bogalusa, Louisiana. November 7, 1968.

. . . Proceedings of the Washington Public Meetings. National Estuarine Pollution Study. Seattle, Washington. July 23, 1968. Aberdeen, Washington. July 25, 1968.

. . . Projects of the Industrial Pollution Control Branch, Washington, D.C.:
U.S. Government Printing Office, 1970.

. . . Water Pollution Problems of Lake Michigan and Tributaries: Actions for
Clean Water (Excerpt), January, 1968.

. . . Water Quality Criteria, report of the National Technical Advisory Committee to the Secretary of the Interior. Washington, D.C.: U.S. Government
Printing Office, 1968.

. . . Upper Potomac River Basin Water Quality Assessment, Chesapeake
Technical Support Laboratory, Middle Atlantic Region. Technical Report
No. 17. Washington, D.C.: U.S. Government Printing Office, 1969.

U.S. Public Health Service. Proceedings of the Third National Conference on
Air Pollution. Washington, D.C. December 12-14, 1966.

. . . Transcript of Conference in the matter of Pollution of Interstate Waters,
Lower Columbia River, Bonneville Dam to Cathlamet, Washington. Second
Session. Portland, Oregon. September 3, 1959.

U.S. Public Health Service, Consumer Protection and Environmental Health
Service, National Air Pollution Control Administration. Control Techniques
for Particulate Air Pollutants. Washington, D.C.: U.S. Government Printing
Office. 1969.

U. S. Public Health Service, Division of Air Pollution. In Quest of Clean Air
for Berlin, New Hampshire. By Paul A. Kenline. 1962.

U.S. Public Health Service, National Center for Air Pollution Control. Technical Report. Lewiston, Idaho; Clarkston, Washington. Air Pollution Abatement Activity. 1967.

VII. State Government Reports and Proceedings

Alabama. Water Pollution Progress Report for the Years 1967-1968. State of
Alabama Water Improvement Commission.

. . . Proceedings of Public Hearing on Water Quality Standards. State of
Alabama Water Improvement Commission. Upper Tombigbee River Basin.
Carrollton, Alabama. November 29, 1966.

. . . Lower Tombigbee River Basin. Vols. I, II. Grove Hill, Alabama.
November 30, 1966.

. . . Chattahoochee River Basin (Upper Portion) Phoenix City, Alabama.
December 21, 1966.

. . . Alabama River Basin. Selma, Alabama. Jan. 10, 1967.

. . . Conecuh, Yellow and Blackwater River Basins. Andalysia, Alabama.
Jan. 17, 1967.

. . . Mobile and Escatawpa River Basins. Vol. I. Mobile, Alabama. Jan. 19, 1967.

California. Air Pollution Control in California. 1969 Annual Report, Air Resources Board. 1970.

Louisiana. Louisiana Air Control Plan, Fiscal Year 1970. Louisiana Air Control Commission.

Maine. Maine Water Resources Plan. Vols. I, II. 1969.

Ohio. Report and Recommendations on Water Quality for Scioto River Basin. Department of Engineering, Ohio Department of Health. 1967.

Tennessee. Public Hearings on Water Uses, Stream Standards and Implementation Plans for Interstate Streams of Tennessee. Vols. II, IV. Tennessee Stream Pollution Control Board. 1967.

Wisconsin. Quest for Clean Waters: Where We Stand in Wisconsin. Progress Report 1. Publication 1101-69. Department of Natural Resources.

. . . Upper Wisconsin River Pollution Investigative Survey. Department of Natural Resources. July, 1970.

. . . Progress Report to the Lake Michigan Enforcement Conference. Department of Natural Resources. March 31, 1970.

. . . Report on an Investigation of the Pollution in the Lower Fox River and Green Bay Made During 1966-1967. Department of Natural Resources. January 4, 1968.

. . . Report on an Investigation of the Pollution of the Menominee River and its Tributaries in Wisconsin Made During 1968. Department of Natural Resources. April 16, 1969.

. . . Report on an Investigation of the Pollution in the Oconto River Drainage Basin made during 1968. Department of Natural Resources. April 13, 1969.

. . . Report on an Investigation of the Pollution in the Upper Wisconsin River Drainage Basin Made During 1963 and early 1964. Department of Natural Resources. June 30, 1964.

Council on Economic Priorities
Economic Priorities Reports and in-depth Studies
Annual Subscription Rates

Individual Subscription*
(1 copy of each Report) $ 25

Sustaining Indiv. Subscription $ 200

Corporate/Institutional/Governmental Subscription
(1 copy of each in-depth study and
up to 10 copies of each Report —
please specify no. desired) $ 350

(Foreign subscription — $ 1.50 add.)

Please make check payable to
The Council on Economic Priorities
456 Greenwich Street
New York, N.Y. 10013

(CEP is a tax-exempt organization.)

Check encl. : _____ Bill me: _____

Name** _____

Address _____

City _____

State _____ Zip _____

  *Student/Clergy rate: $ 7.50
**If Student, specify college and year.